The Penguin Book of

WAR

GREAT MILITARY
WRITINGS

Edited by

JOHN KEEGAN

VIKING

To Matthew

VIKING

Published by the Penguin Group
Penguin Books Ltd, 27 Wrights Lane, London w8 5tz, England
Penguin Putnam Inc., 375 Hudson Street, New York, New York 10014, USA
Penguin Books Australia Ltd, Ringwood, Victoria, Australia
Penguin Books Canada Ltd, 10 Alcorn Avenue, Toronto, Ontario, Canada m4v 3b2
Penguin Books (NZ) Ltd, Private Bag 102902, NSMC, Auckland, New Zealand

Penguin Books Ltd, Registered Offices: Harmondsworth, Middlesex, England

First published 1999
1 3 5 7 9 10 8 6 4 2

Editorial matter copyright © John Keegan, 1999

The Acknowledgements on pages 468–76 constitute an extension of this copyright page

Set in 12/14.5pt Monotype Garamond
Typeset by Rowland Phototypesetting Ltd,
Bury St Edmunds, Suffolk
Printed in England by Clays Ltd, St Ives plc

A CIP catalogue record for this book is available from the British Library

ISBN 0–670–85299–6

Z5669

Penguin Book of War : Great Military
Writings.

CONTENTS

PART II

PART III

EDITOR'S ACKNOWLEDGEMENTS

I would first like to acknowledge the help I have received in compiling this anthology from my son Matthew, to whom the book is dedicated. Next I wish to express my thanks to my assistant, Lindsey Wood, for her help in typing, editing and proofreading. My thanks also go to Al Silverman, a benefactor to me throughout my writing life, who originally proposed the idea for an anthology of military literature, to Eleo Gordon, my editor at Penguin and, as always, to my literary agent, Anthony Sheil. Most of the books from which pages were extracted are in the possession of the library of the Royal Military Academy Sandhurst or the London Library; I thank the librarians and their staff. Finally, as ever, love and thanks to my darling wife Susanne.

The Manor House
Kilmington
Wiltshire
June 1999

INTRODUCTION

The Last of the Gentlemen's Wars was what Major-General J. F. C. Fuller, the British military theorist, entitled his account of the Anglo-Boer War of 1899–1902, in which he had fought as a young officer. The title was apt. Despite their social differences, the Afrikaner farmers of the Transvaal Republic and the Orange Free State and the officers and soldiers of the British regular army opposed against them shared a common military code. Both cared for each other's wounded and prisoners, respected the rights of non-combatants and observed the conventions of parlay and truce. Each side, at least at the outset of hostilities, fought a 'clean' war, face to face, man to man, with rifle or sword on 'equal' terms.

A point of particular, if implicit, agreement in this 'gentlemen's war' was that neither side should employ the services of the African 'natives' who formed the majority of the population in the theatre of war. The Boers, who were fighting in part to defend their privileged status as white men in the black world they ruled, declined in any case to do so. The British, who were protecting an imperial system in which blacks occupied a lower place, also excluded them from the fighting ranks. The reasons for doing so were on both sides, however, more than political. They were social and cultural. The military culture of the European whites – and the Boers remained European by culture despite their long African domicile – was not that of the native peoples. They, the Zulus in particular, had been warriors before their military power was broken by the invaders and had fought with skill and determination. They had not, however, observed the European conventions. The wounded were slaughtered after battle, prisoners too, while the women and children of the defeated might also be killed, if they were not appropriated as slaves or chattels. Victory in war, particularly to the Zulus, gave the winners absolute rights over the lives and properties of the defeated. Little wonder that neither the Boers nor the British wished to reawaken the military habits of the subject peoples.

Differences – and tensions – between the world's military cultures, present and past, are a major theme of this collection. It begins with the Heroic Age of the Greek world, from which comes down to us not only some of the earliest literature of war but also some of the greatest. Homer, in *The Iliad*, describes the events and the personalities of the Trojan War. His account, composed in the eighth century BC, of an episode of the twelfth century, must have been based on folk memory. Both archaeology and scholarship nevertheless attest to the reality of the conflict, while Homer's dramatic verse undoubtedly depicts the heroic style of combat with remarkable veracity.

Heroes represented the societies for which they fought. At the same time, they fought for themselves rather than those who followed them or the wider community beyond the battlefield. Heroes observed a code of honour but not of social responsibility. Heroes fought heroes: that was the essence of the heroic code. It demanded fair fight, face to face, with equal weapons and without the intervention of supporters, who traditionally participated only as spectators. Victory in heroic single combat brought prestige and benefit to the winner's side, since the loser's was diminished by his defeat. It was the peculiar and horrible circumstances of defeat that lent single combat its dramatic quality. The victor was not expected to show mercy. The incapacitated opponent was to be killed without hesitation when he fell, his armour and weapons appropriated as trophies and his body mutilated and exhibited in celebration of the victor's triumph.

Antique heroism is repugnant to the modern world, which associates the heroic ideal with magnanimity in victory and concern for the welfare of the vanquished. It would take many centuries, however, and the implantation of a new ethical culture, that of Christianity, for 'harsh' heroism to give way to the more merciful heroism of the knight of chivalry and the Renaissance gentleman. Until the appearance of those two warrior types, warrior harshness would remain the norm in most societies and most regions of the world.

There was, however, an important and necessary intermediate

stage between the disappearance of the heroic duel as the central act of warfare, and its supersession by collective combat bound by a generally accepted code of laws and practices. That stage was the introduction of drill and discipline on the battlefield. It was closely associated with the rise of political society, in either a monarchical or civic form. Heroic warfare was essentially tribal, a transaction between groups dominated by warriors who held power because they were good at fighting. Such groups were often nomadic or, if settled, existed at a low economic level. Improvements in the techniques of production required a formalization of tribal relationships, either through the institution of kingship or civic assemblies, though the second usually followed the first. Both kingdoms and city states found that a higher level of economic and administrative life required a more structured military system for its defence than primordial warrior ways provided. The result was the formation of armies, monarchical at first in the river lands of the Middle East, then civic in the city states of Greece and Italy.

Without discipline, however, armies are ineffective instruments, needing leadership to order their numbers and concentrate their power. The concentration of power at the highest level achievable with hand-held weapons reached its apogee among the Greek city states of the fifth century BC, where the phalanx, a densely ranked mass of spearmen, won battles by sheer weight and momentum in a few minutes of deadly pushing and stabbing. Since the survival of each of its members depended on maintaining the closest possible bodily contact with neighbours, the phalanx generated an intense spirit of collective effort, which in turn was reflected in the civic life of the Greek states. The flavour of phalanx warfare is transmitted to us through the writings of Thucydides and Xenophon extracted here.

In its heyday, the phalanx was invincible. Like all systems, however, it had the defect of its strength, which was inflexibility. As its enemies learnt to exploit that weakness, competing systems came into being. The most important was that of the rising power of Rome, whose legions preserved the Greeks' solidity on the battlefield but deployed in smaller and looser sub-units armed with throwing as well as stabbing weapons. The legion's manoeuvrability was higher than that of the phalanx, allowing it to outflank as well as confront an enemy formation, and so to win by envelopment rather than sheer weight of numbers.

The legions originated as civic militias. The transformation of the later Roman Republic into an empire, however, changed their social structure, the military duty donated by citizens giving way to long-term service performed by regulars. As a result, the Roman legions came to form the first truly professional army, one that remains the model for all its successors. Yet it preserved an important quality from the tribal past, the ferocity with which it waged war and the pitilessness with which it treated the vanquished. Since the official religion of the Roman Empire was emperor worship, those who opposed the emperor could deserve no mercy and in consequence were to be slaughtered or enslaved. A particular cruelty was reserved for those who rebelled against the emperor's rule. The massacre of the Jewish rebels in Palestine in the first century A D, described by Josephus, exemplifies the fate of impious subjects who fell into the legions' hands.

By the time the end of the Roman Empire in the west came in the fifth century A D, it was a Christian state, as the eastern half, Byzantium, would remain until the fall of Constantinople to the Turks in 1453. The adoption of Christianity as the empire's official religion did not, however, affect the way in which its legions fought; at the end, in any case, the imperial army had become largely barbarian in composition. The rise of Christianity was, nevertheless, to exert a profound influence on the conduct of warfare. So too was Islam, when the proponents of that raw creed emerged from the Arabian desert to conquer Roman Africa, the Near East and Iberia in the seventh and eighth centuries.

The early Christians deplored warfare. It was only after several centuries that their successors recognized pacifism to be irreconcilable with the needs of running a Christian state. They attempted all the same to define moral rules that the warrior should observe. Islam deplored combat between fellow believers, and permitted warmaking against unbelievers only if they actively opposed the faith's spread.

Each religion was thus responsible for introducing into military affairs an entirely new principle, that of fighting, within a strictly defined ethical code, for an idea.

The first outright clash between these two religious entities occurred in the era of the Crusades, expeditions begun at the end of the eleventh century as an attempt to recapture the Holy City of Jerusalem for Christendom. The First Crusade had a successful culmination in the taking of Jerusalem in 1099, and the foundation of several Christian kingdoms in the Near East followed. In turn, that inaugurated three centuries of Muslim–Christian conflict in the Eastern Mediterranean, ending with the expulsion of the last Crusaders to a few island footholds – Rhodes, Cyprus, finally Malta – in the fifteenth century. The long passage of arms had manifold results. One was the adoption by Muslim warriors of the practice of chivalry, in essence both a code of honour and of skill at arms that stratified warfare socially and elevated the mounted warrior – in Christian Europe, the knight – to a dominant place. Cavalry had long commanded battlefields, in Europe since the introduction of the stirrup, an Asian invention, in the ninth century. The rise of the knight, and of his Muslim coeval, had the effect, however, of conferring on him a special military status which resolved into political position also. By the fifteenth century the armoured knight, and his Muslim equivalent, the mameluke, dominated territories from the English Channel to the Persian Gulf.

The era of the mounted warrior was to be short-lived, however. One reason for that was the appearance in the twelfth and thirteenth centuries of hordes of tribal horsemen from the Central Asian steppe, first Mongols, then Turks, who had no truck with chivalry and, by brutal weight of numbers, overthrew states – the Islamic Caliphate and the Byzantine Empire, among them – that put their trust in the concept of soldierly honour as a means of defence. Between 1259 and 1453, the Caliphate and the Eastern Roman Empire were overthrown by Mongols and Turks, as were the Christian Russian kingdoms, whilst the heartland of Christian Europe was brought under lethal threat by the pony-riding nomads. The empire of China, in which a horse aristocracy had long ruled, was also attacked and, for a period, brought under nomad rule. Nomads were, indeed, between the twelfth and fifteenth centuries, to be the most potent military force on the Eurasian land mass.

Their era of power was brought to a close by the second important military development of the pre-modern world, the harnessing of gunpowder to warfare. Gunpowder, probably Chinese in origin, had been used sporadically by armies since the fourteenth century. At the end of the fifteenth, as a propellant for cannon balls, it became a decisive agent in the struggle by rulers within the knightly kingdoms to put down their over-mighty subjects by breaking down castle walls. Knights had been brave but undisciplined opponents of the nomad hordes that menaced the eastern frontier of Christian Europe. Once their power was overthrown, as it was during the sixteenth century, and they and their feudal followers replaced as a military instrument by professional soldiers under royal control, the transition from haphazard to dependable military action by states, whether against domestic rebels or foreign enemies, was assured. Gunpowder made the modern state. States were to use it with terrifying effectiveness against both their domestic and foreign enemies from the sixteenth to nineteenth centuries.

By the seventeenth century, the technology of gunpowder had been extended from that of the large battlefield cannon, served by a crew, to the personal firearm, equipping an individual. Cannon were not thereby superseded. Their importance, indeed, increased, as they became lighter, more mobile and more adaptable to the battlefield. The development of the firearm – the matchlock and the later and handier flintlock musket – transformed, however, the role of the foot soldier whom the cannon had begun by supporting. The musket made the infantryman an instrument of firepower in his own right, particularly when ranked in close formation and drilled to load, aim and fire at the same moment.

The combination of gunpowder and drill proved decisive in beating back the attack of Islam, manifested in the Ottoman Empire, from the frontiers of Christian Europe during the seventeenth century. The Turkish offensive from the Eastern to the Western Mediterranean was checked by their defeat at the Siege of Malta in

1565; their amphibious power was broken at the Battle of Lepanto in 1571. Thereafter, though they continued to stage a series of aggressive land campaigns in the Balkans for another century, the impetus of triumphant Islam was lost. In 1683 the Turks were defeated at the second Siege of Vienna. In 1699, by the Treaty of Carlowitz, they were, for the first time since their seizure of Constantinople in 1453, obliged to cede territory to their Christian enemies and accept Austrian dominance over Hungary, which they had controlled since the previous century.

The Habsburg triumph, deriving from Austrian mastery of the new techniques of gunpowder drill and discipline, swiftly translated into a general European offensive against the non-Christian worlds beyond the Mediterranean and Atlantic. The Spanish, during the course of the sixteenth century, had established their empire in the Americas, through the medium of gunpowder weapons, but the technological backwardness of their Aztec and Inca opponents made their victory a comparatively easy one. The real challenge to the European maritime thrust to world empire was presented when the Spanish, French, English and Dutch venturers met the power of the Moghuls in India and the Manchus in China from the end of the seventeenth century onwards.

China was to prove impervious to European penetration until the nineteenth century, as the nearby Japanese island empire would also. In India, however, and its neighbouring peninsulas and archipelagos, the Dutch, English and French established footholds during the seventeenth century, through the medium of the ocean-going warship mounting heavy cannon, and consolidated their presence by building coastal artillery forts, protecting their trading stations. In the East Indies those outposts soon grew into a Dutch maritime empire. The process took longer in the Indian sub-continent but, by the end of the eighteenth century, it had been repeated there also, most of India belonging or owing fealty to the British, who had triumphed over its decadent Moghul rulers by deploying, against their vast but ill-organized armies, small, disciplined forces of soldiers, drilled to the highest European standards. During the nineteenth century the pattern would be repeated in Africa. The French and British in particular, during the 'scramble' for the continent, would win victory after victory against warrior peoples whose hardihood and bravery proved of no avail, even when they possessed firearms, against European military technique.

Yet there was a paradox at the heart of the conflict between Europe and the world beyond. The power of European arms and military organization could always overcome that of their warrior enemies in any direct conflict between bodies of men in the open. European armies consistently beat Asian and African armies wherever and whenever they met on the battlefield. What the European method of warfare could not defeat, however, was the warrior spirit, even when the European armies engaged were themselves led by warrior officers and contained men of undoubted warrior quality.

The contrast between one method of warfare and another was never better exemplified than in the struggle of Europeans against natives for control of North America. 'Red Indians', as Europeans called them, particularly the Indians of the Great Plains, made a cult of hardihood and courage, even hopeless courage. They were admitted to full manhood in the tribe by rituals of harsh physical suffering. They sustained their reputation as men by resisting, should they fall into enemy hands, indignity and the cruellest torture to the point of death.

In the face of such irrational dedication to the warrior ideal, Europeans found only one recourse to assert their dominance, which was oppression to the point of extermination. During the period of mass settlement of the Great Plains, from 1849 to 1890, white America did indeed move from an attempt to suppress Indian resistance to the incoming of white agriculturalists to near-extermination of the native inhabitants. The programme of settlement eventually succeeded, as sheer weight of numbers determined it should. The breaking of the spirit of Native American warriordom was its necessary concomitant.

European triumphalism would meet an inevitable antithesis. Competition for empire turned Europe in upon itself. By 1910 the era of mass emigration to the New World and to Australasia was over and most of the globe, whether suitable for European settlement or not, had been appropriated by one European government or another, latterly for reasons of national prestige. The limitations of the opportunity to emigrate, when married to disputes and disappointments between governments over the acquisition of colonies, raised tempers at home, particularly between the successful colonial powers, chiefly Britain and France, and the others, particularly Germany and Austria-Hungary, which felt excluded from 'a place in the sun'. Germany and Austria were, however, imperial powers themselves within the European continent, ruling over large minorities of Slav people who sought their own national independence. Deeply suspicious of any action by their neighbours that threatened their imperial integrity, Germany and Austria moved during the first years of the twentieth century to a state of hair-trigger military alert. It was replicated by bordering states. In July 1914 this continental military system led to general war.

The First World War, its course and its outcome, determined the nature of the rest of the century, ensuring that it would be one of almost unrelenting conflict. Defeat, costing two million lives given in vain, embittered Germany and disposed its population to heed Hitler when he promised revenge on coming to power fifteen years after the war's end. Defeat, which gave the Bolsheviks their chance to seize power in Russia, consigned that country to terrible civil war and to a political system dedicated to ideological aggression against others. Defeat, which dismembered Austria, left its former imperial territories as unstable and quarrelsome entities in the no man's land between Bolshevism and emergent Nazism. Defeat, which also overcame the Ottoman Empire, ended Turkish control over the volatile Arab lands, so rendering the region even more unstable than post-war Eastern Europe.

The victory won by Britain and France, at a cost in lives equivalent to that suffered by the vanquished nations, left them uncertain and enfeebled; while their co-victor, the United States, which might have fortified them in the exercise of post-war responsibility, decided after 1918 to withdraw from world affairs altogether. This combination of circumstances – disablement of the great colonial powers, embitterment of their enemies, transformation of Russia into an international malefactor, alienation of the United States from commitment – left the world ripe for upheaval. The likelihood that upheaval would be the outcome was enhanced by the emergence into independence, full, partial, or potential, of states unfitted by their experience under imperial dominance to act without narrow selfishness on the world stage.

Upheaval, as might have been predicted, was indeed the outcome. In 1939 the world descended again into general war, on an even wider scale than in 1914. The loss of life it caused was higher by far than in the First World War – fifty rather than ten million dead, the majority civilian rather than military – while the material destruction that resulted was incomparably greater. The war's aftermath, moreover, left the makings of five decades of local wars, many on a very large scale, in China, India, the Middle East, South-East Asia and Africa. The final destruction of European imperial rule, inaugurated in 1914, released rancours and hostilities which had been kept in check for three centuries and more by maritime colonialism. Some of the ensuing wars took the form of belated conflicts between the European empires and their former subject peoples, as between the French and the Vietnamese, the French and the Algerians, the Dutch and the Indonesians or the Portuguese and the Angolans. Others broke out between colonialized peoples whom imperial government had obliged to live in mutual peace, as in the Indian sub-continent. Others again, as in China, were the result of quarrels over ideologies imported from Europe through contact with the maritime empires.

The strategies, tactics and, above all, the weapons of these wars were almost exclusively European. The Chinese Civil War, culmination of Mao Tse-tung's twenty-year struggle against both Chiang Kai-shek and the Japanese invaders which resulted in the imposition of a form of Bolshevism on China in 1949, was fought with weapons supplied by Russia or captured indirectly from the Americans. Yet the form of warfare practised in China, later in Vietnam and later still by resistance movements in Africa, was not European at all. Europeans, over the course of two thousand years and more, since the emergence of the Greek city states, had evolved a style of conflict that achieved results by face-to-face struggle on a defined battlefield and within a narrow time-frame. The 'battle of decision' had come for them to define the nature of war. Over the centuries, it had served them well, creating stable states and winning empires. Unperceived, however, an alternative style of fighting had always persisted. It derived its validity from the avoidance, whenever possible, of face-to-face combat, of battle at a fixed point, and of observance of the demands of time.

This alternative form of warfare was the obverse of the 'gentlemen's war' around which European civilization had organized itself since the Greeks defined its central values over two millennia ago. Gentlemanly warfare honoured courage in combat to the point of death itself, but also honoured respect for the vanquished and execrated cruelty towards the captured. It deprecated deceit, cunning, expediency and anything that smacked of cowardice, however conducive to eventual victory. Alternative warfare observed no such inhibitions. It could require extraordinary courage of those who fell into the hands of the enemy, resistance to torture in the name of warrior honour, death by ritual combat as cultural custom. It placed, however, no particular virtue on the fight to the death, as, for political and communal reasons, Western warriordom did.

This anthology has been designed to exemplify these contrasting military traditions. Many who read it will understandably be horrified by the record of inhumanity that the testimonies from the non-European world depict. The history of all forms of warfare is, however, essentially inhumane. That should be the message of this collection of warrior words from the European and non-European worlds.

JOHN KEEGAN
January 1999

PART I

This first series of extracts is drawn from the earliest accounts of warfare that have come down to us, but is particularly concerned with war between different cultures – Greeks against Persians, the Romans against the British and the Jews, Muslims against Christians in the Mediterranean world, Europeans against Native Americans. The theme of this section is to illustrate how various are the forms which war may take, and to emphasize that the modern world's understanding of what motivates men and states to war was not necessarily shared by peoples of the past. Modern man sees war as a political or economic necessity. It is important to be reminded that, to our ancestors, war might be seen as a duty owed to the gods, or to the concept of tribal or personal honour, quite as much as to material dictates.

THUCYDIDES
The Melian Dialogue

Thucydides (*c.*460–*c.*400 BC) is regarded as the first modern – and perhaps still the world's greatest – historian, for his history of the Peloponnesian War (431–421 and 421–404). He was not only a chronicler of the conflict but also fought in it, eventually as the commander of an Athenian fleet in the campaign against the Spartan invasion of Thrace in 424–423 BC. For his failure to prevent the Spartans from capturing the city of Amphipolis, though he succeeded in preventing the port of Eion from falling to them, he took exile from Athens and did not return for twenty years. It was in those years that he composed his history, though he appears to have conceived the idea of writing it at the war's outbreak. His account, in eight books, ends at 411. His original intention had been to carry it on to the war's end.

The Peloponnesian War, which fell into two parts, was a contest for mastery in the Greek world between the city states of Sparta (Lacedaemon) and Athens, and their respective allies. Sparta and Athens had together led the other Greeks in the long struggle (499–448 BC) against the Persian emperors' efforts to make them subjects. Victory, however, glorious though it was, brought dissent. The contribution made by Athens to victory was in large measure a maritime one, and led to that city state strengthening its position in the island region of the Aegean that was the gateway to its colonies on the coast of Asia and in the Black Sea. Sparta was a land power, the most military in Greece, with a military aristocracy that repressed its own peasantry and increasingly oppressed its neighbours who had helped in victory over the Persians. When Athens began to fortify its positions on the mainland north of the Peloponnese, to make a better base for its fleet, Sparta felt its land power challenged. When Athens, which despised the undemocratic nature of Spartan government, fell into conflict with Sparta's Peloponnesian neighbours, war broke out. It was a paradoxical conflict. Athens, the centre of the principle of democracy in Greek city-state life, sought to crush the independence of smaller city states which would not join it in the war against Sparta. Sparta, a ruthless dictatorship of warriors over the low-born at home, became the champion of the smaller Peloponnesian states' autonomy.

The philosophical issue came to a head in 416 BC when the island of Melos, south-east of the Peloponnese, was challenged by Athens to declare its position. Historically a Spartan colony, Melos was vulnerable to Athenian naval power. Required to choose their overlord by an Athenian diplomatic mission, the Melians tried to declare neutrality. The Athenians would have none of it. Melos, to them, was a necessary island link in their chain of bases that led from their home base in Greece to the source of their corn supply in Asia Minor.

In 'the Melian Dialogue', Thucydides contrives a debate between the Athenian emissaries and the Melian leaders. It is much more than the report of a diplomatic mission, however; it is widely regarded as an analysis of the universal value of the place of force in relations between states that are guided by self-interest. The Athenians put self-interest first. They tell the Melians that Athens must take possession of Melos if the other islands they have conquered, and which they need for their security, are to remain under their control. The Melians first ask why they cannot merely become friendly neutrals. When told that such neutrality does not satisfy the Athenians' needs, they advance the argument that it would be dishonourable to break with Sparta, which they trust to come to their help if they are threatened. The Athenians reply that force, not trust, is what counts in war, and that honour is a costly position unless it can be defended with arms.

The Melians nevertheless refused to surrender their freedom. War broke out and, as the Athenians had warned, ended in the Melians' utter defeat.

Thomas Hobbes, the seventeenth-century English philosopher of state power, translated Thucydides and used the Melian Dialogue in particular to illustrate the debt the weak owe to the strong. Thucydides's history became, in the nineteenth century, a key text in the education of the British ruling class, at a time when their country's naval power was its principal instrument in the creation of a world empire.

*   *   *

84. The next summer the Athenians made an expedition against the isle of Melos. The Melians are a colony of Lacedaemon that would not submit to the Athenians like the other islanders, and at first remained neutral and took no part in the struggle, but afterwards, upon the Athenians using violence and plundering their territory, assumed an attitude of open hostility. The Athenian generals encamped in their territory with their army, and before doing any harm to their land sent envoys to negotiate. These the Melians did not bring before the people, but told them to state the object of their mission to the magistrates and the council; the Athenian envoys then said:

85. *Athenians*: 'As we are not to speak to the people, for fear that if we made a single speech without interruption we might deceive them with attractive arguments to which there was no chance of replying – we realize that this is the meaning of our being brought before your ruling body – we suggest that you who sit here should make security doubly sure. Let us have no long speeches from you either, but deal separately with each point, and take up at once any statement of which you disapprove, and criticize it.'

86. *Melians*: 'We have no objection to your reasonable suggestion that we should put our respective points of view quietly to each other, but the military preparations which you have already made seem inconsistent with it. We see that you have come to be yourselves the judges of the debate, and that its natural conclusion for us will be slavery if you convince us, and war if we get the better of the argument and therefore refuse to submit.'

87. *Athenians*: 'If you have met us in order to make surmises about the future, or for any other purpose than to look existing facts in the face and to discuss the safety of your city on this basis, we will break off the conversations; otherwise, we are ready to speak.'

88. *Melians*: 'In our position it is natural and excusable to explore many ideas and arguments. But the problem that has brought us here is our security; so, if you think fit, let the discussion follow the line you propose.'

89. *Athenians*: 'Then we will not make a long and unconvincing speech, full of fine phrases, to prove that our victory over Persia justifies our empire, or that we are now attacking you because you have wronged us, and we ask you not to expect to convince us by saying that you have not injured us, or that, though a colony of Lacedaemon, you did not join her. Let each of us say what we really think and reach a practical agreement. You know and we know, as practical men, that the question of justice arises only between parties equal in strength, and that the strong do what they can, and the weak submit.'

90. *Melians*: 'As you ignore justice and have made self-interest the basis of discussion, we must take the same ground, and we say that in our opinion it is in your interest to maintain a principle which is for the good of all – that anyone in danger should have just and equitable treatment and any advantage, even if not strictly his due, which he can secure by persuasion. This is your interest as much as ours, for your fall would involve you in a crushing punishment that would be a lesson to the world.'

91. *Athenians*: 'We have no apprehensions about the fate of our empire, if it did fall; those who rule other peoples, like the Lacedaemonians, are not formidable to a defeated enemy. Nor is it the Lacedaemonians with whom we are now contending: the danger is from subjects who of themselves may attack and conquer their rulers. But leave that danger to us to face. At the moment we shall prove that we have come in the interest of our empire and that in what we shall say we are seeking the safety of your state; for we wish you to become our subjects with least trouble to ourselves, and we would like you to survive in our interests as well as your own.'

92. *Melians*: 'It may be your interest to be our masters: how can it be ours to be your slaves?'

93. *Athenians*: 'By submitting you would avoid a terrible fate, and we should gain by not destroying you.'

94. *Melians*: 'Would you not agree to an arrangement under which we should keep out of the war, and be your friends instead of your enemies, but neutral?'

95. *Athenians*: 'No: your hostility injures us less than your friendship. That, to our subjects, is an illustration of our weakness, while your hatred exhibits our power.'

96. *Melians*: 'Is this the construction which your subjects put on it? Do they not distinguish between states in which you have no concern, and peoples who are most of them your colonies, and some conquered rebels?'

97. *Athenians*: 'They think that one nation has as good rights as another, but that some survive because they are strong and we are afraid to attack them. So, apart from the addition to our empire, your subjection would give us security: the fact that you are islanders (and weaker than others) makes it the more important that you should not get the better of the mistress of the sea.'

98. *Melians*: 'But do you see no safety in our neutrality? You debar us from the plea of justice and press us to submit to your interests, so we must expound our own, and try to convince you, if the two happen to coincide. Will you not make enemies of all neutral Powers when they see your conduct and reflect that some day you will attack them? Will not your action strengthen your existing opponents, and induce those who would otherwise never be your enemies to become so against their will?'

99. *Athenians*: 'No. The mainland states, secure in their freedom, will be slow to take defensive measures against us, and we do not consider them so formidable as independent island powers like yourselves, or subjects already smarting under our yoke. These are most likely to take a thoughtless step and bring themselves and us into obvious danger.'

100. *Melians*: 'Surely then, if you are ready to risk so much to maintain your empire, and the enslaved peoples so much to escape from it, it would be criminal cowardice in us, who are still free, not to take any and every measure before submitting to slavery?'

101. *Athenians*: 'No, if you reflect calmly: for this is not a competition in heroism between equals, where your honour is at stake, but a question of self-preservation, to save you from a struggle with a far stronger Power.'

102. *Melians*: 'Still, we know that in war fortune is more impartial than the disproportion in numbers might lead one to expect. If we submit at once, our position is desperate; if we fight, there is still a hope that we shall stand secure.'

103. *Athenians*: 'Hope encourages men to take risks; men in a strong position may follow her without ruin, if not without loss. But when they stake all that they have to the last coin (for she is a spendthrift), she reveals her real self in the hour of failure, and when her nature is known she leaves them without means of self-protection. You are weak, your future hangs on a turn of the scales; avoid the mistake most men make, who might save themselves by human means, and then, when visible hopes desert them, in their extremity turn to the invisible – prophecies and oracles and all those things which delude men with hopes, to their destruction.'

104. *Melians*: 'We too, you can be sure, realize the difficulty of struggling against your power and against Fortune if she is not impartial. Still we trust that Heaven will not allow us to be worsted by Fortune, for in this quarrel we are right and you are wrong. Besides, we expect the support of Lacedaemon to supply the deficiencies in our strength, for she is bound to help us as her kinsmen, if for no other reason, and from a sense of honour. So our confidence is not entirely unreasonable.'

105. *Athenians*: 'As for divine favour, we think that we can count on it as much as you, for neither our claims nor our actions are inconsistent with what men believe about Heaven or desire for themselves. We believe that Heaven, and we know that men, by a natural law, always rule where they are stronger. We did not make that law nor were we the first to act on it; we found it existing, and it will exist for ever, after we are gone; and we know that you and anyone else as strong as we are would do as we do. As to your expectations from Lacedaemon and your belief that she will help you from a sense of honour, we congratulate you on your innocence but we do not admire your folly. So far as they themselves and their national traditions are concerned, the Lacedaemonians are a highly virtuous people; as for their behaviour to others, much might be said, but we can put it shortly by saying that, most obviously of all people we know, they identify their interests with justice and the pleasantest course with honour. Such principles do not favour your present irrational hopes of deliverance.'

106. *Melians*: 'That is the chief reason why we have confidence in them now; in their own interest they will not wish to betray their own colonists and so help their enemies and destroy the confidence that their friends in Greece feel in them.'

107. *Athenians*: 'Apparently you do not realize that safety and self-interest go together, while the path of justice and honour is dangerous; and danger is a risk which the Lacedaemonians are little inclined to run.'

108. *Melians*: 'Our view is that they would be more likely to run a risk in our case, and would regard it as less hazardous, because our nearness to the Peloponnese makes it easier for them to act and our kinship gives them more confidence in us than in others.'

109. *Athenians*: 'Yes, but an intending ally looks not to the good will of those who invoke his aid but to marked superiority of real power, and of none is this truer than of the Lacedaemonians. They mistrust their own resources and attack their neighbours only when they have numerous allies, so it is not likely that, while we are masters of the sea, they would cross it to an island.'

110. *Melians*: 'They might send others. The sea of Crete is large, and this will make it more difficult for its masters to capture hostile ships than for these to elude them safely. If they failed by sea, they would attack your country and those of your allies whom Brasidas did not reach; and then you will have to fight not against a country in which you have no concern, but for your own country and your allies' lands.'

111. *Athenians*: 'Here experience may teach you like others, and you will learn that Athens has never abandoned a siege from fear of another foe. You said that you proposed to discuss the safety of your city, but we observe that in all your speeches you have never said a word on which any reasonable expectation of it could be founded. Your strength lies in deferred hopes; in comparison with the forces now arrayed against you, your resources are too small for any hope of success. You will show a great want of judgement if you do not come to a more reasonable decision after we have withdrawn. Surely you will not fall back on the idea of honour, which has been the ruin of so many when danger and disgrace were staring them in the face. How often, when men have seen the fate to which they were tending, have they been enslaved by a phrase and drawn by the power of this seductive word to fall of their own free will into irreparable disaster, bringing on themselves by their folly a greater dishonour than fortune could inflict! If you are wise, you will avoid that fate. The greatest of cities makes you a fair offer, to keep your own land and become her tributary ally: there is no dishonour in that. The choice between war and safety is given you; do not obstinately take the worse alternative. The most successful people are those who stand up to their equals, behave properly to their superiors, and treat their inferiors fairly. Think it over when we withdraw, and reflect once and again that you have only one country, and that its prosperity or ruin depends on one decision.'

112. The Athenians now withdrew from the conference; and the Melians, left to themselves, came to a decision corresponding with what they had maintained in the discussion, and answered, 'Our resolution, Athenians, is unaltered. We will not in a moment deprive of freedom a city that has existed for seven hundred years; we put our trust in the fortune by which the gods have preserved it until now, and in the help of men, that is, of the Lacedaemonians; and so we will try and save ourselves. Meanwhile we invite you to allow us to be friends to you and foes to neither party, and to retire from our country after making such a treaty as shall seem fit to us both.'

113. Such was the answer of the Melians. The Athenians broke up the conference saying, 'To judge from your decision, you are unique in regarding the future as more certain than the present and in allowing your wishes to convert the unseen into reality; and as you have staked most on, and trusted most in, the Lacedaemonians, your fortune, and your hopes, so will you be most completely deceived.'

114. The Athenian envoys now returned to the army; and as the Melians showed no signs of yielding the generals at once began hostilities, and drew a line of circumvallation [i.e. surrounded with defensive fortifications] round the Melians, dividing the work among the different states. Subsequently the Athenians returned with most of their army, leaving behind them a certain number of their own citizens and of the allies to keep guard by land and sea. The force thus left stayed on and besieged the place.

115. Meanwhile the Athenians at Pylos took so much plunder from the Lacedaemonians that the latter, although they still refrained from breaking off the treaty and going to war with Athens, proclaimed that any of their people that chose might plunder the Athenians. The Corinthians also commenced hostilities with the Athenians for private quarrels of their own; but the rest of the Peloponnesians stayed quiet. Meanwhile the Melians in a night attack took the part of the Athenian lines opposite the market, killed some of its garrison, and brought in corn and as many useful stores as they could. Then, retiring, they remained inactive, while the Athenians took measures to keep better guard in future.

116. Summer was now over. The next winter the Lacedaemonians intended to invade the Argive territory, but on arriving at the frontier found the sacrifices for crossing unfavourable, and went back again. This intention of theirs made the Argives suspicious of certain of their fellow-citizens, some of whom they arrested; others, however, escaped them. About the same time the Melians again took another part of the Athenian lines which were but feebly garrisoned. In consequence reinforcements were sent from Athens, and the siege was now pressed vigorously; there was some treachery in the town, and the Melians surrendered at discretion to the Athenians, who put to death all the grown men whom they took and sold the women and children for slaves; subsequently they sent out five hundred settlers and colonized the island.

THUCYDIDES
The Final Sea Battle
Syracuse, 413 BC

By 415 BC, at the end of seventeen years of almost continuous fighting in the Peloponnesian War (begun 431), in which the futures of Sparta, Athens and their allies had swayed one way and the other, Athens, under the leadership of the brilliant orator Alcibiades, decided to seek a decision by a widening of the conflict. The objective chosen was Sicily, 'Great Greece', dotted with Greek colonies and regarded on the mainland as the key to the Greek maritime world, which extended from what is today the French Riviera, in the west, to the Black Sea in the east. Syracuse, a major Greek Sicilian city, which had fallen into a local quarrel with an Athenian ally, presented a target, and a fleet of 136 triremes – galleys with three banks of oars, the capital ships of Greek naval warfare – set out and landed 5,000 troops under the city's walls. They won a victory and laid siege. When Sparta sent help to Syracuse, however, the Athenians were unable to complete their siege works and became embroiled in sea fighting in Syracuse Bay, which went against them. When Demosthenes, greatest of Athenian leaders, arrived with reinforcements of troops and ships, he decided that the expedition had been a failure and that retreat was the only safe option. Before the escape could be completed, the Syracusans blockaded the harbour and brought the Athenian fleet to battle. It culminated in a Syracusan victory, the enslavement of the defeated and the execution of Demosthenes. Athens, though it succeeded in sustaining the war for another eleven years, never fully recovered from the disaster.

*   *   *

70. The Syracusans and their allies had already put out with about the same number of ships as before; one detachment guarded the entrance of the harbour, the rest were disposed all round it, so as to attack the Athenians on all sides at once; while the land forces held themselves in readiness at the points at which the vessels might put in to the shore. The Syracusan fleet was commanded by Sicanus and Agatharchus, who had each a wing of the whole force, the Pythen and the Corinthians in the centre. When the Athenians came up to the barrier, the first shock of their charge overpowered the ships stationed there, and they tried to undo the fastenings; after this, as the Syracusans and allies bore down upon them from all quarters, the action spread from the barrier over the whole harbour, and was more obstinately disputed than any preceding. On either side the rowers showed great zeal in bringing up their vessels at the boatswains' orders, and the pilots great skill in manoeuvring, rivalling each other's efforts; once the ships were alongside, the soldiers on board did their best not to let the service on deck be inferior; in short, every man strove to prove himself the first in his particular department. As many ships were engaged in a small compass (never had fleets so large – there were almost two hundred vessels – fought in so narrow a space), the regular attacks with the beak were few, for there was no opportunity to back water or break the line; while the collisions caused by one ship chancing to run foul of another, either in avoiding or attacking a third, were more frequent. So long as a vessel was coming up to the charge the men on the decks rained darts and arrows and stones upon her; but once alongside, the heavy infantry tried to board each other's vessel, and fought hand to hand. In many places it happened, owing to want of room, that a vessel was charging an enemy on one side and being charged herself on another, and that two, or sometimes more, ships had unavoidably got entangled round one, and the pilots had to make plans of attack and defence against several adversaries coming from different quarters; while the huge din caused by the number of ships crashing together not only spread terror, but made the orders of the boatswains inaudible. The boatswains on either side in the discharge of their duty and in the heat of the conflict shouted incessantly orders and appeals to their men; the Athenians they urged to force the passage, and now if ever to show their mettle and make sure of a safe return to their country; to the Syracusans and their allies they cried that it would be glorious to prevent the escape of the enemy, and to win a victory which would bring glory to their country. The generals, on either side, if they saw any vessel in any part of the battle backing ashore without being forced to do so, called out to the captain by name and asked him – the Athenians, whether they were retreating because they thought that they would be more at home on a bitterly hostile shore than on that sea which had cost them so much labour to win; the Syracusans, whether they were flying from the flying Athenians, whom they well knew to be eager to escape by any possible means.

71. Meanwhile the two armies on shore, while victory hung in the balance, were a prey to the most agonizing and conflicting emotions; the Sicilians thirsting to add to the glory that they had already won, while the invaders feared to find themselves in even worse plight than before. The last hope of the Athenians lay in their fleet, their fear for the outcome was like nothing they had ever felt; while their view of the struggle was necessarily as chequered as the battle itself. Close to the scene of action, and not all looking at the same point at once, some saw their friends victorious and took courage, and fell to calling upon heaven not to deprive them of salvation, while others, who had their eyes turned upon the losers, wept and cried aloud, and, although spectators, were more overcome than the actual combatants. Others, again, were gazing at some spot where the battle was evenly disputed; as the strife was protracted without decision, their swaying bodies reflected the agitation of their minds, and they suffered the worst agony of all, ever just within reach of safety or just on the point of destruction. In short, in that one Athenian army, as long as the sea fight remained doubtful, there was every sound to be heard at once, shrieks, cheers, '*We win*,' '*We lose*,' and all the other sounds wrung from a great host in desperate peril; with the men in the fleet it was nearly the same; until at last the Syracusans and their allies, after the battle had lasted a long while, put the Athenians to flight, and with much shouting and cheering chased them in open rout to the shore. The naval force fell back in confusion to the shore, except those who were taken afloat, and rushed from their ships to their camp; while the army, no more with divided feelings, but carried away by one impulse, ran down with a universal cry of dismay, some to help the ships, others to guard what was left of their wall, while the majority began to consider how they should save themselves. Their panic was as great as any of their disasters. They now suffered very nearly what they had inflicted at Pylos; then the Lacedaemonians, besides losing their fleet, lost also the men who had crossed over to Sphacteria; so now the Athenians had no hope of escaping by land, without the help of some extraordinary accident.

72. The sea fight had been severe, and many ships and lives had been lost on both sides; the victorious Syracusans and their allies now picked up their wrecks and dead, and sailed off to the city and set up a trophy, while the Athenians, overwhelmed by their disaster, never even thought of asking leave to take up their dead or wrecks, but wished to retreat that very night. Demosthenes, however, went to Nicias and gave it as his opinion that they should man the ships they had left and make another effort to force their passage out next morning; saying that they had still left more ships fit for service than the enemy, the Athenians having about sixty remaining as against less than fifty of their opponents'. Nicias was quite of his mind; but when they wished to man the vessels, the sailors, who were so utterly overcome by their defeat as no longer to believe in the possibility of success, refused to go on board.

XENOPHON

The Battle of Cunaxa and Death of Cyrus

The Peloponnesian War, which ended in a Spartan triumph in 404 BC, left Greece full of workless veterans, who were ready to take service as mercenaries under whoever would hire them.

In 401 BC, Cyrus, the younger brother of Artaxerxes II, Emperor of Persia, decided to make an attempt to seize the throne and raised an army in Asia Minor, of which he was satrap, or subordinate ruler. Among the 13,000 Greeks he persuaded to join him was Xenophon, an Athenian who had come to admire the victorious Spartans but also saw in Cyrus the type of military aristrocrat by whom he aspired to be led. Under the command of Clearchus, the Spartan general of the 10,000 hoplites, or heavy infantry, of the Greek contingent, Xenophon took part in the march from Sardis, on the Aegean coast of Asia Minor, to Babylon, where at Cunaxa the army of Cyrus met that of Artaxerxes.

Xenophon's account of the battle is an important source for our understanding of the nature of Greek warfare. Traditionally, Greek armies were weak in cavalry, of which the Persians had an abundance. The hoplites, supported by their light infantry or peltasts, depended for success when faced by horsemen or charioteers on their ability to keep formation, stand firm and deliver a disabling charge at the critical moment. It was the spectacle of their dense, heavily armoured ranks and the sound of their battle cry, the ululation or *eleleu*, which unnerved their opponents. At Cunaxa, Greek discipline and fervour for combat overcame Persian numbers, as it had done time and again in the wars between the enormous Asiatic empire and the tiny peninsular cluster of Greek city states.

*   *   *

It was already the middle of the morning, and they had nearly reached the place where Cyrus intended to halt, when Pategyas, a Persian and a good friend of Cyrus, came into sight, riding hard, with his horse in a sweat. He immediately began to shout out, in Persian and Greek, to everyone in his way that the King with a great army in order of battle was approaching. There was certainly considerable confusion at this point, for the Greeks and everyone else thought that he would be upon them before they could form up in position. Cyrus leapt down from his chariot, put on his breastplate, mounted his horse and took hold of his javelins. He gave orders for all the rest to arm themselves and to take up their correct positions. This was done readily enough. Clearchus was on the right wing, flanked by the Euphrates, next to him was Proxenus and then the other Greeks, with Menon holding the left wing of the Greek army. As to the native troops, there were about a thousand Paphlagonian cavalry stationed with Clearchus and also the Greek peltasts on the right; Ariaeus, Cyrus's second-in-command, was on the left with the rest of the native army. Cyrus and about six hundred of his personal cavalry in the centre were armed with breastplates, and armour to cover the thighs. They all wore helmets except for Cyrus, who went into the battle bare-headed. All their horses had armour covering the forehead and breast; and the horsemen also carried Greek sabres.

It was now midday and the enemy had not yet come into sight. But in the early afternoon dust appeared, like a white cloud, and after some time a sort of blackness extending a long way over the plain. When they got nearer, then suddenly there were flashes of bronze, and the spear points and the enemy formations became visible. There were cavalry with white armour on the enemy's left and Tissaphernes was said to be in command of them. Next to them were soldiers with wicker shields, and then came hoplites with wooden shields reaching to the feet. These were said to be Egyptians. Then there were more cavalry and archers. These all marched in tribes, each tribe in a dense oblong formation. In front of them, and at considerable distances apart from each other, were what they called the scythed chariots. These had thin scythes extending at an angle from the axles and also under the driver's seat, turned toward the ground, so as to cut through everything in their way. The idea was to drive them into the Greek ranks and cut through them.

But Cyrus was wrong in what he said at the time when he called together the Greeks and told them to stand their ground against the shouting of natives. So far from shouting, they came on as silently as they could, calmly, in a slow, steady march.

At this point Cyrus himself with his interpreter Pigres and three or four others rode up and shouted out to Clearchus, telling him to lead his army against the enemy's centre, because that was where the King was. 'And if we win there,' he said, 'the whole thing is over.' Clearchus saw the troops in close order in the enemy's centre and he heard too from Cyrus that the King was beyond the Greek left (so great was the King's superiority in numbers that he, leading the centre of his own army, was still beyond Cyrus's left), but in spite of this he was reluctant, from fear of encirclement, to draw his right wing away from the river. He replied to Cyrus, then, that he would see to it that things went well.

While this was going on the Persian army continued to move steadily forward and the Greeks still remained where they were, and their ranks filled up from those who were continually coming up. Cyrus rode by some way in front of the army and looked along the lines both at the enemy and at his own troops. Xenophon, an Athenian, saw him from his position in the Greek line and, going forward to meet him, asked if he had any orders to give. Cyrus pulled in his horse and said: 'The omens are good and the sacrifices are good.' He told him to tell this to everyone, and while he was speaking he heard a noise going along the ranks and asked what the noise was. Xenophon told him that it was the watchword now being passed along the ranks for the second time. Cyrus wondered who had given the word and asked what it was, and Xenophon told him it was 'Zeus the Deliverer and Victory.' On hearing this Cyrus said: 'Then I accept the word. Let it be so,' and with these words he rode away to his own position in the field.

By now the two armies were not more than between six and eight hundred yards apart, and now the Greeks sang the paean and began to move forward against the enemy. As they advanced, part of the phalanx surged forward in front of the rest and the part that was left behind began to advance at the double. At the same time they all raised a shout like the shout of 'Eleleu' which people make to the War God, and then they were all running forward. Some say that to scare the horses they clashed their shields and spears together. But the Persians, even before they were in range of the arrows, wavered and ran away. Then certainly the Greeks pressed on the pursuit vigorously, but they shouted out to each other not to run, but to follow up the enemy without breaking ranks. The chariots rushed about, some going through the enemy's own ranks, though some, abandoned by their drivers, did go through the Greeks. When they saw them coming the Greeks opened out, though one man stood rooted to the spot, as though he was at a racecourse, and got run down. However, even he, they said, came to no harm, nor were there any other casualties among the Greeks in this battle, except for one man on the left wing who was said to have been shot with an arrow.

Cyrus was pleased enough when he saw the Greeks winning and driving back the enemy in front of them, and he was already being acclaimed as king by those who were with him; but he was not so carried away as to join in the pursuit. He kept the six hundred cavalry of his personal bodyguard in close order, and watched to see what the King would do, as he was sure that his position was in the Persian centre. Indeed, all Persian commanders are in the centre of their own troops when they go into battle, the idea being that in this way they will be in the safest spot, with their forces on each side of them, and that also if they want to issue any orders, their army will receive them in half the time. The King, too, on this occasion was in the centre of his army, but was all the same beyond Cyrus's left wing. Seeing, then, that no frontal attack was being made either on him or on the troops drawn up to screen him, he wheeled right in an outflanking movement.

Then Cyrus, fearing that the King might get behind the Greeks and cut them up, moved directly towards him. With his six hundred he charged into and broke through the screen of troops in front of the King, routed the six thousand, and is said to have killed their commander, Artagerses, with his own hand. But while they turned to flight, Cyrus's own six hundred lost their cohesion in their eagerness for the pursuit, and there were only a very few left with him, mostly those who were called his 'table companions'. When left with these few, he caught sight of the King and the closely formed ranks around him. Without a moment's hesitation he cried out, 'I see the man,' charged down on him, and struck him a blow on the breast which wounded him through the breastplate, as Ctesias the doctor says – saying also that he dressed the wound himself.

But while he was in the very act of striking the blow, someone hit him hard under the eye with a javelin. In the fighting there between Cyrus and the King and their supporters, Ctesias (who was with the King) tells how many fell on the King's side. But Cyrus was killed himself, and eight of the noblest of his company lay dead upon his body. It is said that when Artapatas, the most trusted servant among his sceptre-bearers, saw Cyrus fall, he leapt from his horse and threw himself down on him. Some say that the King ordered someone to kill him on top of Cyrus, others that he drew his scimitar and killed himself there. He had a golden scimitar, and used to wear a chain and bracelets and the other decorations like the noblest of the Persians; for he had been honoured by Cyrus as a good friend and a faithful servant.

XENOPHON
A Plundering Expedition

Cyrus died in the battle of Cunaxa, leaving his 10,000 Greek mercenaries isolated in the heart of the Persian Empire, 1,500 miles from their starting place at Sardis. They refused to surrender and Artaxerxes, knowing he could not defeat them, offered to provide them with guides to find their way home. He also provided them with an escort under Tissaphernes, the general against whom Cyrus had originally raised his revolt. When the chance offered, Tissaphernes managed to separate the commanders of the Ten Thousand from their men and deliver them to Artaxerxes, who executed them. He expected that their leaderless followers would then capitulate. New leaders emerged, however, foremost among them Xenophon, who – in the *Anabasis* – was to write the history of their retreat through hostile territory to the Black Sea coast. The march across 1,000 miles of the rugged Caucasus lasted five months, during which the Ten Thousand supplied themselves by plundering the inhabitants, who organized a running fight against the interlopers. When the Ten Thousand, reduced by privation and casualties, eventually reached the coast in 400 BC, the vanguard raised what was to become the legendary cry, 'The sea! The sea!'

They remained short of supplies and, from their base at Trapezus (modern Trabzon in Turkey) continued to plunder the surrounding countryside. In the following passage Xenophon describes their methods. The *Anabasis*, because it showed that a fierce and determined army of hoplites could live off the country and brave any Persian army that opposed it, is often held to have inspired Alexander the Great's successful expedition to conquer the Persian Empire (336–323 BC).

*   *   *

In the end it was no longer possible to get provisions from close at hand so as to return to camp on the same day. Xenophon therefore took some of the people of Trapezus as guides, and led half the army against the Drilae, leaving the other half behind to guard the camp – a necessary measure, since the Colchians, driven out as they had been from their own homes, had collected in large numbers and were occupying dominating positions on the heights. The people of Trapezus failed to lead the Greeks to places where it was easy to get provisions, since they were on friendly terms with the inhabitants of such places. However, they were willing enough to show them the way to the country of the Drilae, from whom they had suffered harm. It was difficult and mountainous ground, and the people were the most warlike of all the inhabitants of the Black Sea coastline.

As the Greeks advanced into the interior, the Drilae fell back, first setting fire to all their settlements which seemed to them indefensible. Thus there was nothing for the Greeks to take except for an odd pig or ox or other animal that had escaped the fire. They had one position which was their capital, and there they all gathered together. There was a tremendously deep ravine all round the place, and the roads into the fortification were difficult to get at.

The peltasts had got about a half a mile ahead of the hoplites, and, crossing over the ravine and seeing a lot of cattle and other booty, they attacked the fortifications. A number of spearmen, who had come out for plunder, followed after them, so that there were more than two thousand who crossed over the ravine. However, they were not able to take the place by assault, which was not surprising, as there was a broad ditch round it with the earth thrown up to form a rampart, and with a palisade on top of the rampart and wooden towers erected at frequent intervals. They therefore tried to retreat, but the enemy pressed hard upon them. Being unable, then, to get back, as the way down from the fortifications to the ravine was only broad enough for people to go down it in a single file, they sent a messenger to Xenophon, who was at the head of the hoplites. The messenger said that the place was full of all kinds of supplies. 'But,' he said, 'we cannot take it, because the enemy come out and attack us, and our return route is a difficult one.'

On receiving this information Xenophon advanced to the ravine and ordered the hoplites to halt there. He himself with the captains crossed over and examined the position to see whether it would be better to withdraw the troops who had crossed already, or, on the assumption that the place could be taken, to bring the hoplites across too. It seemed that it would be impossible to withdraw without considerable loss of life: the captains were of the opinion that they could take the place; and Xenophon agreed with them, relying also on the results of the sacrifices, for the soothsayers had indicated that there would be a battle, but the final result of the expedition would be successful. He therefore sent the captains back to bring the hoplites across, and stayed where he was himself. He brought all the peltasts back from the ditch and forbade any of them to engage in long-range fighting. When the hoplites arrived, he ordered each captain to form up his company in the way in which he thought his men would fight best; for the captains who were continually competing with each other in doing brave deeds, were now next to each other. They did as they were told, and Xenophon then ordered all the peltasts to advance with their javelins at the ready, and the archers to have their arrows fitted to the string, as they would both have to discharge their weapons as soon as he gave the signal. He told the light troops to have their wallets full of stones, and sent reliable people to see that these orders were obeyed.

When everything was ready, and the captains and lieutenants and other officers who considered themselves just as good men as their immediate superiors, had all taken up their positions, they were all actually in sight of each other, since, because of the nature of the ground, they were in a crescent-shaped formation. Then, after they had sung the paean and the trumpet had sounded, the hoplites raised the war cry and charged, with the missiles all being hurled together – spears, arrows, stones from slings, and a lot also thrown by hand, together with firebrands, which some people used in the attack. Under this weight of weapons the enemy were forced to abandon the palisade and the towers. This gave Agasias the Stymphalian an opportunity. He stripped off his armour and climbed up, only wearing his tunic. Others then helped each other up or climbed up by themselves and, to all appearance, the position was won. The peltasts and light infantry rushed inside, each man making off with what booty he could. Xenophon, however, stood by the gates and kept back as many of the hoplites as he could outside, since there were fresh enemy troops coming into sight in strong positions on the high ground. Before much time had gone by, there was a shout from inside, and people came running out, some carrying booty, with here and there a wounded man among them. There was a crush around the gates and, in reply to questions, the men who were being driven out said that there was a citadel inside and a large enemy force which kept charging down from it and falling on the Greeks who were inside.

Xenophon then ordered Tolmides the herald to proclaim that those who wanted plunder had permission to go in. A lot of men surged forward and, as the ones who were pushing their way in forced back those who were being driven out, they shut the enemy up again inside the citadel. Everything outside the citadel was plundered and the Greeks brought the booty out of the gates. The hoplites took up position, some at the palisade, and some on the path that led up to the citadel. Xenophon and the captains then considered whether it was possible to capture the citadel. If they could, it would mean that they could get away safely: otherwise it appeared that it would be a very difficult business to retire. After considering the matter, they came to the conclusion that the position was absolutely impregnable, and therefore made their dispositions for the retreat.

Each man pulled up the stakes in the palisade opposite him. They sent back those who were unfit for action or carrying booty, together with most of the hoplites; and the captains kept with them the men in whom each had special confidence. When they began their retreat, large numbers of the enemy, armed with shields, spears, greaves and Paphlagonian helmets, charged out at them from inside, and others climbed on to the roofs of the houses on each side of the way leading up to the citadel. Thus it was not safe even to drive them back by the gates leading there, since they threw down great pieces of timber and so made it awkward for them either to stay where they were or to retreat. The approach of darkness increased their alarm.

However, while they were still fighting and still doubtful what to do next, some god showed them a way of saving themselves. One of the houses on the right, through someone or other's action, suddenly caught fire. When this house collapsed, the enemy fled from the row of houses on the right. Xenophon, by a stroke of luck, saw what had happened, and gave the order to set fire to the houses on the left as well. These, being of wood, were soon blazing, and so the enemy fled from these houses too. Now the only enemy force which still caused trouble was the one in front, and it was obvious that these intended to fall upon them on their way out of the town and down into the ravine. Xenophon then ordered all those who were out of range of the missiles to carry wood into the space between them and the enemy. When enough wood had been brought, they set fire to it and also set fire to the houses next to the rampart, so that the enemy might have their attention attracted there. In this way they managed with difficulty to retreat from the place under the protection of a fire between them and the enemy. The whole city was burned to the ground – houses, towers, palisade and everything else except the citadel.

Next day the Greeks marched back, taking their supplies with them. They were apprehensive about the return journey to Trapezus, as the road was steep and narrow; and so they pretended to set an ambush. A Mysian in the army, who was called Mysus too, took ten Cretans with him and stayed behind in some wooded country, pretending that he was trying to keep out of enemy observation. The shields of the party, which were of brass, kept on flashing into sight. The enemy observed all this and were alarmed just as if it was a real ambush, and meanwhile the army carried out the descent. When Xenophon considered that they had got far enough away, he signalled to the Mysian to run back at full speed. So he and his men started up and ran for it. The Cretans (who saw that they were being overtaken in the race) jumped down from the road into a wood and, rolling over and over among the undergrowth, got away safe. The Mysian ran down the road and shouted for help. Soldiers came to his aid and picked him up wounded. Then the rescue party retreated step by step and, being shot at by the enemy, with some of the Cretan archers shooting back at the enemy. So they all got back safely to the camp.

JULIUS CAESAR
First British Expedition

The Greeks, who invented the face-to-face battle to the death and, under Alexander the Great, initiated the greatest campaign of conquest the world had yet seen, eventually succumbed to the power of Rome, an Italian city state which had learnt Greek methods of infantry fighting but had succeeded in avoiding both the profitless small-scale quarrelling that eventually exhausted the Greek world, and the over-ambitious long-range campaigning that destroyed Alexander's empire. The Romans had a gift of incorporating their defeated neighbours into a widening Latin community. Once they began to campaign beyond the limits of the Italian peninsula, they also adopted a cautious, step-by-step pattern of conquest, quite at variance with Alexander's megalomaniac engorgement of one territory after another. By the middle of the second century BC, Greece had effectively become a Roman possession. By its end, Rome had begun its expansion into Gaul (modern France), the opening of a long drawn-out campaign that was to culminate in the conquest of Gaul by Julius Caesar in 58–51 BC.

In 55 BC Caesar decided to extend his offensive to include the Britons, whom he believed provided support to and a refuge for his Gaulish enemies. He assembled two legions and a fleet of ships and landed near Dubris (modern Dover) to bring the British to battle. The legion, a force of up to 10,000 troops, mainly infantry but containing elements of other arms including cavalry and engineers, provided the model of organization on which all modern military formations are based. Its long-service soldiers, commanded by a corps of professional experts, the centurionate, regularly defeated any barbarian army it met, if circumstances suited its style of fighting. In 55 BC, and in the following year, Caesar found he was fighting too far from his base in Gaul, which had not yet been fully secured. He therefore withdrew and Britain was not to be conquered until the invasion of the Emperor Claudius in 43 AD. Caesar's account of the first battle provides, nevertheless, a striking picture of Roman military methods in the field.

*   *   *

A fleet of about eighty ships, which seemed adequate for the conveyance of two legions, was eventually commissioned and assembled, together with a number of warships commanded by the chief of staff, officers of general rank, and auxiliary commanders. At another port, some eight miles higher up the coast, were eighteen transports which had been prevented by adverse winds from joining the main fleet at Boulogne: these were allotted to the cavalry. The remainder of the army under Sabinus and Cotta was sent on a punitive expedition against the Menapii and those cantons of the Morini which had not been represented in the recent delegation. Another general officer, Publius Sulpicius Rufus, was ordered to guard the harbour with a force that seemed large enough for that purpose.

Arrangements were now complete, the weather was favourable, and we cast off just before midnight. The cavalry had been ordered to make for the northern port, embark there, and follow on; but they were rather slow about carrying out these instructions, and started, as we shall see, too late. I reached Britain with the leading vessels at about 9 a.m., and saw the enemy forces standing under arms all along the heights. At this point of the coast precipitous cliffs tower over the water, making it possible to fire from above directly on to the beaches. It was clearly no place to attempt a landing, so we rode at anchor until about 3.30 p.m., awaiting the rest of the fleet. During this interval I summoned my staff and company commanders, passed on to them the information obtained by [the reconnaissance of] Volusenus, and explained my plans. They were warned that, as tactical demands, particularly at sea, are always uncertain and subject to rapid change, they must be ready to act at a moment's notice on the briefest order from myself. The meeting then broke up: both wind and tide were favourable, the signal was given to weigh anchor, and after moving about eight miles up channel the ships were grounded on an open and evenly shelving beach.

The natives, however, realized our intention: their cavalry and war chariots (a favourite arm of theirs) were sent ahead, while the main body followed close behind and stood ready to prevent our landing. In the circumstances, disembarkation was an extraordinarily difficult business. On account of their large draught the ships could not be beached except in deep water; and the troops, besides being ignorant of the locality, had their hands full: weighted with a mass of heavy armour, they had to jump from the ships, stand firm in the surf, and fight at the same time. But the enemy knew their ground: being quite unencumbered, they could hurl their weapons boldly from dry land or shallow water, and gallop their horses which were trained to this kind of work. Our men were terrified: they were inexperienced in this kind of fighting, and lacked that dash and drive which always characterized their land battles.

The warships, however, were of a shape unfamiliar to the natives; they were swift, too, and easier to handle than the transports. Therefore, as soon as I grasped the situation I ordered them to go slightly astern, clear of the transports, then full speed ahead, bringing up on the Britons' right flank. From that position they were to open fire and force the enemy back with slings, arrows, and artillery. The manoeuvre was of considerable help to the troops. The Britons were scared by the strange forms of the warships, by the motion of the oars, and by the artillery which they had never seen before: they halted, then fell back a little; but our men still hesitated, mainly because of the deep water.

At this critical moment the standard-bearer of the Tenth Legion, after calling on the gods to bless the legion through his act, shouted: 'Come on, men! Jump, unless you want to betray your standard to the enemy! I, at any rate, shall do my duty to my country and my commander.' He threw himself into the sea and started forward with the eagle. The rest were not going to disgrace themselves; cheering wildly they leaped down, and when the men in the next ships saw them they too quickly followed their example.

The action was bitterly contested on both sides. But our fellows were unable to keep their ranks and stand firm; nor could they follow their appointed standards, because men from different ships were falling in under the first one they reached, and a good deal of confusion resulted. The Britons, of course, knew all the shallows: standing on dry land, they watched the men disembark in small parties, galloped down, attacked them as they struggled through the surf, and surrounded them with superior numbers while others opened fire on the exposed flank of isolated units. I therefore had the warships' boats and scouting vessels filled with troops, so that help could be sent to any point where the men seemed to be in difficulties. When everyone was ashore and formed up, the legions charged: the enemy was hurled back, but pursuit for any distance was impossible as the cavalry transports had been unable to hold their course and make the island. That was the only thing that deprived us of a decisive victory.

The natives eventually recovered from their panic and sent a delegation to ask for peace, promising to surrender hostages and carry out my instructions. These envoys brought with them Commius, who, it will be remembered, had preceded us to Britain. When he had landed and was actually delivering my message in the character of an ambassador he had been arrested and thrown into prison. Now, after their defeat, the natives sent him back: in asking for peace they laid the blame for this outrage upon the common people and asked me to overlook the incident on the grounds of their ignorance. I protested against this unprovoked attack which they had launched after sending a mission to the Continent to negotiate a friendly settlement, but agreed to pardon their ignorance and demanded hostages. Some of these were handed over at once, others, they said, would have to be fetched from a distance and would be delivered in a few days. Meanwhile they were ordered to return to their occupations on the land, and chieftains began to arrive from the surrounding districts, commending themselves and their tribes to my protection. Peace was thus concluded.

Late on the fourth day after our landing in Britain the eighteen transports with cavalry on board had sailed from the northern port with a gentle breeze; but as they neared the British coast and were within sight of the camp a violent storm had blown up, and none of them could hold their course. Some had been driven back to the point of embarkation; others, in great peril, had been swept down channel, westwards, towards the southernmost part of the island. Notwithstanding the danger, they had dropped anchor, but now shipped so much water that they were obliged to stand out to sea as darkness fell and return to the Continent.

It happened to be full moon that night; and at such times the Atlantic tides are particularly high, a fact of which we were ignorant. The result was that the warships, which had been beached, became waterlogged: as for the transports riding at anchor, they were dashed one against another, and it was impossible to manoeuvre them or to do anything whatever to assist. Several ships broke up, and the remainder lost their cables, anchors and rigging. Consternation naturally seized the troops, for there were no spare ships in which they could return and no means of refitting. It had been generally understood, too, that we should winter in Gaul, and consequently no arrangements had been made for winter food supplies in Britain.

The British chieftains at my headquarters sized up the situation and put their heads together. They knew we had no cavalry and were short of grain and shipping; they judged the weakness of our forces from the inconsiderable area of the camp, which was all the smaller because we had brought no heavy equipment; and they decided to renew the offensive. Their aim was to cut us off from food supplies and other material and to prolong the campaign until winter. They were confident that if the present expeditionary force were wiped out or prevented from returning, an invasion of Britain would never again be attempted. Accordingly they renewed their vows of mutual loyalty, slipped away one by one from our camp, and secretly reassembled their forces from the countryside.

JOSEPHUS
The Horrors of the Siege

In 66 AD the Jews of the Roman province of Judaea (most of modern Israel) rose in revolt against the rule of the Emperor Nero. Revolt had been and would remain uncommon within the Roman world. That of Boudicca in Britain in 61 AD, and of the Jews in 66, were exceptional; most of the peoples subjected to Roman rule after conquest accepted its benefits. The Jews, however, were an exceptional people, the empire's only mono-theists. The early emperors refrained from interfering in their religious practices. With the rise of emperor worship, however, a conflict between their demands to be venerated as gods and Jewish insistence on the uniqueness of the God of the Bible threatened. The command of the Emperor Gaius (37–41 AD) that a statue of himself be erected in the Temple at Jerusalem, a demand that was the ultimate sacrilege to the Jews, was averted only by the Emperor's murder by his enemies in Rome. Religious differences rumbled on in subsequent reigns. In 66 AD the Emperor's representative in Judaea, the procurator Gessius Florus, confiscated part of the Temple treasure and put down the Jews' protests with force. Rebellion quickly spread throughout the whole province. It was effectively repressed, until the only centre of resistance remained the walled city of Jerusalem.

The siege laid to it in 69 AD is described by Flavius Josephus, a Jewish official of the Roman administration who had abandoned his own people and later rose to wealth and high political rank within the Roman imperial system. His account of the siege is one of the best surviving descriptions of the attack and defence of fortified places in the ancient world. It contains anomalies, particularly vast exaggerations of the numbers of casualties, a common failing of ancient historians. It is also partisan, lauding his own part in the operations, though he is honest enough to reveal that the Jews regarded him as a traitor, particularly for the atrocities he had committed at Jotapata, earlier in the war. He may also exaggerate the divisions within the city, and the extent of the quarrels between Simon, leader of the resistance, and his followers. The story of dissent, common in sieges, is nevertheless credible. So, too, is his account of conflict over food, secretion of wealth and robbery of the weak by the strong, who, under the abnormality of siege conditions, turned to crime, in a short-sighted attempt to profit from their isolation from the outside world.

Many of the places mentioned by Josephus – the pool of Siloam, the brook of Kidron, the Temple Mount – are still identifiable at Jerusalem. The city eventually fell to Titus, the Roman commander, son of the new Emperor Vespasian, in September 70. The Jews outside the city, which was laid waste in the aftermath of the siege, continued their resistance. The last rebels, who occupied the mountain stronghold of Masada, were overcome in 73.

*   *   *

The Romans had begun work on the 12th May (69 A D), but they only completed the platforms on the 29th, after seventeen days of continuous toil; for all four were of vast size. One, facing Antonia, was raised by the Fifth Legion opposite the middle of the Quince Pool; another, built by the Twelfth, was thirty feet away. A long way from these, to the north of the City, was the work of the Tenth, near the Almond Pool; forty-five feet from this the Fifteenth built theirs by the High Priest's Monument. But from within the City John tunnelled through the ground near Antonia, supporting the galleries with wooden props, and by the time the engines were brought up he had reached the platforms and left the works without solid support. Next he carried in faggots daubed with pitch and bitumen and set them alight, so that as soon as the props were burnt away the entire tunnel collapsed, and with a thunderous crash the platforms fell into the cavity. At once there arose a dense cloud of smoke and dust as the flames were choked by the debris; then when the mass of timber was burnt away a brilliant flame broke through. This sudden blow filled the Romans with consternation; the ingenuity of the Jews plunged them into despondency; as they had felt sure that victory was imminent the shock froze their hope of success even in the future. To fight the flames seemed useless, for even if they did put them out their platforms were already swallowed up.

Two days later Simon's forces assaulted the other two platforms; for the Romans had brought up their Batterers on this side and were already rocking the wall. Tephthaeus, who came from Garis in Galilee, and Megassarus, a servant of Queen Mariamne, accompanied by a man from Adiabene (the son of Nabataeus) nicknamed because of a disability Ceagiras ('Cripple'), picked up firebrands and rushed at the engines. In the whole course of the war the City produced no one more heroic than these three, or more terrifying. They dashed out as if towards friends, not massed enemies; they neither hesitated nor shrank back, but charged through the centre of the foe and set the engines on fire. Pelted with missiles and thrust at with swords on every side, they refused to withdraw from their perilous situation till the weapons were ablaze. When the flames were already shooting up the Romans came running from the camps to the rescue. But the Jews advanced from the wall to stop them, grappling with those who attempted to quench the flames and utterly disregarding their personal danger. The Romans tugged at the Batterers while the wicker covers blazed: the Jews, surrounded with the flames, pulled the other way, and seizing the red-hot iron would not leave go of the Rams. From the engines the fire spread to the platforms, outstripping the defenders. Meantime the Romans were enveloped in flames, and despairing of saving their handiwork began to withdraw to their camps. The Jews pressed them hard, their numbers constantly swelled by reinforcements from the City, and emboldened by their success attacked with the utmost violence till they actually reached the Roman fortifications and engaged the defenders.

There is an armed picquet, periodically relieved, which occupies a position in front of every Roman camp and is subject to the very drastic regulation that a man who retires, no matter what the circumstances, must be executed. These men, preferring death with honour to death as a penalty, stood their ground, and their desperate plight shamed many of the runaways into making a stand. Quick-loaders were set up on the wall to drive off the mass of men that poured out of the City without the slightest thought for their own safety. These grappled with all who stood in their path, falling recklessly upon the Roman spears and flinging their very bodies against the foe. It was less by actions than by supreme confidence that they gained the advantage, and it was Jewish audacity rather than their own casualties that made the Romans give ground.

At this crisis Titus arrived from Antonia, to which he had withdrawn to choose a site for more platforms. He expressed the utmost contempt for the soldiers, who after capturing the enemy's walls were in danger of losing their own, and were enduring a siege themselves through letting the Jews out of prison to attack them! Then he put himself at the head of a body of picked men and tried to turn the flank of the enemy, who although assailed from the front wheeled round to meet this new threat and resisted stubbornly. In the confusion that followed, blinded by the dust and deafened by the uproar, neither side could distinguish friend from foe. The Jews stood firm, not so much through prowess now as through despair of victory; the Romans were braced by respect for the honour of their arms, especially as Caesar [that is, Titus] was in the forefront of danger. The struggle would probably have ended, such was the fury of the Romans, with the capture of the whole mass of Jews, had they not forestalled the crisis of the battle by retreating to the City. With their platforms destroyed the Romans were downhearted, having lost the fruits of their prolonged labours in a single hour; many indeed felt that with conventional weapons they would never take the City.

Titus held a council of war. The more sanguine spirits were for bringing the whole army into action in a full-scale assault. Hitherto only a fraction of their forces had been engaged with the enemy; if they advanced en masse the Jews would yield to the first onslaught, overwhelmed by the rain of missiles. The more cautious urged either that the platforms should be reconstructed, or that abandoning these they should merely blockade the City and prevent the inhabitants from making sallies or bringing in food, leaving them to starve and refraining from combat. For there was no battling with despair, when men desired only to die by the sword and so escape a more horrible fate. Titus himself thought it unwise to let so large a force remain idle, while there was no point in fighting those who were certain to destroy each other. To throw up platforms was a hopeless task with timber so scarce, to prevent sallies still more hopeless. For to form a ring of men round so big a City and over such difficult terrain was impracticable, and highly dangerous in view of sudden attacks. The known paths might be blocked, but the Jews would contrive secret ways out, driven by necessity and knowing the ground. Again, if provisions were smuggled in the siege would be prolonged still further, and he was afraid the lustre of his triumph would be dimmed by its slowness in coming. Given time anything could be accomplished, but reputations were won by speed. If he was to combine speed with safety he must build a wall round the entire City. That was the only way to block all egress and force the Jews to abandon their last hope of survival and surrender the City. If they did not, hunger would make them easy victims. For he would not wait for things to happen, but would resume construction of the platforms when resistance had been weakened. If anyone thought the task too great to carry out, he must remember that little tasks were beneath the dignity of Rome, and that without hard work nothing great could be achieved, unless by a miracle.

Having thus convinced the generals Titus ordered them to divide up the work between their units. An inspired enthusiasm seized the soldiers, and when the circuit had been marked out there was competition not only between legions but even between cohorts. The private was eager to please his decurion, the decurion his centurion, the centurion his tribune; the tribunes were ambitious for the praise of the generals; and of the rivalry between the generals Caesar himself was judge. He personally went round several times every day to inspect the work. Starting at the Assyrians' Camp, where his own quarters were, he took the wall to the New City below, and from there through the Kidron to the Mount of Olives. Then he bent the line towards the south, enclosing the Mount (as far as the rock called The Dovecot) and the next eminence, which overhangs the valley near Siloam. From there he went in a westerly direction down into Fountain Valley, then up by the tomb of Ananus the high priest, embracing the hill where Pompey's camp had been. Then turning north he passed the village called The House of Peas, and rounding Herod's Tomb went east till he finished up at his own camp, the starting-place. The wall measured 4½ miles, and outside were built on thirteen forts with a combined circumference of over a mile. Yet the whole task was completed in three days, though it might well have taken months – the speed passed belief. Having surrounded the City with this wall and garrisoned the forts, Titus himself took the first night watch and went the rounds; the second he entrusted to Alexander; the third was assigned to the legion commanders. The guards drew lots for periods of sleep, and all night long they patrolled the intervals between the forts.

The Jews, unable now to leave the City, were deprived of all hope of survival. The famine became more intense and devoured whole houses and families. The roofs were covered with women and babes, the streets full of old men already dead. Young men and boys, swollen with hunger, haunted the squares like ghosts and fell wherever faintness overcame them. To bury their kinsfolk was beyond the strength of the sick, and those who were fit shirked the task because of the number of the dead and uncertainty of their own fate; for many while burying others fell dead themselves, and many set out for their graves before their hour struck. In their misery no weeping or lamentation was heard; hunger stifled emotion; with dry eyes and grinning mouths those who were slow to die watched those whose end came sooner. Deep silence enfolded the City, and a darkness burdened with death. Worse still were the bandits, who broke like tomb-robbers into the houses of the dead and stripped the bodies, snatching off their wrappings, then came out laughing. They tried the points of their swords on the corpses, and even transfixed some of those who lay helpless but still alive, to test the steel. But if any begged for a sword-thrust to end their sufferings, they contemptuously left them to die of hunger. Everyone as he breathed his last fixed his eyes on the Temple, turning his back on the partisans he was leaving alive. The latter at first ordered the dead to be buried at public expense as they could not bear the stench; later, when this proved impossible, they threw them from the walls into the valleys. When in the course of his rounds Titus saw these choked with dead, and a putrid stream trickling from under the decomposing bodies, he groaned, and uplifting his hands called God to witness that this was not his doing.

While such were the conditions in the City the Romans were exuberant, for none of the partisans sallied out now that they too were despondent and hungry. There was an abundance of corn and other necessaries from Syria and the neighbouring provinces, and the soldiers delighted to stand near the wall and display their ample supplies of food, by their own abundance inflaming the hunger of the enemy. But when suffering made the partisans no more ready to submit, Titus took pity on the remnant of the people, and in his anxiety to rescue the survivors again began constructing platforms, though it was difficult to get timber. Round the City it had all been cut down for the previous works, and the soldiers had to collect new supplies from more than ten miles away. Concentrating on Antonia, from four directions they raised platforms much bigger than the earlier ones. Caesar made the round of the legions, speeding the work and showing the bandits they were in his hands. But they alone seemed to have lost all sense of remorse, and making a division between soul and body acted as if neither belonged to them. For their souls were as insensitive to suffering as their bodies to pain — they tore the carcase of the nation with their fangs, and filled the prisons with the defenceless.

Simon actually put Matthias, who had made him master of the City, to death by torture. The son of Boethus and the scion of the chief priests, he enjoyed the absolute trust and respect of the people. When the masses were being roughly handled by Zealots whom John had already joined, he persuaded the people to accept Simon's aid, having made no pact with him but expecting no mischief from him. When, however, Simon arrived and got the City into his power, he treated Matthias as an enemy like the rest, and the furtherer of his cause as a mere simpleton. Matthias was brought before him and accused of favouring the Romans, and without being allowed to defend himself, was condemned to die with three of his sons. The fourth had already made his escape to Titus. When Matthias begged to be killed before his children, pleading for this as a favour because he had opened the gates to Simon, the monster ordered him to be killed last. So his sons were murdered before his eyes and then his dead body was thrown on to theirs, in full view of the Romans. Such were the instructions that Simon had given to Ananus, son of Bagadates, the most brutal of his henchmen; and he mockingly enquired whether Matthias hoped for assistance from his new friends. Burial of the bodies was forbidden. After their deaths an eminent priest named Ananias, son of Masbalus, and the clerk of the Sanhedrin [the Supreme Council of Jerusalem], Aristaeus, whose home was Emmaus, together with fifteen distinguished citizens, were put to death. Josephus's father was kept under lock and key, and an edict forbade anyone in the City to associate with him through fear of betrayal. Any who condoled with him were executed without trial.

Seeing all this Judas, son of Judas, a subordinate whom Simon had entrusted with command of a tower, partly through disgust at these brutal murders but chiefly with an eye to his own safety, collected the ten most reliable of his men. 'How long,' he asked, 'shall we endure these horrors? What hope of survival have we if we remain loyal to a scoundrel? We are starving already and the Romans have almost got in. Simon is betraying his best friends and is likely soon to jump on us; but the word of the Romans can be trusted. So come on! Let us surrender the wall and save ourselves and the City! Simon has lost hope already! It won't hurt him if he gets his deserts a bit sooner.' This argument convinced the ten, and at stand-to he sent off the rest in different directions to avoid discovery. Three hours later he shouted from his tower to the Romans, but some of them were scornful, others mistrustful, the majority uninterested: in any case the City would soon fall into their lap. Titus advanced with his heavy infantry towards the wall, but Simon stole a march on him, occupied the tower first, arrested the men, executed them before the eyes of the Romans, and threw their mutilated bodies over the wall.

At this time, as he went round making yet another appeal, Josephus was struck on the head by a stone and fell to the ground unconscious. Seeing him fall the Jews ran out, and would have dragged him into the City had not Caesar promptly sent men to protect him. While they fought Josephus was picked up, knowing little of what was going on. The partisans thought they had disposed of the man they hated most, and whooped for joy. When the report spread through the City the survivors of the populace were overcome with despair, believing that they had really lost the man with whose help they hoped to desert. When Josephus's mother was told in prison that her son was dead, she said to the guards that she had foreseen this ever since Jotapata; while he was alive she might as well have had no son. Privately she lamented to her maids that this was the only result of bringing children into the world – she would not even bury the son whom she had expected to bury her. But the false report neither grieved her nor cheered the bandits for long. Josephus soon recovered from the blow, and went forward to shout that it would not be long before they paid the penalty for wounding him, and to implore the people to trust him. His reappearance brought new hope to the common folk, to the partisans' consternation.

Some of the deserters, seeing no other way, promptly jumped from the wall. Others advanced as if to battle armed with stones, then fled to the Romans. Their fate was worse than if they had stayed in the City, and the hunger they had left behind was, as they discovered, less lethal than the plenty the Romans provided. They arrived blown up by starvation as if by dropsy, then stuffed their empty bellies non-stop till they burst – except for those who were wise enough to restrain their appetites and take the unaccustomed food a little at a time. Those who escaped this danger fell victims to another disaster. In the Syrian camp one deserter was caught picking gold coins out of his excreta. As I mentioned, they swallowed coins before leaving, because they were all searched by the partisans, and there was a great deal of gold in the City. In fact it fetched less than half the old price. But when the trick was discovered through one man, the rumour ran round the camps that the deserters were arriving stuffed with gold. The Arab unit and the Syrians cut open the refugees and ransacked their bellies. To me this seems the most terrible calamity that happened to the Jews: in a single night nearly two thousand were ripped up.

When Titus learnt of this atrocity he was on the point of surrounding the perpetrators with his cavalry and shooting them down. But far too many were involved; in fact those to be punished far outnumbered their victims. Instead he summoned both the auxiliary and the legionary commanders, some of whose men were accused of participating, and spoke angrily to both groups. Was it possible that some of his own soldiers did such things on the off chance of gain, and had no respect for their own weapons that were made of silver and gold? The Arabs and Syrians, serving in a war that was not the concern of fashion, ended by letting Romans take the blame for their blood-thirsty butchery and their hatred for the Jews; for some of his own soldiers shared their evil reputation. The foreigners therefore he threatened to punish with death if any man was caught after this committing such a crime. The legionary commanders he instructed to ferret out suspected offenders and bring them before him.

But avarice, it seems, scorns every penalty and an extraordinary love of gain is innate in man, nor is any emotion as strong as covetousness. At other times these passions are kept within bounds and overawed by fear. But it was God who condemned the whole nation and turned every means of escape to their destruction. So what Caesar forbade with threats was still done to the deserters in secret, and the refugees, before the rest noticed them, were met and murdered by the foreign soldiers, who looked round in case any Roman saw them, then ripped them up and pulled the filthy money out of their bowels. In few, however, was any found, the majority being victims of an empty hope. Fear of this fate caused many of the deserters to return.

John, when there was nothing left that he could extort from the people, turned to sacrilege and melted down many of the offerings in the Temple and many of the vessels required for services, basins, dishes, and tables, not even keeping his hands off the flagons presented by Augustus and his consort. For the Roman emperors honoured and adorned this shrine at all times. But now this Jew stole even the gifts of foreigners, telling his companions that they need not hesitate to use God's property for God's benefit, and that those who fought for the Sanctuary were entitled to live on it. Accordingly he emptied out the sacred corn and oil which the priests kept in the Inner Temple to pour on burnt offerings, and shared them out to the crowd, who without a qualm swallowed a pailful or smeared it on themselves. I cannot refrain from saying what my feelings dictate. I think that if the Romans had delayed their attack on these sacrilegious ruffians, either the ground would have overwhelmed it, or lightning would have destroyed it like Sodom. For it produced a generation far more godless than those who perished thus, a generation whose mad folly involved the nation in ruin.

But why should I describe these calamities one by one? While they were happening Mannaeus, the son of Lazarus, fled to Titus and told him that through a single gate which had been entrusted to him 115,880 corpses had been carried out between the day the Romans pitched their camp near the City – April 14th – and the first of July. All these were the bodies of paupers. Though he was not himself in charge he had to pay the expenses out of public funds, so was obliged to keep count. The rest were buried by their own kin, who merely brought them out and threw them clear of the City. After Mannaeus many distinguished citizens deserted, and these reported that in all 600,000 pauper bodies had been thrown out at the gates: of the others the number was unknown. When it was no longer possible to carry out the penniless, they said, the corpses had been heaped up in the biggest houses and the doors locked. The price of corn was fantastically high, and now the City was walled round and they could not even gather herbs, some were in such dire straits that they raked the sewers and old dunghills and swallowed the refuse they found there, so that what once they could not bear to look at now became their food.

When the Romans heard of all this misery they felt pity: the partisans, who saw it with their own eyes, showed no regrets but allowed these things to come upon them too; for they were blinded by the doom that was closing in on the City and on themselves.

USĀMAH IBN-MUNQIDH
An Arab-Syrian Gentleman

The preaching of a new monotheistic religion by the Arab Prophet Muhammad (died 632) initiated the most spectacular campaign of conquest the Western world had seen since that of Alexander the Great in the fourth century BC. Between 633 and 732, the Arabs, professing the faith of Islam that Muhammad had founded, created an empire that stretched from Afghanistan in the east to the Pyrenees in the west. A consequence of the Arab victories was that Christian power was extinguished not only in its historic centre in the Holy Land, but also in Egypt, Syria, North Africa and Spain, while the Christian empire of Byzantium, the surviving element of the Roman Empire with its capital at Constantinople, was brought under continuous pressure in Asia Minor. In 1071 the Seljuk Turks, a Central Asian nomadic horse people who had recently converted to Islam, defeated the Byzantine army at Manzikert, on Asia Minor's border with the Caucasus mountains, and overran Asia Minor, threatening Constantinople and Christian Southern Europe.

The Byzantine Emperor Alexius I, secular leader of Eastern Christianity, was driven by the military crisis resulting from Manzikert ('the Dreadful Day') to appeal to Pope Urban II for Western help in 1095. Despite the opening of what would prove a definitive breach between the Latin Christianity of the West (Roman Catholicism) and the Greek Christianity of the East (Orthodoxy) in the Great Schism of 1054, Urban decided to respond to Alexius's appeal. At the Synod of Clermont he therefore preached a call to the Christian knights of the West to mount against Muslim power a military campaign that would become known as the First Crusade.

The Crusade departed in 1096, reached Constantinople but then, falling out with the Emperor Alexius, declined to join him in the recapture of Asia Minor and pressed on to seize the Holy Land and Jerusalem, which had become the Crusade's popular target. The Holy City fell to the Crusaders on 18 July 1099. In the aftermath of the conquest, the leaders of the Crusade set themselves up as rulers in the Holy Land and its adjoining territories, establishing the Kingdom of Jerusalem, the Counties of Tripoli and Edessa and the Principality of Antioch.

These 'Latin Kingdoms' came under immediate counter-attack from the neighbouring regions of the Islamic Caliphate, with seats at Cairo and Baghdad, and from the Muslim Turks. Islam was a warrior society, whose military class practised a style of warfare which, in its emphasis on courage, skill at arms, and horsemanship, closely resembled the chivalry of the Western knight. Usāmah Ibn-Munqidh was a representative of the Muslim knightly class. His description of the attack on the Crusader fortress of Kafartāb – Kafr Tab, south-west of Aleppo in Syria – fascinates both because of its depiction of the Muslim idea of military honour, and also because the details of siege warfare revealed exactly follow those of the Jews in the defence of Jerusalem more than a thousand years earlier. Mining (and counter-mining) with the object of burning away timber props supporting the excavation, was originally an Assyrian invention of the first millennium BC. It would nevertheless continue in use until gunpowder became available to siege engineers in the sixteenth century AD.

* * *

Another illustration I witnessed in the year 509 [Islamic calendar; 1131 AD]. My father (may Allah's mercy rest upon his soul!) had set out at the head of the army to join the Isbāslār Bursuq ibn-Bursuq (may Allah's mercy rest upon his soul!), who had arrived on an expedition ordered by the Sultan. Bursuq commanded a huge army including a large number of amirs, among whom were the Amir-al-Juyush [commander of the armies] Uzbech the lord of al-Mawsil, Sunqur Dirāz the lord of al-Rahabah, the Amir Kundughadi, al-Hājib al-Kabir [Grand Chamberlain] Baktimur, Zanki ibn-Bursuq (who was a veritable hero), Tamirak, Ismāʻil al-Bakji, and others. They camped before Kafartāb, in which were the two brothers of Theophile at the head of the Franks [Crusaders], and attacked it. The troops from Khurāsān entered the trench and began to dig an underground tunnel. Convinced that they were on the point of perdition, the Franks set the castle on fire. The roofs were burned and fell upon the horses, beasts of burden, sheep, pigs and captives – all of whom were burned up. The Franks remained clinging to the walls at the top of the castle.

It occurred to me to enter the underground tunnel and inspect it. So I went down in the trench, while the arrows and stones were falling on us like rain, and entered the tunnel. There I was struck with the great wisdom with which the digging was executed. The tunnel was dug from the trench to the barbican. On the sides of the tunnel were set up two pillars, across which stretched a plank to prevent the earth above it from falling down. The whole tunnel had such a framework of wood that extended as far as the foundation of the tower. The tunnel was narrow. It was nothing but a means to provide access to the tower. As soon as they got to the tower, they enlarged the tunnel in the wall of the tower, supported it on timbers and began to carry out, a little at a time, the splinters of stone produced by boring. The floor of the tunnel, on account of the dust caused by the digging, was converted into mud. Having made the inspection, I went on without the troops of Khurāsān recognizing me. Had they recognized me, they would not have let me off without the payment of a heavy tribute.

They then began to cut dry wood and stuff the tunnel with it. Early the next morning they set it on fire. We had just at that time put on our arms and marched, under a great shower of stones and arrows, to the trench in order to make an onslaught on the castle as soon as its tower tumbled over. As soon as the fire began to have its effect, the layers of mortar between the stones of the wall began to fall. Then a crack was made. The crack became wider and wider and the tower fell. We had assumed that when the tower would fall we should be able to enter as far as our enemy. But only the outer face of the wall fell, while the inner wall remained intact. We stood there until the sun became too hot for us, and then returned to our tents after a great deal of damage had been inflicted on us by the stones which were hurled against us.

After resting until noontime, there set out all of a sudden a footman from our ranks, singlehanded and carrying his sword and shield. He marched to the wall of the tower which had fallen, and the sides of which had become like the steps of a ladder, and climbed on it until he got as far as its highest point. As soon as the other men of the army saw him, about ten of them followed him hastily in full armour and climbed one after the other until they got to the tower, while the Franks were not conscious of their movements. We in turn put on our armour in our tents and advanced. Many climbed the tower before all our army had wholly arrived.

The Franks now turned upon our men and shot their arrows at them. They wounded the man who was first to climb. So he descended. But the other men continued to climb in succession until they stood facing the Franks on one of the tower walls between two bastions. Right in front of them stood a tower the door of which was guarded by a cavalier in full armour carrying his shield and lance, preventing entrance to the tower. On top of that tower were a band of Franks, attacking our men with arrows and stones. One of the Turks climbed, under our very eyes, and started walking towards the tower, in the face of death, until he approached the tower and hurled a bottle of naphtha on those who were on top of it. The naphtha flashed like a meteor falling upon those hard stones, while the men who were there threw themselves on the ground for fear of being burnt. The Turk then came back to us.

Another Turk now climbed and started walking on the same wall between the two bastions. He was carrying his sword and shield. There came out to meet him from the tower, at the door of which stood a knight, a Frank wearing double-linked mail and carrying a spear in his hand, but not equipped with a shield. The Turk, sword in hand, encountered him. The Frank smote him with the spear, but the Turk warded off the point of the spear with his shield and, notwithstanding the spear, advanced towards the Frank. The latter took to flight and turned his back, leaning forward, like one who wanted to kneel, in order to protect his head. The Turk dealt him a number of blows which had no effect whatsoever, and he went on walking until he entered the tower.

Our men proved too numerous and too strong for the enemy. So the latter delivered the castle, and the captives came down to the tents of Bursuq ibn-Bursuq.

Among those who were assembled in the large tent of Bursuq ibn-Bursuq in order to set for themselves a price for their liberty, I recognized that same man who had set out with his spear against the Turk. He, who was a sergeant, stood up and said, 'How much do ye want from me?' They said, 'We demand six hundred dinārs.' He pooh-poohed them, saying, 'I am a sergeant. My stipend is two dinārs a month. Wherefrom can I get you six hundred dinārs?' And saying this, he went back and sat among his companions. And he was huge in size. Seeing him, the Amir al-Sayyid al-Sharif, who was one of the leading amirs, said to my father (may Allah's mercy rest upon his soul!), 'O my brother, seest thou what manner of people these are? In Allah we seek refuge against them.'

By the decree of Allah (worthy of admiration is He!) our army departed from Kafartāb to Dānith and were surprised to meet early Tuesday morning, the twenty-third of Rabi' II [the fourth month of the Muslim calendar, corresponding roughly to April], the army of Antioch. The capitulation of Kafartāb took place on Friday, the twelfth of Rabi' II. The Amir al-Sayyid (may Allah's mercy rest upon his soul!) was killed, together with a large body of Muslims.

My father (may Allah's mercy rest upon his soul!) with whom I had parted at Kafartāb returned [to Kafartāb] after the army had been defeated. We were still at Kafartāb guarding it with the intention of rebuilding it; for the isbāslār [castellan] had delivered it into our hands. We were bringing out the captives, each two chained to one man from Shayzar. Some of them had half of their bodies burned and their legs remained. Others were dead by fire. I saw in what befell them a great object lesson. We then left Kafartāb and returned to Shayzar in the company of my father (may Allah's mercy rest upon his soul!), who had lost all the tents, loads, mules, camels and baggage he had, and whose army was dispersed.

JEAN FROISSART
The Battle of Crécy, 26 August 1346

The Battle of Crécy, with that of Agincourt (1415), was one of the two crucial English victories over France in the Hundred Years War (1337–1453). The kingdoms of France and England were bound together in a complex relationship. The Kings of England, as Dukes of Aquitaine in south-western France, were the sovereign French king's feudal vassals. As sovereigns of England, a strong, centralized kingdom, which France was not, they had ambitions, however, to consolidate their French possessions and challenge their French overlords for mastery in France. In 1338 Edward III of England actually proclaimed himself King of France, a title he and his successors were to sustain for over a century. He found allies in Flanders and Brittany and began a war in mainland France in 1337. Philip VI of France responded by invading Aquitaine (Gascony) in 1346. To distract the French army from the south-west, Edward made a landing in Normandy, near Cherbourg, in July of that year and marched on Paris. Finding themselves increasingly isolated in hostile territory, the English then struck out across the river Seine towards the Channel coast, in the hope of finding safety in Flanders or of recrossing the Channel. Cornered by a larger and pursuing French army, which had its base in Paris, they eventually took up a position at Crécy, near Abbeville on the river Somme (near where a British army would fight a major battle in defence of France against the Germans in 1916, during the First World War). The French army contained a large contingent of mercenary Genoese crossbowmen, masters of one of the most effective weapons of late medieval warfare; the crossbow was to be challenged and bested by the longbow of the English and Welsh archers of Edward's army in the battle that followed.

*   *   *

There is no man, unless he had been present, that can imagine or describe truly the confusion of that day, especially the bad management and disorder of the French, whose troops were out of number. What I know, and shall relate in this book, I have learned chiefly from the English, and from those attached to Sir John of Hainault, who was always near the person of the King of France. The English, who, as I have said, were drawn up in three divisions, and seated on the ground, on seeing their enemies advance, rose up undauntedly and fell into their ranks. The Prince's [Edward, Prince of Wales, known as 'the Black Prince', Edward III's eldest son] battalion, whose archers were formed in the manner of a portcullis, and the men-at-arms in the rear, was the first to do so. The Earls of

Northampton and Arundel, who commanded the second division, posted themselves in good order on the Prince's wing to assist him if necessary.

You must know that the French troops did not advance in any regular order, and that as soon as their King came in sight of the English his blood began to boil, and he cried out to his marshals, 'Order the Genoese forward and begin the battle in the name of God and St Denis.' There were about 15,000 Genoese crossbowmen; but they were quite fatigued, having marched on foot that day six leagues, completely armed and carrying their crossbows, and accordingly they told the Constable they were not in a condition to do any great thing in battle. The Earl of Alençon hearing this, said, 'This is what one gets by employing such scoundrels, who fall off when there is any need for them.' During this time a heavy rain fell, accompanied by thunder and a very terrible eclipse of the sun; and, before this rain, a great flight of crows hovered in the air over all the battalions, making a loud noise; shortly afterwards it cleared up, and the sun shone very bright; but the French had it in their faces, and the English on their backs. When the Genoese were somewhat in order they approached the English and set up a loud shout, in order to frighten them; but the English remained quite quiet and did not seem to attend to it. They then set up a second shout, and advanced a little forward; the English never moved. Still they hooted a third time, advancing with their crossbows presented, and began to shoot. The English archers then advanced one step forward, and shot their arrows with such force and quickness, that it seemed as if it snowed. When the Genoese felt these arrows, which pierced through their armour, some of them cut the strings of their cross-bows, others flung them to the ground, and all turned about and retreated quite discomfited.

The French had a large body of men-at-arms on horseback to support the Genoese, and the King, seeing them thus fall back, cried out, 'Kill me those scoundrels, for they stop up our road without any reason.' The English continued shooting, and some of their arrows falling among the horsemen, drove them upon the Genoese, so that they were in such confusion, they could never rally again.

In the English army there were some Cornish and Welsh men on foot, who had armed themselves with large knives, these advancing through the ranks of the men-at-arms and archers, who made way for them, came upon the French when they were in this danger, and falling upon earls, barons, knights, and squires, slew many, at which the King of England was exasperated. The valiant King of Bohemia was slain there; he was called Charles of Luxembourg, for he was the son of the gallant king and emperor, Henry of Luxembourg, and, having heard the order for the battle, he inquired where his son the Lord Charles was; his attendants answered that they did not know, but believed he was fighting. Upon this, he said to them, 'Gentlemen, you are all my people, my friends, and brethren at arms this day; therefore, as I am blind, I request of you to lead me so far into the engagement that I may strike one stroke with my sword.' The knights consented, and in order that they might not lose him in the crowd, fastened all the reins of their horses together, placing the King at their head that he might gratify his wish, and in this manner advanced towards the enemy. The Lord Charles of Bohemia, who already signed his name as King of Germany, and bore the arms, had come in good order to the engagement; but when he perceived that it was likely to turn out against the French he departed. The King, his father, rode in among the enemy, and he and his companions fought most valiantly; however, they advanced so far that they were all slain, and on the morrow they were found on the ground with all their horses tied together.

The Earl of Alençon advanced in regular order upon the English, to fight with them, as did the Earl of Flanders in another part. These two lords with their detachments, coasting, as it were, the archers, came to the Prince's battalion, where they fought valiantly for a length of time. The King of France was eager to march to the place where he saw their banners displayed, but there was a hedge of archers before him: he had that day made a present of a handsome black horse to Sir John of Hainault, who had mounted on it a knight of his, called Sir John de Fusselles, who bore his banner; the horse ran off with the knight and forced his way through the English army, and when about to return, stumbled and fell into a ditch and severely wounded him; he did not, however, experience any other inconvenience than from his horse, for the English did not quit their ranks that day to make prisoners: his page alighted and raised him up, but the French knight did not return the way he came, as he would have found it difficult from the crowd. This battle, which was fought on Saturday, between La Broyes and Crécy, was murderous and cruel; and many gallant deeds of arms were performed that were never known: towards evening, many knights and squires of the French had lost their masters, and wandering up and down the plain, attacked the English in small parties; but they were soon destroyed, for the English had determined that day to give no quarter, nor hear of ransom from anyone.

Early in the day some French, Germans, and Savoyards had broken through the archers of the Prince's battalion, and had engaged with the men-at-arms; upon this the second battalion came to his aid, and it was time they did so, for otherwise he would have been hard pressed. The first division, seeing the danger they were in, sent a knight off in great haste to the King of England, who was posted upon an eminence near a windmill. On the knight's arrival he said, 'Sir, the Earl of Warwick, the Lord Stafford, the Lord Reginald Cobham, and the others who are about your son, are vigorously attacked by the French, and they entreat that you will come to their assistance with your battalion, for if numbers should increase against him, they fear he will have too much to do.' The King replied, 'Is my son dead, unhorsed, or so badly wounded that he cannot support himself?' 'Nothing of the sort, thank God,' rejoined the knight, 'but he is in so hot an engagement that he has great need of your help.' The King answered, 'Now, Sir Thomas, return to those that sent you, and tell them from me not to send again for me this day, nor expect that I shall come, let what will happen, as long as my son has life; and say that I command them to let the boy win his spurs, for I am determined, if it please God, that all the glory of this day shall be given to him, and to those into whose care I have entrusted him.' The knight returned to his lords and related the King's answer, which mightily encouraged them, and made them repent they had ever sent such a message.

It is a certain fact that Sir Godfrey de Harcourt, who was in the Prince's battalion, having been told by some of the English that they had seen the banner of his brother engaged in the battle against him, was exceedingly anxious to save him; but he was too late, for he was left dead on the field, and so was the Earl of Aumarle, his nephew. On the other hand, the Earls of Alençon and Flanders were fighting lustily under their banners with their own people; but they could not resist the force of the English, and were there slain, as well as many other knights and squires, who were attending on, or accompanying, them.

The Earl of Blois, nephew to the King of France, and the Duke of Lorraine, his brother-in-law, with their troops, made a gallant defence; but they were surrounded by a troop of English and Welsh, and slain in spite of their prowess. The Earl of St Pol, and the Earl of Auxerre, were also killed, as well as many others. Late after vespers, the King of France had not more about him than sixty men, every one included. Sir John of Hainault, who was of the number, had once remounted the King, for his horse had been killed under him by an arrow: and seeing the state he was in, he said, 'Sir, retreat whilst you have an opportunity, and do not expose yourself so simply; if you have lost this battle, another time you will be the conqueror.' After he had said this he took the bridle of the King's horse and led him off by force, for he had before entreated him to retire. The King rode on until he came to the castle of La Broyes, where he found the gates shut, for it was very dark: he ordered the Governor of it to be summoned, who, after some delay, came upon the battlements, and asked who it was that called at such an hour. The King answered, 'Open, open, Governor, it is the fortune of France.' The Governor hearing the King's voice immediately descended, opened the gate, and let down the bridge; the King and his company entered the castle, but he had with him only five barons: Sir John of Hainault, the Lord Charles of Montmorency, the Lord of Beaujeu, the Lord of Aubigny, and the Lord of Montfort. It was not his intention, however, to bury himself in such a place as this, but having taken some refreshments, he set out again with his attendants about midnight, and rode on under the direction of guides, who were well acquainted with the country, until about daybreak he came to Amiens, where he halted. This Saturday the English never quitted their ranks in pursuit of anyone, but remained on the field guarding their position and defending themselves against all who attacked them. The battle ended at the hour of vespers, when the King of England embraced his son and said to him, 'Sweet son, God give you perseverance: you are my son; for most loyally have you acquitted yourself; you are worthy to be a sovereign.' The Prince bowed very low, giving all honour to the King, his father. The English during the night made frequent thanksgivings to the Lord for the happy issue of the day; and with them there was no rioting, for the King had expressly forbidden all riot or noise.

On the following day, which was Sunday, there were a few encounters with the French troops; however, they could not withstand the English, and soon either retreated or were put to the sword. When Edward was assured that there was no appearance of the French collecting another army, he sent to have the number and rank of the dead examined. This business was entrusted to Lord Reginald Cobham and Lord Stafford, assisted by three heralds to examine the arms, and two secretaries to write down the names. They passed the whole day upon the field of battle, and made a very circumstantial account of all they saw: according to their report it appeared that 80 banners, the bodies of 11 princes, 1,200 knights, and about 30,000 common men were found dead on the field.

JEHAN DE WAVRIN
A French Knight's Account of Agincourt

This vivid account of one of the most famous battles of the late Middle Ages brings the story of the Hundred Years War seventy years onward from Crécy. In the intervening period, the English had won another famous victory at Poitiers (1356), conducted a vicious guerrilla campaign in the interior which provoked a successful French counter-offensive, and eventually entered into a truce which it was agreed at the Treaty of Paris (1396) was to last for thirty years. Ensuing civil war in France, however, prompted the young and vigorous Henry V of England to intervene on the side of the faction challenging the authority of King Charles VI of France. Henry landed in Normandy in August 1415, but became involved in the siege of the port of Harfleur, which delayed his advance. Deciding after the capture of the city that the approach of winter precluded continuing the campaign, he set off on an overland march towards the port of Calais, from which he planned to return to England by the short sea route. Charles VI had, however, gathered an army to cut off his escape and on 24 October Henry found his path barred at the village of Agincourt, just short of his objective. The two armies spent a wet and miserable night in the open and next morning prepared for battle. The field was bordered on each side by thick belts of woodland. In the gap between, the English archers hammered a barricade of pointed stakes into the ground, while the knights armed for combat. The French, who outnumbered the English by 30,000 to 9,000, were confident of victory and had spent much of the night drinking. The English, in sober mood, prayed for divine assistance in their desperate straits. Henry heard Mass three times while he awaited the clash of arms. Then he led his army forward, on foot. What followed was to prove one of the most extraordinary reversals of fortune in the history of hand-to-hand fighting with edged weapons.

*   *   *

Then on the morning of the next day, that is to say, Friday, St Crispin's Day, the 25th of October 1415, the Constable and all the other officers of the King of France, the Dukes of Orléans, Bourbon, Bar, and Alençon, the Counts of Eu, Richemont, Vendôme, Marle, Vaudemont, Blaumont, Salines, Grampré, Roussy, Dampmartin, and generally all the other nobles and warriors armed themselves and issued from their bivouac; and then it was ordered by the Constable and marshals of the King of France that three battalions should be formed . . .

When the battalions of the French were thus formed, it was grand to see them; and as far as one could judge by the eye, they were in number fully six times as many as the English. And when this was done the French sat down by companies around their banners, waiting the approach of the English, and making their peace with one another; and then were laid aside many old aversions conceived long ago; some kissed and embraced each other, which it was affecting to witness; so that all quarrels and discords which they had had in time past were changed to great and perfect love. And there were some who breakfasted on what they had. And these Frenchmen remained thus till nine or ten o'clock in the morning, feeling quite assured that, considering their great force, the English could not escape them; however, there were at least some of the wisest who greatly feared a fight with them in open battle. Among the arrangements made on the part of the French, as I have since heard related by eminent knights, it happened that, under the banner of the Lord of Croy, eighteen gentlemen banded themselves together of their own choice, and swore that when the two parties should come to meet they would strive with all their might to get so near the King of England that they would beat down the crown from his head, or they would die, as they did; but before this they got so near the said King that one of them with the lance which he held struck him such a blow on his helmet that he knocked off one of the ornaments of his crown. But not long afterwards it only remained that the eighteen gentlemen were all dead and cut to pieces; which was a great pity; for if every one of the French had been willing thus to exert himself, it is to be believed that their affairs would have gone better on this day. And the leaders of these gentlemen were Louvelet de Massinguehem and Garnot de Bornouille . . .

The French had arranged their battalions between two small thickets, one lying close to Agincourt, and the other to Tramecourt. The place was narrow, and very advantageous for the English, and, on the contrary, very ruinous for the French, for the said French had been all night on horseback, and it rained, and the pages, grooms, and others, in leading about the horses, had broken up the ground, which was so soft that the horses could only with difficulty step out of the soil. And also the said French were so loaded with armour that they could not support themselves or move forward. In the first place they were armed with long coats of steel, reaching to the knees or lower, and very heavy, over the leg harness, and besides plate armour also most of them had hooded helmets; wherefore this weight of armour, with the softness of the wet ground, as has been said, kept them as if immovable, so that they could raise their clubs only with great difficulty, and with all these mischiefs there was this, that most of them were troubled with hunger and want of sleep. There was a marvellous number of banners, and it was ordered that some of them should be furled. Also it was settled among the said French that every one should shorten his lance, in order that they might be stiffer when it came to fighting at close quarters. They had archers and crossbowmen enough, but they would not let them shoot, for the plain was so narrow that there was no room except for the men-at-arms.

Now let us return to the English. After the parley between the two armies was finished and the delegates had returned, each to their own people, the King of England, who had appointed a knight called Sir Thomas Erpingham to place his archers in front in two wings, trusted entirely to him, and Sir Thomas, to do his part, exhorted every one to do well in the name of the King, begging them to fight vigorously against the French in order to secure and save their own lives. And thus the knight, who rode with two others only in front of the battalion, seeing that the hour was come, for all things were well arranged, threw up a baton which he held in his hand, saying 'Nestrocq' ['Now strike'], which was the signal for attack; then dismounted and joined the King, who was also on foot in the midst of his men, with his banner before him. Then the English, seeing this signal, began suddenly to march, uttering a very loud cry, which greatly surprised the French. And when the English saw that the French did not approach them, they marched dashingly towards them in very fine order, and again raised a loud cry as they stopped to take breath.

Then the English archers, who, as I have said, were in the wings, saw that they were near enough, and began to send their arrows on the French with great vigour. The said archers were for the most part in their doublets, without armour, their stockings rolled up to their knees, and having hatchets and battleaxes or great swords hanging at their girdles; some were barefooted and bareheaded, others had caps of boiled leather, and others of osier, covered with harpoy or leather.

Then the French, seeing the English come towards them in this fashion, placed themselves in order, everyone under his banner, their helmets on their heads. The Constable, the Marshal, the admirals, and the other princes earnestly exhorted their men to fight the English well and bravely; and when it came to the approach the trumpets and clarions resounded everywhere; but the French began to hold down their heads, especially those who had no bucklers, for the impetuosity of the English arrows, which fell so heavily that no one durst uncover or look up. Thus they went forward a little, then made a little retreat, but before they could come to close quarters, many of the French were disabled and wounded by the arrows; and when they came quite up to the English, they were, as has been said, so closely pressed one against another that none of them could lift their arms to strike their enemies, except some that were in front, and these fiercely pricked with the lances which they had shortened to be more stiff, and to get nearer their enemies.

The French had formed a plan which I will describe, that is to say, the Constable and Marshal had chosen ten or twelve hundred men-at-arms, of whom one party was to go by the Agincourt side and the other on that of Tramecourt, to break the two wings of the English archers; but when it came to close quarters there were but six score left of the band of Sir Clugnet de Brabant, who had the charge of the undertaking on the Tramecourt side. Sir William de Saveuse, a very brave knight, took the Agincourt side, with about three hundred lances; and with two others only he advanced before the rest, who all followed, and struck into these English archers, who had their stakes fixed in front of them, but these had little hold in such soft ground. So the said Sir William and his two companions pressed on boldly; but their horses stumbled among the stakes, and they were speedily slain by the archers, which was a great pity. And most of the rest, through fear, gave way and fell back into their vanguard, to whom they were a great hindrance; and they opened their ranks in several places, and made them fall back and lose their footing in some land newly sown; for their horses had been so wounded by the arrows that the men could no longer manage them. Thus, by these principally and by this adventure, the vanguard of the French was thrown into disorder, and men-at-arms without number began to fall; and their horses feeling the arrows coming upon them took to flight before the enemy, and following their example many of the French turned and fled. Soon afterwards the English archers, seeing the vanguard thus shaken, issued from behind their stockade, threw away their bows and quivers, then took their swords, hatchets, mallets, axes, falcon-beaks and other weapons, and, pushing into the places where they saw these breaches, struck down and killed these Frenchmen without mercy, and never ceased to kill till the said vanguard which had fought little or not at all was completely overwhelmed, and these went on striking right and left till they came upon the second battalion, which was behind the advance guard, and there the King personally threw himself into the fight with his men-at-arms. And there came suddenly Duke Anthony of Brabant, who had been summoned by the King of France, and had so hastened for fear of being late, that his people could not follow him, for he would not wait for them, but took a banner from his trumpeters, made a hole in the middle of it, and dressed himself as if in armour; but he was soon killed by the English. Then was renewed the struggle and great slaughter of the French, who offered little defence; for, because of their cavalry above mentioned, their order of battle was broken; and then the English got among them more and more, breaking up the two first battalions in many places, beating down and slaying cruelly and without mercy; but some rose again by the help of their grooms, who led them out of the mêlée; for the English, who were intent on killing and making prisoners, pursued nobody.

And then all the rearguard, being still on horseback, and seeing the condition of the first two battalions, turned and fled, except some of the chiefs and leaders of these routed ones. And it is to be told that while the battalion was in rout, the English had taken some good French prisoners.

And there came tidings to the King of England that the French were attacking his people at the rear, and that they had already taken his sumpters [pack animals] and other baggage, which enterprise was conducted by an esquire named Robert de Bornouille, with whom were Rifflart de Plamasse, Yzembart d'Agincourt, and some other men-at-arms, accompanied by about six hundred peasants, who carried off the said baggage and many horses of the English while their keepers were occupied in the fight, about which robbery the King was greatly troubled; nevertheless he ceased not to pursue his victory, and his people took many good prisoners, by whom they expected all to become rich, and they took from them nothing but their head armour.

At the hour when the English feared the least there befell them a perilous adventure, for a great gathering of the rearguard and centre division of the French, in which were many Bretons, Gascons, and Poitevins, rallied with some standards and ensigns, and returned in good order, and marched vigorously against the conquerors of the field. When the King of England perceived them coming thus he caused it to be published that every one that had a prisoner should immediately kill him, which those who had any were unwilling to do, for they expected to get great ransoms for them. But when the King was informed of this he appointed a gentleman with two hundred archers whom he commanded to go through the host and kill all the prisoners, whoever they might be. This esquire, without delay or objection, fulfilled the command of his sovereign lord, which was a most pitiable thing, for in cold blood all the nobility of France was beheaded and inhumanly cut to pieces, and all through this accursed company, a sorry set compared with the noble captive chivalry, who when they saw that the English were ready to receive them, all immediately turned and fled, each to save his own life. Many of the cavalry escaped; but of those on foot there were many among the dead.

When the King of England saw that he was master of the field and had got the better of his enemies he humbly thanked the Giver of victory, and he had good cause, for of his people there died on the spot only about sixteen hundred men of all ranks, among whom was the Duke of York, his great-uncle, about whom he was very sorry. Then the King collected on that place some of those most intimate with him, and inquired the name of a castle which he perceived to be the nearest; and they said, 'Agincourt.' 'It is right then,' said he, 'that this our victory should for ever bear the name of Agincourt, for every battle ought to be named after the fortress nearest to the place where it was fought.'

When the King of England and his army had been there a good while, waiting on the field, and guarding the honour of the victory more than four hours, and no one whatever, French or other, appeared to do them injury, seeing that it rained and evening was drawing on, he returned to his quarters at Maisoncelles. And the English archers busied themselves in turning over the dead, under whom they found some good prisoners still alive, of whom the Duke of Orléans was one; and they carried the armour of the dead by horse loads to their quarters. And they found on the field the Duke of York and the Earl of Oxford, whom they carried into their camp; and the French did little injury to the said English, except in the matter of these two.

When evening came the King of England, being informed that there was so much baggage accumulated at the lodging places, caused it to be proclaimed everywhere with sound of trumpet that no one should load himself with more armour than was necessary for his own body, because they were not yet wholly out of danger from the King of France. And this night the corpses of the two English princes, that is to say, the Duke of York and the Earl of Oxford, were boiled, in order to separate the bones and carry them to England; and this being done, the King further ordered that all the armour that was over and above what his people were wearing, with all the dead bodies on their side, should be carried into a barn or house, and there burned altogether; and it was done according to the King's command.

Next day, which was Saturday, the King of England and his whole army turned out from Maisoncelles, and passed through the scene of slaughter, where they killed all the French that they found still living, except some that they took prisoners; and King Henry stood there, looking on the pitiable condition of those dead bodies, which were quite naked, for during the night they had been stripped as well by the English as by the peasantry.

ANDREW WHEATCROFT
The Fall of Constantinople

By the middle of the fifteenth century the Ottoman Turks, originally horse nomads from the Central Asian Steppe, had conquered almost all the territory of the once great Eastern Roman Empire, confining Constantine XI, the last Byzantine emperor, within his capital at Constantinople. The city, which stands on the European shore of the Bosphorus, the channel dividing Europe from Asia, is built on two peninsulas separated by the inlet of the Golden Horn. The city proper, surrounding the 'Great Church' of Sancta Sophia, occupies the southernmost peninsula and was defended, as it still is, by the monumental Walls of Theodosius.

Constantinople had withstood many sieges, by Avars, Arabs and Bulgars, and once before by the Ottomans. In previous emergencies, however, the Byzantine emperors had been able to call upon reserves from their surviving possessions in Europe and Asia to beat off the attackers, and to seek assistance from fellow Christian states. By the middle of the fifteenth century, however, Christian Europe, divided against itself and locked in conflict with Muslim power in Spain and the Balkans, had little help to offer, while Constantinople had become a Christian island in an Islamic sea. After the Ottoman defeat of the

Polish-Hungarian 'Last Crusade' to the Black Sea coast in 1444, Constantinople survived in effect on the Ottomans' indulgence, dependent on them even for food supply.

In February 1453 a new Ottoman leader, Mehmed II ('the Conqueror'), decided to terminate the anomaly of the Eastern Roman Empire's existence. Assembling an army of 80,000 and a siege train of 70 heavy cannon, supported by a large fleet, he proceeded to impose a close blockade. Venice, which was attempting to sustain its own island empire in Turkish waters, sent a fleet to lend assistance but it was soon driven off. By April the Turks had isolated the city and were battering its massive walls with twelve monster guns, some cast on the spot. The Byzantine emperor had only 10,000 soldiers to defend the perimeter, formed by 10 miles of wall and the water barrier of the Golden Horn. The key to the defence was a great iron chain stretched across the mouth of the Golden Horn, blocking the entrance to the inlet and denying access to the weaker northern shore of the city. As long as it held, the garrison was just sufficient in strength to man the fortifications. Should the Turks find a way into the Golden Horn, the city's defence was undermined. Andrew Wheatcroft's description of Mehmed's solution to the problem of the siege is a brilliant account of a great military operation. It is also an epitaph on the extinction of the power of Rome, which had made Europe a millennium and a half earlier and which still remains, culturally, intellectually and politically, the dominant influence of the life of the conti-nent. The fall of Constantinople – 'New Rome' – seemed indeed to contemporaries to threaten the survival of their world, and its eventual subordination to an Asiatic civilization of which Mehmed the Conqueror was the personification.

*　　*　　*

On the night of 18 April, Mehmed launched his first full-scale assault on the weak points in the walls. The fighting was desperately hard, but the defenders threw back all the assaults. Not a single Turk crossed over the low earthworks plugging the gaps in the great walls.

The sultan [Mehmed] determined on another approach. On the day following the assault on the land walls, the Turkish fleet attempted to break the iron chain protecting the Golden Horn. They failed. Two days later, four Genoese ships ran through the blockade to safety in the Golden Horn. Spectators crowded the sea walls of the city to watch this gladiatorial combat between the four tall vessels with the cross of Christ emblazoned on their sails and the hundreds of smaller Turkish craft milling around them like insects. The sultan rode along the shore close to the Golden Horn to watch his ships triumph over the Christians. As the combat went against the Turks, he could be clearly seen from the city walls.

Spurring his horse into the water, splashing through the shallows until his fine robes were soaked through, his shouts of encouragement turned to threats and curses as his seamen failed before his eyes.

The failure of the first assaults by sea and land made it clear to Mehmed that the city would not yield easily. After the Christians had held off the initial assault, there were murmurs in the Turkish camp that Constantinople would never be taken. The sultan responded by tightening his grip around the city. The iron chain still protected the mouth of the Golden Horn, so Mehmed decided to outflank it. In the greatest secrecy he created a rough slipway from the Bosphorus up over the steep heights of Galata to the Golden Horn, a distance of some seven stadia (1,400 yards). This slipway was made of planks and tree trunks laid side by side, sometimes roped together where the incline was steepest. All the planks were greased with sheep's fat and oil. At the same time his carpenters built large cradle-like sledges on which the keel of a ship would rest. On the morning of 23 April, eighty ships were dragged out of the water on to the cradles; then, with hundreds of soldiers and teams of oxen straining at the ropes attached to the forward corners of the sledges, they were dragged up the greased slipway. As the ships were hauled by brute force up the steep slope, their crews sat at their oars and the sails were unfurled, giving the impression that they were sailing up the hill and on to the plateau above. The commanders even ran along between the ranks of oarsmen whipping them and shouting at them to row harder. As the chronicler put it,

'It was a strange spectacle and unbelievable in the telling except to those who actually did see it – the sight of ships borne along the mainland as if sailing on the sea.' But it was real enough. Within a few hours a Turkish fleet was slithering down the hillside into the waters of the Golden Horn. Now the Turks could attack the city on every side.

By the end of the second week of the siege a stalemate had been reached between attackers and defenders. The guns battered away at the walls, and the defenders attempted to repair the breaches. Christian and Turkish ships skirmished indecisively in the restricted waters of the Golden Horn. Both sides tried to undermine their opponents' morale. The Turks impaled some captive sailors in full sight of the triple walls, so the Christians responded by hanging all their Turkish captives, and dangling them from the battlements opposite the place where the sailors' bodies lay swollen in the heat. The stench of rotting flesh from the hundreds of corpses overlay both the city and the Turkish camp.

There were dangers for both sides in this stalemate. Disunity threatened the Christians. The garrison was made up of many discordant nationalities, and, as the siege continued and conditions became harsher, violent tensions rose between Greeks and Italians, and between traditional enemies like the Venetians and the Genoese. The Turkish side was just as riven by factions, with the sultan and his closest supporters facing the now open opposition of the great families of Anatolia. The sultan's will was absolute so long as he was successful, but if Mehmed failed to capture the city he would be deposed. Once deposed, he could then expect a visit from the executioner, who would strangle him in time-honoured Turkish fashion.

As April moved into May, both sides tried every trick to break the deadlock. The Turks attacked the boom, and launched repeated attacks on the walls. At midnight on 12 May 50,000 Turks poured towards a gap battered in the wall at the weaker, northern, end of the fortifications. The situation seemed desperate for the Christians, and 'most of us believed', the Venetian surgeon Nicolo Barbaro wrote in his diary, 'that they would capture the city.' But, miraculously, this Turkish attack was driven off, as were all the Turks' other assaults from land and sea. When the Turks attempted to mine underneath the walls, the Christians dug counter-mines; fourteen mining attempts were made, and all were frustrated. The stalemate continued.

Both Turks and Christians began to look for omens of heavenly intervention. The Christians were dismayed when a statue of the Virgin fell to the ground during a procession through the streets. An unseasonal cloud of mist settled over the city, and was immediately taken as an omen that God had withdrawn his light from his people. The sultan Mehmed was similarly despondent when he saw a mysterious shaft of sunlight which seemed to suffuse the great church of St Sophia (Aye Sofya) with a golden glow; this he interpreted as a sign of divine intervention on the side of the city. As the siege stretched into its sixth week, with no sign of a relieving Christian fleet from the west, the defenders came to rely more and more on a miracle. They realized that, as each day passed, their reserves of powder and shot, as well as food and other supplies, were dwindling. Their confidence, likewise, was ebbing away. The shaft of light which so disturbed the sultan was taken within Constantinople as the fulfilment of an ancient prophecy that the city would perish in a pillar of fire, God having abandoned his people.

[The Emperor] Constantine and the Genoese commander Giustiniani read other signs as they surveyed the Turkish camp from the walls of the city. There was great evidence of activity in the Turkish camp and on board the ships in the Golden Horn. The cannons were being manoeuvred so as to concentrate their fire on the temporary stockades and fortifications with which Giustiniani's soldiers had filled the gaping holes in the walls. Turkish irregulars were to be seen collecting branches and vines, both of which were used to make scaling ladders and fascines (large bundles of brushwood). The fascines would be piled one on top of the other, against the rampart and the stockades, to give attacking troops a foothold. The alert Christian commanders also noticed the sultan in the dawn of Monday 28 May as he rode slowly around the walls and along the line of the Golden Horn, surveying the city for new points of weakness. All the evidence pointed to a major attack in prospect.

Before noon, Constantine's spies in the Turkish camp had confirmed his surmise. They told him that Mehmed had summoned all the commanders to his tent at sunset. The Christian war council concluded that the great assault would not be long delayed.

The Turkish plan was to attack along the whole line of the walls, by land and sea simultaneously. The bashi-bazouks [Turkish irregulars, noted for their brutality] would go first. Then the Anatolian feudal levies were to attack on the southern sector of the land walls, while the janissaries [regular infantry] were to be massed in the centre opposite the most damaged section of the fortifications, which was the weak point in the Christian defence. The Turkish ships were to mount a concerted assault all along the sea walls. The flotilla in the Golden Horn was to shower the defenders with spears and arrows, to prevent any reinforcements being drawn from the sea walls to the main battle. The Turkish ships cruising endlessly back and forth in the Sea of Marmara were loaded with soldiers and scaling ladders ready to exploit any weakening of the Christian defence.

After sunset on 28 May an unnatural silence fell over the Turkish camp, broken only by the sound of the evening prayers. The Venetian Barbaro saw how both Turk and Christian now put their trust in God, 'each side having prayed to its God, we to ours and they to theirs, the Lord Almighty with his Mother in Heaven decided that They must be avenged in this battle of the morrow for all the sins committed.' In Constantinople it was now the eve of St Theodosia. In her church on the Golden Horn a curious air of normality prevailed. The church was filled, as for centuries past, with an abundance of roses and wild flowers. The worshippers began gathering as normal, although it was noted they were now mostly women: the bulk of the men were serving on the walls.

But this year there was a Turkish fleet anchored within hailing distance of the church, and plainly making ready for battle.

The congregation's prayers were now for the deliverance of the city. Throughout the day the icons and holy relics of the city had been brought out from their shrines, paraded through the streets and up on to the walls to invoke the aid of God and His saints. The churches tolled their bells, and the words of the Kyrie Eleison were chanted endlessly. At sunset, the emperor and the notables formed a long procession to the great church of St Sophia. Before he entered the tall bronze doors of the church, the emperor Constantine turned and spoke to his people for the last time: 'The Turks have their artillery, their cavalry, their hordes of soldiers. We have Our God and Saviour.'

As the night drew on, the Turkish camp came to life. Cannon and huge war machines were dragged closer to the walls. Their fire was pinpointed on the weak points. Long lines of bashi-bazouks came forward carrying fascines to fill the ditch in front of the rampart. Many of these 'expendable' irregular soldiers were killed by the Christian soldiers firing from the walls, but more came forward to replace them, all apparently oblivious to the hail of missiles around them. About half-past one in the morning, the sultan, now satisfied with the preparations, ordered his war banner to be unfurled. At this signal, the word went out along the line: begin the main attack.

In the centre of the line, Giustiniani and his soldiers had agreed to man the stockade and the Outer Wall. Behind them the gates through to the city were locked. Giustiniani and his men were trapped between the advancing Turks and their own fortifications. They had to win or die. Through the darkness they heard an explosion of sound, of drums beating, a clamour of cymbals and cries of 'Allah, Allah,' as the bashi-bazouks rushed forward in a confused mass. Flares were lit, and cast a flickering light over the pulsing Turkish horde below them. The Christians poured a devastating fire down on to the irregulars battering against the rampart. The fiercest attack came along the valley of the Lycus. Like Giustiniani and his men, the bashi-bazouks were trapped and unable to retreat, because Mehmed had placed his janissaries behind them, with orders to kill any who fell back from the walls. After two hours, the sultan judged that the irregulars had sapped the strength of the defence and allowed them to retire.

The Anatolians and the janissaries pushed forward over the bodies of the fallen bashibazouks. Where the irregulars had wavered under fire, the janissaries advanced file by file in a solid disciplined mass. When one fell, another took his place. But even they could find no chink in the Christian defence. Some succeeded in lodging their scaling ladders against the stockade, and a few even reached the top, but none survived for more than a few seconds. The sultan, who had led his crack troops to the edge of the ditch, finally ordered them to withdraw. As they fell back, the Turkish guns began to batter away again at the defences, drowning out the church bells now tolling over the city. For four hours, until just before dawn, it seemed that a miracle might be accomplished. Every Turkish assault had been thrown back. The temporary stockades of wood and earth had proved an effective barrier – as good as the walls themselves. But the unrelenting pressure of the attack eventually uncovered a weak point. At the northern end of the walls, close to the Golden Horn and the imperial palace, a small side gate called the Kerkoporta had been used to launch night attacks on the Turkish lines. After the last raid, it had not been properly secured. At the height of the battle, some alert Turks noticed that it was open, and burst through. They were eventually killed, and the gate was shut, but not before they had pulled down the Christian flags from the towers and raised Turkish banners in their place. Roughly at the same time, in the heart of the battle, a chance musket shot wounded Giustiniani. He decided to leave the battle. The emperor begged him to stay, for he had inspired the defence, but he refused. Constantine reluctantly unlocked the gate through the Great Wall, and Giustiniani was carried away by his men-at-arms to a Genoese ship in the Golden Horn. The simultaneous loss of their leader and some of their best soldiers dismayed the remaining Christians. Some shouted that the Turks had taken the city, as they saw the enemy banners flying over the towers near the Kerkoporta, and Constantine galloped off along the space between the walls to rally the troops on the northern flank. Rumours spread that the Turks had broken through the walls, and the fighting around the stockade slackened. This sudden weakening was noticed by the sultan, who was close to the fiercest part of the battle. He pushed the janissaries back into the attack, promising riches and fame to the first man into the city. This time a huge janissary named Hassan clambered up the stockade and held the defenders at bay while more Turkish soldiers joined him. Eventually the Christians cut him down, but the defensive line had now been breached. Once the janissaries had secured a foothold on the Christian side of the stockade, their discipline began to overwhelm the exhausted defenders. They pressed the Christians back to the Great Wall. There the janissaries' scimitars rose and fell remorselessly. Soon a heaving pile of dead, dying and wounded Christians was piled up shoulder high at the foot of the Great Wall.

The tenor of the battle changed immediately. The Turks clambered up the Great Wall unopposed, and broke open the Military Gate of St Romanus; long lines of janissaries trotted through into the defenceless city. Within a few minutes, the Turks had penetrated the walls in two other places and were pouring through into the city beyond. Dawn was breaking, and along the sea walls the Turks saw their flags replacing the Christian banners on the towers and high buildings to the north. They pressed their attack forward with increased zeal.

The defenders began to abandon their posts, running home to protect their families. The Turkish scaling ladders were no longer pushed away, and the attack triumphed in every sector.

The emperor Constantine preferred to die with his city rather than survive as a captive. He threw away his helmet with the imperial eagle, and his emblazoned surcoat, and sought an anonymous death in the heart of the fighting. The city finally fell: as Edward Gibbon expressed it, 'after a siege of fifty-three days, that Constantinople, which had defied the power of Chosroes [the Persians], the Chagan [the Avars], and the caliphs [the Arabs], was irretrievably subdued by the arms of Mohammed the Second.'

The consequences of the Turkish victory were horrifying. In April, with his army gathered before the walls, the sultan had summoned the city to surrender and had guaranteed the lives and property of all within. When the offer was turned down, he promised his soldiers, in accordance with the customs of warfare, that they should make free with the city for three days after its capture. Their passions were already inflamed, both by the prospect of plunder and by a desire for revenge. The Turkish troops had suffered from the long siege almost as much as the Christian defenders, and they had not forgotten the bodies of Turkish prisoners hanging from the towers and battlements. The chronicler Kritovoulos tells how the janissaries and other soldiers killed 'without rhyme or reason', because they had been roused by the 'taunts and curses' hurled at them from the walls all through the siege. Once they had entered the city, 'they killed so as to frighten all the City and to terrorize and enslave all by the slaughter.'

DIEGO HURTADO DE MENDOZE
Such Botching, Disorder and Chaos

Christian Europe's attempt to re-establish a foothold in the Islamicized East, during the two centuries of the Crusades (1095–1291), was in part caused and simultaneously matched by a continuing Muslim offensive into the European continent itself. In 1354 the Ottoman Turks – the Seljuk Turks' victory over the Eastern Roman (Byzantine) Emperor at Manzikert in 1071 having largely destroyed Christian power in Asia Minor – established their first foothold in Europe by crossing the Straits of the Dardanelles and capturing Gallipoli. It provided a base from which they proceeded to conquer Greece, Bulgaria and Serbia. In 1453 they extinguished the Eastern Roman Empire by the capture of Constantinople and began advances that threatened Hungary and the capital of the Holy Roman Empire at Vienna. Meanwhile, however, Christian Europe had begun to strike back against the other centre of Muslim power in Europe, Spain, which had fallen to the Arabs in 712.

Small Christian footholds had survived in northern Spain, for which a campaign of reconquest began in the tenth century. It culminated at the end of the fifteenth, when the union, in 1479, of the crowns of the Spanish kingdoms of Castile and Aragon, in the persons of Queen Isabella and King Ferdinand, led to final victory over the last Muslim (Moorish) possessions in the south. Many Moors chose to leave Spain and return to North Africa, from where the Muslim invasion had been launched. Those who remained, known to the Spaniards as the Moriscos and largely concentrated in the southern region of Granada, accepted Christianity in return for being allowed to retain certain cultural and legal privileges.

In practice, while conforming outwardly to Christianity, many Moriscos remained Muslim at heart. Their Christian overlords acquiesced none the less in the fiction of the Moriscos' conformity, and protected them against the centralizing power of the Spanish kingdom. At the beginning of the sixteenth century, however, the growing power of the Ottoman Turks in the Eastern Mediterranean confronted the Christian powers of the Western Mediterranean – Venice, the Papal States and Spain in particular – with a new challenge. The conflict culminated in the Ottoman Turks' offensive against the island of Malta, at the meeting point between the eastern and western halves of the inland sea. It was held by the last of the Crusading orders, the Knights of St John. They succeeded in holding the Ottomans at bay. In the aftermath of their victory in 1565, the government of Spain decided to bring the Moriscos under pressure to conform absolutely to Christian Orthodoxy. They refused to do so and broke into revolt in 1568. What followed is known as the War of Granada. The Spanish government of Philip II had used the riches won by the conquistadors of the Americas to raise the first truly modern army. By 1568, however, it was so deeply committed to defending the Spanish possessions elsewhere in Europe, particularly against the Protestant rebels of the Netherlands, that it had no efficient soldiers to put down the revolt in Granada. The account that follows is a graphic description of what happens when an undisciplined army, largely motivated by the hope of personal enrichment, takes the field.

*   *   *

If I stand back from my narrative and try to account for this disastrous defeat and ask how it was that a nation as vigorous as ours, whose soldiers are accustomed to undergo the greatest hardships without wavering in their duty for, amongst us, loyalty is a point of honour and we are vain of our military honour which in war is not the least important factor – if I just stand back and ask myself how, in this campaign, we could comport ourselves so unmartially and so unvalorously, I find myself thinking of the numerous well-disciplined and famous armies with which I myself have served, or observed. Some of these were led by the Emperor Charles [Charles V, Holy Roman Emperor from 1519 to 1556], one of the best soldiers the world has seen for many centuries; others by his rival, King Francis I of France, a man who was no less bold or experienced; but there has never been a more formidable host than the one which the Emperor's son, our current sovereign, King Philip II of Spain, took quite recently onto the field of Durlan to defend the Estates of Flanders against Francis I's son, Henri II of France. Never have I myself seen nor have I heard tell of an army that was better armed, better disciplined, more completely equipped in every particular than was that one. The provisioning was superb. The artillery were first class. There was superabundance of munitions. The private soldiers were of the best. There was an ample seasoning of courtiers and gentleman volunteers. The generals, the captains, the officers were all of very high quality.

The result was that we had a resounding success. We were able to dictate a peace treaty of which the whole world has had to take stock. By its terms our client, Duke Filibert, was restored to his dukedom of Savoy, something that was extraordinarily difficult to achieve in that French power lies so much closer to Savoy than ours does.

What a contrast to the rabble we fielded in Granada! Never in all my life have I witnessed such botching, such disorder, such chaos in the commissariat, such wanton squandering of money and time, or come across a soldiery so cowardly, so greedy, so lacking in perseverance or so profoundly ill disciplined.

The causes, in my view, are to be found right at the beginning of the war when the Marquess of Mondéjar had had to rely on adventurers and on levies raised by the city councils in order to put the rebellion down.

These feeble, greedy robbers believed that the enemy was neither well armed nor in the least formidable. They came rushing out of their houses without leaders and they were never properly formed into regiments. Their homes were close by and their one desire was to get back to them as quickly as they could with as much booty as they could carry.

They were a raw rabble who came to war with no experience of soldiering, and a raw rabble they remained, leaving the army as ignorant of what soldiering is about as they had been on the day that they had entered it.

Now when the Marquess of Mondéjar was still in charge, this scandalous state of affairs was very largely concealed. The marquess was a man of high character and he was immensely thorough and hard working. He knew the country and he knew the kind of rabble on which he was going to have to rely. He gathered round him a core of trusted volunteers who stiffened and goaded the levies. He himself never rested. He was here, there and everywhere, whenever there was a crisis, whatever the hour of the day or night, however far away or difficult to get to. He led from the front.

But after his enemies had succeeded in getting him dismissed, disaster followed on disaster. The enemy grew from strength to strength and our forces daily became weaker and worse. Panic, which is contagious, spread from one panicky soldier to the next and that is the worst thing that can happen to an army in wartime. There was no attempt to share out prize money. Each kept for himself that which he had taken and attempted at once to get home with it by the shortest route. Often enough men were cut down with their arms too full of loot to defend themselves. If there was no plunder to be had, they either refused to march or, so soon as they had found out, deserted and scuttled home.

This was a mountain war. There was not much to eat. There was very little proper equipment. The men had to sleep on the ground. There was no wine for them to drink. Their only pay was their keep.

Often, for weeks on end, they saw very little or no money. Once they found that their expectations of enormous plunder were largely illusory, they lost interest and ceased to want to fight. Poor, hungry, impatient, they fell ill and died or were killed as they tried to desert to their homes. The truth is they were quite useless. They preferred anything at all to the hardships of war and would not fight, except for plunder.

I must speak now of their captains and officers. Many of these grew tired of reprimanding and punishing and attempting to control their soldiers and gave themselves over to the vices of the rank and file. Whole camps took their tone from such men.

This, though, is not the whole picture, for amongst the levies who were sent by the cities there were indeed men who kept themselves well reined in, either because they were well born and they felt it was incumbent on them to behave as gentlemen, or because they were honest men and they would have been ashamed of themselves had they behaved like their fellows. Then, too, there were the contingents who were sent by the great lords. These were picked men, highly disciplined companies, many of whose members could have been officers, had they not elected to serve in the ranks because they wished to serve in a high cause and prove themselves in hand-to-hand combat against the Moor.

Captains, officers, soldiers – all were of the best and these were spirited, highly disciplined soldiers ready to undertake any venture, however perilous.

It was men such as these who, in the end, were responsible for our victory. Amongst them the city of Granada's own volunteers call for high praise. I cannot end these reflections without saying that it is my considered judgement that if we want to avoid a similar damaging experience next time it becomes necessary to call on the city councils for levies, we must prepare for the future.

FRANCESCO BALBI DI CORREGGIO
The Siege of Malta

The Ottoman Turks' capture of Constantinople in 1453 was the critical preliminary to their land offensive into Southern Europe and North Africa and their maritime conquest of the Eastern Mediterranean in the early modern age. By the end of the fifteenth century they had completed their conquest of Serbia, Bosnia, Albania and Greece. Under Suleiman the Magnificent (Sultan from 1520 to 1566), they extended their empire into Syria and Egypt, incorporated into it the Christian kingdom of Hungary, reduced most of the Venetian islands in the Aegean and, in 1529, laid siege to Vienna, Habsburg capital of the Holy Roman Empire. They also conducted major campaigns into Persia, secured footholds in North Africa, invaded Southern Italy and raided into the Western Mediterranean.

It seemed, by mid-century, that no Christian power could stand against that of the Muslim Turks. Not only were they the standard-bearers of a spiritually dynamic Islam, they were also the exemplars of a new style of warfare, based on mastery of contemporary high technology and on the organization of a new sort of professional army, the janissaries. The janissaries, recruited by the enslavement of Christian boys from the Balkans, converted to Islam and bound to long military apprenticeship, were soldiers pure and simple. Their life was war and, under the green banner of the Ottoman Sultan, who in 1517 had also appointed himself Caliph, or head of the religion of Islam, they appeared to Christian Europe, by the middle of the sixteenth century, to be invincible.

By 1565 the only barrier remaining between the Turks' stronghold in the Eastern Mediterranean and their newly acquired outposts in the Western Mediterranean was the island of Malta, owned and garrisoned by the Crusading Order of

St John. The order had begun as a religious community of hospitallers in Jerusalem. During the Crusades, its knights had become soldiers and, after their expulsion from the Holy Land, had stoutly defended the island of Rhodes (1522) and Libyan Tripoli (1551) against the Turkish aggression. Their possessions reduced to Malta, granted them by the Habsburg Emperor Charles V after their expulsion from those places and in recognition of their stalwart Christian militancy, the knights girded themselves in 1565 to withstand a final Ottoman onslaught. The Turks landed on the island in May of that year with 30,000 troops, to oppose a force of 500 knights, supported by 8,500 Maltese and foreign mercenaries. The order's Grand Master was Jean de la Valette. Under his inspired leadership, the defence of the fortifications around the Grand Harbour was sustained for five months, until the Ottomans abandoned the siege and withdrew in defeat. The knights' successful defence, won at the cost of about 6,000 casualties to the Turks' 24,000, checked the expansion of Islamic power into the Western Mediterranean for good. The Order of St John, which exists to this day, annually celebrates a Victoria Mass in September to commemorate the triumph.

*   *   *

Tuesday, 7 August 1565

On the seventh [of August], one hour before daylight, we saw that all the Turks on Cortin had commenced to move on Saint Michael's, and those from the fleet were being conveyed in boats from Marsa-M'xett to Is-Salvatur. This was a sign that the Turks would make an assault that day, as it turned out.

At daybreak a general assault was made on Saint Michael's as well as the Post of Castile, with so much shouting, beating of drums and blaring of trumpets that would have caused wonder had we not experienced it before.

The strength of the assailants on Saint Michael's was 8,000 and those on the Post of Castile 4,000. They attacked simultaneously as was their plan and as we had anticipated. But, when they left their trenches to come to the assault we were already at our posts, the hoops alight, the pitch boiling: in fact, all the materials for our defence were ready for action, and, when they scaled the works they were received like men who were expected.

The assaults on this day were most daring and well fought on both sides with great bitterness and much bloodshed. The greatest effort was made against the Post of Colonel Robles and that of Bormla, where Don Bernardo de Cabrera commanded. These Posts were the most vulnerable because of being so levelled they seemed easiest to gain. Here most of the fighting of the day took place, the Turks throwing in their main force. It was here that the greatest havoc was done amongst them both from the effect of the incendiary missiles and the fire from the traverses of these two Posts which faced each other and supported one another, bringing a deadly cross-fire to bear on the enemy who hoped to enter by this locality.

During the action their artillery did not fire as usual so as not to risk hitting their own men who were in close formation and very exposed. As on all other occasions, we fought behind cover having been made wiser by our losses in the past.

Although the assault on Saint Michael's was most severe, the attack on the Post of Castile was not less determined. In fact, the situation here became so serious that a knight of the Habit, a man of position, went to the Grand Master (who was in the square with the reserves, waiting to be called where most required) and said to him: 'Your Lordship, come to the relief of Castile because the Turks are coming in.'

Unmoved, the Grand Master turned to his knights and said: 'Let us all go and die there, for this is the day.' Having said this, with admirable courage, he took his helmet from a page and his pike from another, and went towards the Post of Castile followed by all the reserves.

When he arrived at the gate leading to the menaced position, the Prior of Champagne, the Bailiff of Eagle, the Conservator La Motta, Captain Romegas, and Commander Saquenville tried to dissuade him from going to the place of danger, but, to their sorrow, he insisted, and even wanted to mount the angle of the Cavalier of

Castile on which the Turks had already gained a footing, but this they would not allow him to do because it was so very exposed to the enemy artillery of both Is-Salvatur and Karkara; so he went to the low battery of Claramonte pike in hand, like a common soldier; but, when he looked up and saw the Post of the Spur of Buonainsegna full of Turks he took an arquebuse from a soldier, and, pointing it towards the enemy, called out: 'There boys, there.'

At this sign, all of us who were in this low battery aimed upwards at the enemy and fired as fast as we could, while those of ours who were above pelted them so furiously with incendiary missiles and stones as to force them to retire with heavy loss.

When the principal knights saw that we were no longer in danger they persuaded the Grand Master to retire from the place where he stood, surrounded by more than twenty dead. He consented to withdraw, like a good captain who knew that, after God, our salvation depended on his life; but he was far from retiring to rest for he did not go further than the gate of the inner works, where he stopped.

It always pained him to see any of our dead although he dissembled his feelings and when anyone was killed he praised him so as to infuse courage into the others.

On this day His Lordship was wounded in the leg, but this did not cause him to relax his duties although his leg was in bandages.

During this assault the Imperial Standard of the Sultan of Turkey was seen on the walls of Castile at the Post of Buonainsegna. It had the white tail of a horse with many tassels. We threw many lines with hooks so as to take it and at last caught it. With our pulling one way and the enemy the other, the knob which surmounted it fell off and in this way they saved their standard; but not before many of its gold and silver tassels had been burnt by our fire.

The assault lasted nine hours, from daybreak until after noon, during which time the Turks were relieved by fresh troops more than a dozen times, while we refreshed ourselves with drinks of well-watered wine and some mouthfuls of bread, for so great was the care which his most illustrious Lordship had for us, that, seeing he could not relieve us by fresh men (like the Turks) because he had so few, he cheered us in this manner. Since he could not provide fresh men he had given orders that, on days when assaults were made, many bottles of watered wine and bread should be freely supplied at all the Posts which were engaged. He had also ordered that many barrels of salt water should be kept at all the Posts so as to afford relief to those who suffered from burns, and were it not for all these thoughtful provisions no human endurance could have withstood the fury and pertinacity of the Turks who were so many and we so few.

Victory was ours again but it was due to Divine agency rather than to human effort, for the enemy had intended this to be their final assault and no man who could fight had been left behind in the camp or with the fleet.

As to ourselves, in spite of all the help and encouragement which the Grand Master afforded us, not one could stand on his legs from fatigue or wounds. Many of ours were killed. But the Lord came to our aid in the following manner.

When these assaults had lasted fully nine hours, one might say that our Lord inspired our cavalry who were in the City with their horses. On that day they went out as usual, and, as they did not see a single Turk anywhere, they pushed on as far as the Marsa where they became aware of the great danger their Order was in. Not knowing how else they could help because they were barely a hundred horse and as many infantry, they made an onslaught on the sick and other non-combatants who were there, killing as many of them as were found, and shouting 'Victory and relief' the whole time.

Some Turks from the fleet who were stationed on the promontory of Saint Elmo were the first to notice this commotion going on at the Marsa, and, forming themselves into a squadron, set off in good order to the Marsa. The Turks who were attacking the Post of Castile and Saint Michael's noticed the movement of this squadron, but, seeing that it had not advanced more than a hundred paces when it turned about and made for the fleet with all haste, they halted and abandoned the attack.

At the same time the news reached the Commander of the land forces that those who had been left behind at the Marsa had all been killed and the tents plundered. This news spread to the trenches where it grew until it was said that strong reliefs for us had arrived, and that if they did not retire in time they would all lose their heads. This false rumour had such an effect on the enemy that they all retired from their trenches waiting for orders from the Pasha or any of their officers. The first to leave were those facing the Post of Castile, and, on emerging from the ditch, came under the fire of our arquebuses at the Post of Auvergne, and many were so killed. When those who were attacking Saint Michael's saw the flight of their men on the promontory of Saint Elmo they hesitated; but when, soon after, their wounded came in and exaggerated our force a thousand-fold, they raced each other out of their trenches and none of their officers could stop them.

This sudden retirement of the Turks bewildered us, for we did not know its reason. We thought that it might be due to some discord between their different elements, as often happens in war, or that some renegades had sailed away with part of their fleet, or, again, that some relief force of ours had landed and was close on the enemy.

We soon learned that the Sicilians, from their Post, had been the first to detect our cavalry at the Marsa fighting sword in hand and had immediately reported it to the Grand Master, who sent a vedette to the top of the clock tower to verify the information.

The Sicilians then gave a shout of 'Victory and relief'. This shout was passed on from Post to Post, and while it cheered us it put panic into the hearts of the Turks, and there was not one of them who remained in the trenches or one of ours who did not mount on the parapets.

Meanwhile the Turks had not discovered what had happened. Mustapha Pasha formed up his men and marched on Santa Margarita in good order so as to take advantage of that position and its guns against all eventualities. On arrival he halted and waited for reliable information as to what was really going on.

The Turks soon became aware that a mere handful of men had baulked them of such a great victory, putting them into a state of panic and disorder. They advanced, with their flags unfurled, towards our men; but scouts gave timely warning. Each horseman mounted a foot soldier behind him and in perfect order, without losing a single man, retired to the old City after having killed many of the enemy, and saved the Order and all the besieged.

The humiliation of the Pashas and all the Turks was great when they realized that so few men had done them so much harm and put such a fear into them. Mustapha Pasha's resentment centred on Piali, because, he said, if after forming his men on the promontory of Saint Elmo he had marched on the Marsa the demoralization of his own men would not have happened. In any case, if he did not advance there was no need to retire in such great haste and disorder. Piali replied that he had received reports of a Christian landing in great force, and, under such circumstances, it was his duty to save the fleet, knowing that the Sultan thought more of it than of armies such as this one; and having said this to Mustapha he left him to digest it.

Judging by the haste with which the Turks removed their dead more than 2,000 must have been killed before Saint Michael's, and their wounded (as we ascertained later) were twice that number. Before the Post of Castile more than 200 of their most distinguished men died; among them being the Greek Ochali (El Louk Aly). On our side we had sixty killed, but the wounded exceeded that number. The following are the names of the more prominent among them.

Ensign Muñatones. It was said that one of our soldiers shot him in the right hand through carelessness.

Commander (Francisco) Torrellas. Commander (Antonio) Fuster.

Commander (Gabriel) Serralta (? Ceralta) and Don Jorge Fabellon. (Not otherwise identified as a knight.)

All these knights of the Habit were wounded.

Those who greatly distinguished themselves at Saint Michael's on this day were:

Colonel Melchor Robles. The Prior of Hungary (Don Vincenzo Caraffa).

Don Francisco de Bargas Manrique (de Burgues Manrique). Hernando de Heredia.

Captain Martelo (Antonio Martelli). Don Juan Mascon.

Don Bernardo de Cabrera. The knight Adorno (Gregorio Adorno).

Juan Burato (Giovanni Buratto or Buralto, a postulant).

Among the soldiers were sergeant Chaparro; the lieutenant of Martelli, Silvestro del Testo; Giulio Crudeli, his ensign; Mathias de Ribera; and sergeant Chacon; Giromimo, a slave of Commander Fortunio (Ramon Fortuyn), who was wounded by a grappling iron while fighting on a parapet; and Bartholomé (from Majorca), a retainer of Commander Serralta, and other retainers of knights. Many of the Maltese also fought valiantly.

Those who distinguished themselves at the Post of Castile were the most valiant Lord Jean de La Valette, most worthy Grand Master of the Venerable Hospital of Jerusalem, our Commander-in-Chief during the siege.

Commander (Pedro) Buonainsegna. Commander Chencho Gascon (Gencio, or Lorenzo Guasconi). Don Juan de Mendoca. Don Vasco de Acuña. Captain Romegas (Mathurin d'Aulx de Lescout, *dit* Romegas). Commander Pierre de Giou. Commander Sacambila (Louis de Mailloc Saquenville). (Esteban) Claramonte. Zaportella. (There were three of the name.) Don Rodrigo Maldonado. Fraguo (Don Alonso del Frago),

and many other knights of the Habit, soldiers of fortune, and retainers of knights. Among the soldiers, the Grand Master's tailor, a Maltese named Marco, fought very well. Vincenzo Cigala, a clerk; Nicolò Rodio; Master Juan Oliver; Lorencio Puche from Majorca; Mendoza; and the knight Peri Juan Alegre (not a knight of Saint John and not otherwise identified), took part in the fighting at the Post of Claramonte.

The knights who took no part in this fighting yet rendered good service, for, by order of their Grand Master they guarded the other Posts which were very thinly held on account of the greater part of the knights and soldiers being called to reinforce the menaced positions.

I cannot pass over in silence the valour shown during the whole of the siege by all the young knights – of all nations – who took part in the defence with great pluck and readiness and fought in the most dangerous places, filling the gaps made by the fallen with as much courage as if they had been old soldiers. They were undaunted by the many and horrible deaths which they continually witnessed, and, as I am not acquainted with the names of all I prefer not to mention those I know lest I should appear to be partial, which I certainly am not.

As soon as all the Turks had retired from Saint Michael's Colonel Robles, in the presence of all, went on his knees and gave thanks to our Lord for the great victory which it had pleased Him to grant us; and he sent a request to the Grand Master to have a Te Deum sung at San Lorenzo because we had been given one of the greatest triumphs which Christians had ever achieved. The bearer of this message had no need to deliver it because he found the Grand Master in the church of San Lorenzo giving thanks to our Lord, as he always did after a Turkish retirement, and the Te Deum invoked by Robles was being sung with great solemnity. When it was over a procession was made, and if it was not so imposing as those usually made by this Order, the tears of many men and women demonstrated its devotion.

The knights who came to our aid with their cavalry were Commander Vincenzo Anastagi (because Boisbreton had been wounded and his place taken by Lugny) . . .

Meanwhile some were not wanting who, affecting great solicitude for the safety of the Grand Master, and because they saw the breaches were very wide and believed we could not hold out, advised him to retire within Saint Angelo with the best part of the Order because there he could wait in greater security until relief should come. This counsel became known to the soldiers and each one spoke freely of what he thought of these advisers.

When the Grand Master heard this, ready as he was to be the first to die for his Order, he caused all the relics and all that was of value to be taken to Saint Angelo, and, in order to dispel any doubt, he ordered the bridge to be removed, thus making it clear to all that there should be no retirement and that we should defend the Birgu or die in the attempt.

Before the Turks made their last assault they had already reached the entrance to the ditch of the Post of Castile by means of well-covered trenches. The entrance to this ditch was defended by the casemate of the Post of Auvergne with eight gun emplacements, four above and four in the ditch, but, owing to its small size, it could not mount heavy artillery. When the Turks reached the entrance to the ditch and saw the harm which they suffered whenever they attacked or during retirement and that their batteries could not cover them or touch the casemate, they determined to erect a trench at the mouth of the ditch which would allow them to go in and out and which should be strong enough to withstand the fire from the guns of the casemate. They kept to the right of the battery at the spur of the cavalier [earthwork] where it came under the fire from Karkara which already reached nearly as far as the middle ditch. (It is not clear whether it was the fire from Karkara or the new trench which reached nearly as far as the middle ditch.) They dug till they reached the outer revetment of the contrefosse without our being able to molest them although we were always throwing stones and firebowls at them from our parapet. It was on this parapet that Captain Esteban Calderon, a knight of the Habit, lost his life while looking over to see what they were doing.

When the Turks reached this wall they made an opening in it, and through this they began to throw earth and fascines into the ditch. When the earth was heaped to the level of this opening in the wall they pushed it further into the ditch. They did all this under the fire of a small gun which Captain Romegas had mounted for the purpose and which he himself aimed and fired all day long, doing the Turks much harm. As the gun emplacement was very confined, and so as to prevent the carriage from going to pieces from the effects of the recoil, Romegas ordered a buffer of hawsers to be put behind it.

When the Turks had thus completed three parts of their trench, their position behind it became secure and they finished it to perfection according to their design.

While these operations by the Turks were taking place, the Grand Master was informed that it was impossible to prevent them from completing the trench and, when finished, it would be greatly to our disadvantage. As a counter-measure he ordered a tunnel to be dug to the right of this trench inside the walls at the point of the cavalier, whence it was intended to oppose those who entered the ditch or to serve as a sally port whence we could go out and destroy their trench. Unfortunately, when our tunnel was completed it was found that the sap head was commanded by two pieces of the battery at Karkara which had bombarded the traverses of the Post of Don Rodrigo Maldonado. For this reason we did not proceed with the opening of the tunnel but left it as it was lest it should be of use in some other way; inside the works, however, other defences were erected in which were many devices for our defence.

INGA CLENDINNEN
Aztecs

The wars between Islam and Christianity were certainly a conflict of cultures; different as they were, however, those cultures shared much in common: belief in a single God, omnipotent yet benevolent, and in the duty to obey God's law, revealed in written form and interpreted by His earthly representative. Christians and Muslims found it easy to hate each other; but they did so because each regarded the other as heretics, perverters of the true monotheistic faith. The strictly pious of both religions, moreover, were troubled by the violence into which their differences drew them. Christians accepted that the taking of human life was a wrong that could be justified only by pursuit of some higher good. Muslims shared a similar view, acknowledging a duty of protection to unbelievers who submitted to Islamic authority, even if they persisted in religious error.

The Spaniards who crossed the Atlantic at the end of the fifteenth century quickly encountered a culture whose religion resembled theirs in no way whatsoever. The Aztecs of Mexico had created a civilization which, in its power, wealth and sophisticated political structure, greatly impressed the invaders.

Its religion, and the warfare that served it, both baffled and disgusted them, for the Aztec gods were in no way benevolent, requiring a daily sacrifice of human victims to placate them. The qualities which Christians and Muslims alike sought to achieve in the practice of their ideal of the religious life – humility, charity and ultimately self-sacrifice – were not those the Aztecs believed their gods wanted at all. Blood sacrifice was the means of pleasing their deities, whose appetite for human suffering and death was insatiable.

Only in one respect did the religion of the Aztecs and those of the monotheists of the Old World resemble each other, and that was in respect for the nobility of death freely accepted for a religious purpose. Even so there were fundamental differences. Christians venerated believers who preferred an unresisting death to denying God. Muslims commemorated warriors who died in the holy war against the infidel. Aztecs thought the most noble form of human behaviour was to co-operate bravely in a ritual form of combat that would lead, slowly and painfully, to their own killing. Inga Clendinnen, an Australian historian, describes Aztec ritual combat in an account that embraces a description of a wider encounter between Old and New World cultures uncomprehending in the extent of their dissimilarity.

*   *   *

The killing of selected warrior captives, usually accompanied by torture, was unremarkable among Amerindians, as the Huron 'burning' of the Seneca warrior indicates [this is described in the introduction to the Ragueneau extract, pp. 92–3], but the process presumably varied in accord with different understandings of war and the consequent relationship between captor and captive. In Tenochtitlán notable captives, or those taken in a major campaign, were presented before the idol of Huitzilopochtli and then displayed at the royal palace before Moctezoma, while speeches were made on the death they would die. The warrior from a Nahua city participant in Mexica understandings of war was particularly cherished, being tended by stewards in the local temple and constantly visited, adorned, and admired by his captor and the captor's devoted entourage of local youths. Such a man presented for death before Huitzilopochtli's shrine crowning the great temple pyramid ideally leapt up the steps shouting the praises of his city. (That act of courage might have been made easier by the great bulk of the pyramid, which loomed so huge that a man at the base or on the long climb upwards could not see what awaited him.) Some, we are told, faltered on the stairs, and wept or fainted. They were dragged up by the priests. But, for most, pulque [alcoholic drink made from agave], anger, pride, or the narrowing existential focus of their days somehow got them through.

Mexica combat at its best was a one-to-one contest of preferably close-matched combatants, with one predestined to triumph, one to die. Given the fated outcome, and given the warrior obligation to seek and embrace the 'flowery death' on the field of battle or the killing stone, no shame need attach to defeat. The captive was in a deep sense the reflex of his captor, who accordingly took a tense and proprietary interest in that final performance. The quality of his own courage would be on public trial there.

Such prized captives were preferably offered at the festival of Tlacaxipeualiztli, the 'Feast of the Flaying of Men', on what the Spaniards thought of as the 'gladiatorial stone', to die after having engaged in combat with a sequence of selected Mexica warriors. The victim was tethered by the waist to a rope fastened to the centre of a round stone, about waist high, a metre and a half wide, and elevated in its turn on a platform about the height of a man. The 'display' element was made explicit by the procession of 'gods' (high priests in the regalia of their deities) who formally took their places around the small round stage. The tethered victim was given a long draught of pulque, and most ceremoniously presented with weapons: four pine cudgels for throwing, and a war club, the club being studded not with the usual shallow flint blades but with feathers. He then had to fight up to four leading Mexica warriors armed with bladed clubs, who fought from the platform, so giving the captive the advantage of height – an equivocal advantage, as we will see.

Despite the combat theme, the conditions so carefully constructed in the 'gladiatorial' encounter bore slight resemblance to ordinary battle. The combat with each warrior was presumably timed, so there was pressure on the Mexica warrior to perform at maximum. The victim, elevated above his opponent and released from the inhibition against killing which prevailed on the battlefield, could whirl his heavy club and strike at the head of his antagonist with unfamiliar freedom. The Mexica champions were also presented with a temptingly easy target. The victim could be disabled and brought down with one good blow to the knee or ankle, as on the battlefield. But such a blow would simultaneously abort the spectacle and end their glory, so the temptation had to be resisted. Their concern under these most taxing and public circumstances was rather to give a display of the high art of weapon handling: in an exquisitely prolonged performance to cut the victim delicately, tenderly with those narrow blades, to lace the living skin with blood (this whole process was called 'the striping'). Finally, the victim, a slow-carved object lesson of Mexica supremacy, exhausted by exertion and loss of blood, would falter and fall, to be dispatched by the usual heart excision.

Throughout all this the captor, who had nurtured his captive with such care and pride, watched his mirrored self on public display. His warrior at last dead, the heart burnt in the eagle vessel in homage to Huitzilopochtli, the head removed for use in a priestly dance and then skewered on the appropriate skull rack, the cadaver carried to his home calpulli, the captor was given a gourd fringed with quetzal feathers and filled with the blood drawn from the welling chest cavity to carry through the city, daubing the blood on the mouths of the stone idols in all the temples. Then he returned to his own ward temple to flay and dismember the body, and to distribute the limbs in the conventional way. Later again, he watched while his kin, summoned to his home household, ate a small ritual meal of maize stew topped by a fragment of the dead warrior's flesh, as they wept and lamented the likely fate of their own young warrior. For that melancholy 'feast' the captor put off his glorious captor's garb, and was whitened, as his dead captive had been, with the chalk and feathers of the predestined victim.

The captor himself did not eat the flesh, saying, 'Shall I perchance eat my very self?' He had earlier, we are told, addressed his captive as his 'beloved son', and was addressed in turn as 'beloved father'. A surrogate 'uncle' had supported the captive through his last combat, offering him his draught of pulque, sacrificing quail on his behalf, and wailing for him after his death. There has been a tendency to take the invocation of kin terms as indicative of a particular emotional response, but that claim seems ill-founded: there was slight tenderness in the Huron's slow killing of the Seneca prisoner [see p. 92], for all the mutual use of kin terminology. Neither do we see any trace of grief for the victim in the Mexica ritual: the tears shed are shed for the victor, and his putative fate. I have written elsewhere on the ambivalence of the privileges attaching to the honour of offering one's captive on the gladiatorial stone, and the acuteness of the psychological manipulations which blurred the boundaries of self, as the two identities were juxtaposed and overlaid. The offering warrior was projected into a terrible and enduring intimacy with his victim: having proudly tended and taunted him through the days and weeks of his captivity, and watched his own valour measured in the captive's public display, he had seen life leave the young body and its pillaging of heart, blood, head, limbs, and skin. Then he had lent out the flayed skin to those who begged the privilege, and pulled it on over his own body as it went through its slow transformations: tightening and rotting on the living flesh; corrupting back into the earth from which it had been made. Powerful emotions must have been stirred by these extravagant and enforced intimacies with death, and more with the decay and dissolution of the self, but there is no indication that pity or grief for the victim were among them.

What of the victim? It was clearly essential for reasons sacred and secular that the warriors tethered to the stone should fight, and fight well; the spectacle and the value of the offering would collapse should they whimper and beg for a quick death. There must always have been an element of risk here, but most captives seem to have performed adequately, and some magnificently. There could have been no individual bargaining. The warrior's life had been forfeit from the moment of his submission on the field of battle, or at least from the cutting of his warrior scalp lock. How, then, were they persuaded to fight?

In view of the uninhibited triumphing over comrades in Mexica warrior houses, I would guess warrior victims were often enough teased into anger and so to high performance, especially as 'wrath' was identified as the elevated state in which a warrior was suffused by sacred power. The victim was also more subtly conditioned. He had been presented by his captor to the people in a sequence of different regalia over the preceding four days at the pyramid of Xipe Totec, 'Our Lord the Flayed One', so coming to know the place where he was to die. He had practised the routines: on each occasion he had been forced to engage in mock combat, and then to submit to a mock heart excision, the 'heart' being made of unsoftened maize kernels. On his last night of life he kept vigil with his captor. His scalp lock was cut at midnight, marking his social death as warrior: he would fight not in his warrior regalia but in the whitened chalk and feathers of the sacrificial victim. It was as designated victim that he watched other men from his people, men he had known when they were alive, fight and fall on the stone, until it was his turn for his last display of maximum skill and valour. If he died well, his name would be remembered and his praises sung in the warrior houses of his home city.

The 'rehearsals' – the garments changed again and again, the mock combats at the stone, the mock heart excisions – doubtless reduced the individual's psychological capacity to resist as he was led step by step down a narrowing path. We will see that same technique of conditioning by familiarization used on non-warrior victims. The pulque given the gladiator came late, and I suspect its effect was more psychological than physiological as he took the taste of the sacred drink into his mouth. But the best guarantee was the co-operation which came from common understandings. For such public deaths victims were preferably taken in a special kind of war: the 'Flowery Wars' initiated by the first Moctezoma. These were battles staged by mutual arrangement between the three cities of the Triple Alliance, and the three ultramontane provinces of Tlaxcala, Huejotzingo, and Cholula, solely for the mutual taking of prisoners worthy of sacrificial death. The men who fought in the Flowery Wars were men of the highest rank, and they fought against matched opponents. Their capture was in a sense a selection by the god, and perhaps borne the more stoically for that. The finest demonstrations on the gladiatorial stone depended on agreement as to the nature and the necessity of the performance itself.

If few warrior captives died under such intense scrutiny, some suffered crueller fates, and there no co-operation was assumed. Victims destined for the singularly agonizing death required for the celebration of the Fire God were tightly bound before they were cast into the fire, to be hooked out, still living but badly burned, and dispatched by the usual heart excision . . . What we see in the handling of warrior victims is a pragmatic and finely adjusted balance between direct physical control (those bound victims cast into the fire), coercion, and psychological conditioning and reward. That they would die was unproblematical; it was the manner of their deaths which required management.

GEORGE PEELE
(?1558–1596)
Farewell to Arms
To Queen Elizabeth

His golden locks time hath to silver turned;
O time too swift, O swiftness never ceasing!
His youth 'gainst time and age hath ever spurned,
But spurned in vain; youth waneth by increasing:
Beauty, strength, youth, are flowers but fading seen;
Duty, faith, love, are roots, and ever green.

His helmet now shall make a hive for bees;
And, lovers' sonnets turned to holy psalms,

A man-at-arms must now serve on his knees,
And feed on prayers, which are age's alms:
But though from court to cottage he depart,
His saint is sure of his unspotted heart.

And when he saddest sits in homely cell,
He'll teach his swains this carol for a song:
'Blest be the hearts that wish my sovereign well,
Curst be the souls that think her any wrong'.
Goddess, allow this agèd man his right,
To be your beadsman now, that was your knight.

FATHER PAUL RAGUENEAU
An Attack by Iroquois Warriors

The Spanish and Portuguese conquest of Central and South America in the sixteenth century was followed by the penetration of North America in the seventeenth by the Dutch, English and French. In what would become Canada, the principal motive for venture into the interior was the search for fur. Fur-bearing animals had been almost exterminated in Europe by that date but the demand for fur – particularly beaver skin for hats – was enormous. In pursuit of beaver, largely trapped by native tribes and traded to Europe through middlemen, the French, who dominated the business, moved ever further up the St Lawrence River into the Great Lakes as territory was hunted out. Their initial monopoly was disputed both by the English of New England and by the Dutch in what is now New York state, and local conflict was the consequence. The native Americans, however, also joined in the fur wars, those of the Iroquois federation seeking in particular to create a barrier between the European settlements nearer the sea and the Indian territories on and beyond the Great Lakes to which the 'fur frontier' was, under pressure of hunting, retreating. The Iroquois, by imitation, were themselves attempting to become middlemen, particularly between the Algonquin and the Huron, who occupied the fur-bearing regions, on the one hand, and the French, on the other.

War between the North American natives of the Great Lakes was as cruel as that between the much more sophisticated Aztecs of Mexico, particularly in its demand that captives should play a co-operative role in the ritual torture that led to their deaths. Horrified observers of these ceremonies were the French Jesuits who had taken the conversion of Redskin America to Christianity as one of the many, and worldwide, missions that fell to them in the seventeenth century. One of them, a missionary with the Huron in 1637, describes (in another piece from Inga Clendinnen's work quoted in a previous extract) their treatment of a prisoner from the Seneca, a hostile tribe. 'He was a man of fifty, who had briefly been adopted into the family of a Huron chief, until it was decided that the wounds he had suffered in combat unfitted him for incorporation into the tribe. On an appointed night, and after a sequence of feasting in which he had joined, he was brought into the Huron council house where eleven fires had been lit. It was filled with people, the young warriors, who had been equipped with burning torches, having been warned to torture him slowly so that he would survive until daylight. The Huron chief, having announced that the prisoner would die by fire, and how his body would then be divided, had him brought in. "Now he began to run a circuit around the fires, again and again, while everyone tried to burn him as he passed; he shrieked like a lost soul; the whole cabin resounded with cries and yells. Some burned him, some seized his hands and snapped bones, others thrust sticks through his ears, still others bound his wrists with cords, pulling at each end with all their might, so as to cut flesh and crush bone." The torture continued throughout the night, the victim being revived when he fainted and given food and shown kindness. When he could, he addressed his tormentors in kinship terms, was answered as kin, and, when he resumed the circuit of the fires, sang his warrior songs. Eventually, at dawn, still conscious, he was taken outside the cabin, tied to a post, burned to death with heated axe-heads and dismembered. His feet, hands and head were then distributed to those to whom they had been promised.'

The Jesuit priest Paul Ragueneau describes an Iroquois raid on their Huron enemies during the fur war of 1642–53. Some of the defeated Huron would have suffered the fate members of their tribe had inflicted upon the Seneca.

*   *   *

In consequence of the bloody victories obtained by the Iroquois over our Hurons at the commencement of the spring of last year, 1649, and of the more than inhuman acts of barbarity practiced toward their prisoners of war, and the cruel torments pitilessly inflicted on Father Jean de Brébeuf and Father Gabriel Lallemant – terror having fallen upon the neighboring villages – all the inhabitants dispersed. These poor, distressed people forsook their lands, houses, and villages, in order to escape the cruelty of an enemy whom they feared more than a thousand deaths. Many, no longer expecting humanity from man, flung themselves into the deepest recesses of the forest, where, though it were with wild beasts, they might find peace. Others took refuge upon some frightful rocks that lay in the midst of a great Lake nearly four hundred leagues in circumference – choosing rather to find death in the waters, or from the cliffs, than by the fires of the Iroquois. A goodly number having cast in their lot with the people of the Neutral Nation, and with those living on the Mountain heights, whom we call the Tobacco Nation, the most prominent of those who remained invited us to join them, rather than to flee so far away.

This was exactly what God was requiring of us – that, in times of dire distress, we should flee with the fleeing, accompanying them everywhere; that we should lose sight of none of these Christians, although it might be expedient to detain the bulk of our forces wherever the main body of fugitives might decide to settle down.

We told off certain of our Fathers, to make some itinerant Missions – some, in a small bark canoe, for voyaging along the coasts, and visiting the more distant islands of the great Lake, at sixty, eighty, and a hundred leagues from us; others to journey by land, making their way through forest depths and scaling the summits of mountains.

But on each of us lay the necessity of bidding farewell to that old home of Sainte Marie – to its structures, which, though plain, seemed, in the eyes of our poor Savages, master-works of art; and to its cultivated lands, which were promising us an abundant harvest.

It was between five and six o'clock, on the evening of the fourteenth of June, that a part of our number embarked in a small vessel we had built. I, in company with most of the others, trusted myself to some logs, fifty or sixty feet in length, which we had felled in the woods, and dragged into the water, binding all together, in order to fashion for ourselves a sort of raft that should float on that faithless element. We voyaged all night upon our great Lake, by dint of arms and oars; and we landed without mishap, after a few days, upon an island, where the Hurons were awaiting us, and which was the spot we had fixed upon for a general reunion, that we might make of it a Christian island.

The Hurons who were awaiting us on that Island, called the Island of Saint Joseph, had sown there their Indian corn; but the Summer droughts had been so excessive that they lost hope of their harvest, unless Heaven should afford them some favoring showers. On our arrival they besought us to obtain this favor for them; and our prayers were granted that very day.

These grand forests, which, since the Creation of the world, had not been felled by the hand of any man, received us as guests; while the ground furnished to us, without digging, the stone and cement we needed for fortifying ourselves against our enemies. In consequence, thank God, we found ourselves very well protected, having built a small fort according to military rules, which, therefore, could be easily defended, and would fear neither the fire, the undermining, nor the escalade [assault with ladders] of the Iroquois.

Moreover, we set to work to fortify the village of the Hurons, which was adjacent to our abode. We erected for them bastions, which defended its approaches – intending to put at their disposal the strength, the arms, and the courage of our Frenchmen.

The War had already made its ravages, not only in the devastation which occurred in the preceding Winter, but in the number of massacres which happened all through the Summer, on the mainland in the vicinity of this Island. But that nothing might be lacking in the miseries of an afflicted people, all the days and nights of Winter were but nights of horror, passed in constant fear and expectation of a hostile party of Iroquois, of whom tidings had been received; these (it was said) were to come to us to sweep this Island, and to exterminate, with us, the remnants of a nation drawing to its end.

In the Mountains, the people of which we name the Tobacco Nation, we have had, for some years past, two missions; in each were two of our Fathers. The one nearest to the enemy was that which bore the name of Saint Jean; its principal village, called by the same name, contained about five or six hundred families. It was a field watered by the sweat of one of the most excellent Missionaries who had dwelt in these regions, Father Charles Gamier – who was also to water it with his blood, since there both he and his flock have met death, he himself leading them even unto Paradise.

The day approaching in which God would make a Church triumphant of that which, up to that time, had always been in warfare, and which could bear the name of a Church truly suffering, we received intelligence of it, toward the close of the month of November, from two Christian Hurons, escaped from a band of about three hundred Iroquois, who told us that the enemy was still irresolute as to what measures he would take – whether against the Tobacco Nation, or against the island on which we were. Thereupon, we kept ourselves in a state of defence, and detained our Hurons, who had purposed taking the field to meet that enemy.

At the same time we caused the tidings to be speedily conveyed to the people of the Tobacco Nation, who received it with joy, regarding that hostile band as already conquered, and as occasion for their triumph. They resolutely awaited them for some days; then, wearying because victory was so slowly coming to them, they desired to go to meet it – at least, the inhabitants of the village of St Jean, men of enterprise and valor. They hastened their attack, fearing lest the Iroquois should escape them, and desiring to surprise the latter while they were still on the road. They set out on the fifth day of the month of December, directing their route toward the place where the enemy was expected. But the latter, having taken a roundabout way, was not met; and, to crown our misfortune, the enemy, as they approached the village, seized upon a man and woman who had just come out of it. They learned from these two captives the condition of the place, and ascertained that it was destitute of the better part of its people. Losing no time, they quickened their pace that they might lay waste everything, opportunity so greatly favoring them.

It was on the seventh day of the month of last December, in the year 1649, toward three o'clock in the afternoon, that this band of Iroquois appeared at the gates of the village, spreading immediate dismay, and striking terror into all those poor people – bereft of their strength and finding themselves vanquished; when they thought to be themselves the conquerors. Some took to flight; others were slain on the spot. To many, the flames, which were already consuming some of their cabins, gave the first intelligence of the disaster. Many were taken prisoners, but the victorious enemy, fearing the return of the warriors who had gone to meet them, hastened their retreat so precipitately, that they put to death all the old men and children, and all whom they deemed unable to keep up with them in their flight.

It was a scene of incredible cruelty. The enemy snatched from a Mother her infants, that they might be thrown into the fire; other children beheld their Mothers beaten to death at their feet, or groaning in the flames – permission, in either case, being denied them to show the least compassion. It was a crime to shed a tear, these barbarians demanding that their prisoners should go into captivity as if they were marching to their triumph. A poor Christian Mother, who wept for the death of her infant, was killed on the spot, because she still loved, and could not stifle soon enough her Natural feelings.

Father Charles Gamier was, at that time, the only one of our

Fathers in that mission. When the enemy appeared, he was just then occupied with instructing the people in the cabins he was visiting. At the noise of the alarm, he went out, going straight to the Church, where he found some Christians. 'We are dead men, my brothers,' he said to them. 'Pray to God, and flee by whatever way you may be able to escape. Bear about with you your faith through what of life remains; and may death find you with God in mind.' He gave them his blessing then left hurriedly, to go to the help of souls. A prey to despair, not one dreamed of defence. Several found a favorable exit for their flight; they implored the Father to flee with them, but the bonds of Charity restrained him. All unmindful of himself, he thought only of the salvation of his neighbor. Borne on by his zeal, he hastened everywhere, either to give absolution to the Christians whom he met, or to seek, in the burning cabins, the children, the sick, or the catechumens, over whom, in the midst of the flames, he poured the waters of Holy Baptism, his own heart burning with no other fire than the love of God.

It was while thus engaged in holy work that he was encountered by the death which he had looked in the face without fearing it, or receding from it a single step. A bullet from a musket struck him, penetrating a little below the breast; another, from the same volley, tore open his stomach, lodging in the thigh, and bringing him to the ground. His courage, however, was unabated. The barbarian who had fired the shot stripped him of his cassock, and left him weltering in his blood, to pursue the other fugitives.

This good Father, a very short time after, was seen to clasp his hands, offering some prayer; then, looking about him, he perceived at a distance of ten or twelve paces, a poor dying Man – who, like himself, had received the stroke of death, but had still some remains of life. Love of God, and zeal for Souls, were even stronger than death. Murmuring a few words of prayer, he struggled to his knees, and, rising with difficulty, dragged himself as best he might toward the sufferer, in order to assist him in dying well. He had made but three or four steps, when he fell again, somewhat heavily. Raising himself for the second time, he got, once more, upon his knees and strove to continue on his way; but his body, drained of its blood, which was flowing in abundance from his wounds, had not the strength of his courage. For the third time he fell, having proceeded but five or six steps. Further than this we have not been able to ascertain what he accomplished – the good Christian woman who faithfully related all this to us having seen no more of him, being herself overtaken by an Iroquois, who struck her on the head with a war hatchet, felling her upon the spot, though she afterward escaped. The Father, shortly after, received from a hatchet two blows upon the temples, one on either side, which penetrated to the brain. His body was stripped and left, entirely naked, where it lay.

Two of our Fathers, who were in the nearest neighboring mission, received a remnant of these poor fugitive Christians who arrived all out of breath, many of them all covered with their own blood. The night was one of continual alarm, owing to the fear, which had seized all, of a similar misfortune. Toward the break of day, it was ascertained from certain spies that the enemy had retired. The two Fathers at once set out, that they might themselves look upon a spectacle most sad indeed, but nevertheless acceptable to God. They found only dead bodies heaped together, and the remains of poor Christians – some who were almost consumed in the pitiable remains of the still-burning village; others deluged with their own blood; and a few who yet showed some signs of life, but were all covered with wounds – looking only for death, and blessing God in their wretchedness. At length, in the midst of that desolated village, they descried the body they had come to seek; but so little cognizable was it, being completely covered with its blood, and the ashes of the fire, that they passed it by. Some Christian savages, however, recognized their Father who had died for love of them. They buried him in the same spot on which their Church had stood, although there no longer remained any vestige of it, the fire having consumed all.

WILLIAM DUNBAR
Braddock at the Monongahela

The struggle for mastery in North America between the English and the French, begun at the end of the seventeenth century (King William's War), was by the middle of the eighteenth century moving to a climax. The British – as they should be described after the union of the crowns of England and Scotland in 1707 – were by mid-century demographically by far the strongest power in the continent, their settlers in New England, the Middle Atlantic colonies and the South outnumbering the French of New France (Canada) by a million to fifty thousand. The geographical position of the French was, however, dominant. Not only did they control the St Lawrence river, 'the great highway into the continent', they also occupied the Great Lakes and, by an overland route hinging on the future metropolis of Chicago, extended their power to the upper reaches of the Mississippi, from which, through a chain of forts, it descended to the Gulf of Mexico at New Orleans. They were also pushing eastward from the Mississippi–Missouri basin into the complex of river systems which joins it via the waterways of the Ohio and the Tennessee. That region, known to Americans as the 'Old North West', was in mid-century a no man's land. Through the river system, the French had easy access to it, and so the opportunity to occupy the western descents from the Appalachian mountain chain, which confined the British to the Atlantic coastal region. The British, well aware of French ambitions, were meanwhile seeking to break across the Appalachians from Virginia, set up posts in the Old North West and to bring French America under attack from a new direction.

In 1754 a British military expedition, led by Colonel George Washington, invaded the area, with the object of building a fort at the confluence of the Allegheny and Monongahela rivers. On arrival he found that the French had already built a fort on his objective (modern Pittsburgh) and he was forced to withdraw. In the following year, however, a much stronger expedition of troops sent from Britain, under the command of General Edward Braddock, arrived in Virginia, crossed the Appalachians and set out for the Monongahela. On 9 July, after a long and difficult march, it was ambushed by the French just short of their objective. Braddock's 2,000 men, who included Americans as well as British, suffered a bloody defeat at the hands of 900 French troops and their Indian allies, the survivors escaping as best they could through the wilderness. The victory of the French demonstrated their superiority in forest fighting. It was, however, their last American victory. In the following year, at the outbreak of the Seven Years War, they became embroiled in a campaign which would lead to the total destruction of New France in 1763.

*   *   *

On Wednesday the 9th Inst, We were advanced within 9 miles of Fort du Quesne, & in order to reach it were to pass the Monongahela in 2 different places. by 2 in the Morning Col: Gage with the 2 Companies of Grenadiers, to wch I belonged, with 150 Men besides was ordered with 2 six pounders, to cross the River, & cover the March of the General wth the Rest of the Army. This We executed witht any disturbance from the Enemy, and when we had yet possession of the Bank of the sd crossing, we were remained drawn up, till the general came with the rest of the Army, & passed the River in a Column. The Ground from thence to the French Fort, we were told was pretty. good, & the woods open, but all upon the ascent [.] Col: Gage was then ordered with his advanced Party to march on, and was soon followed by the general. We had not marched above 800 yards from the River, when we were allarmed by the Indian Hollow [i.e. holloa], & in an instant, found ourselves attacked on all sides, their methods, they immediately seise a Tree, & are certain of their Aim, so that before the Genl came to our assistance, most of our advanced Party were laid sprawling on the ground. our Men unaccustomed to that way of fighting, were quite confounded, & behaved like Poltrons, nor could the examples, nor the Intreaties of their officers prevail with them, to do any one [what was ordered]. This they denied them, when we begged of them not to throw away their fire, but to follow us with fixed Bayonets, to drive them from the hill & trees, they never minded us, but threw their fire away in the most confused manner, some in the air, others in the ground, & a great many destroyed their own Men & officers. When the General came up to our assistance, men were seized with the same Pannic, & went into as much disorder, some Part of them being 20 deep. The officers in order to remedy this, advanced into the front, & soon became the mark of the Enemy, who scarce left one, that was not killed or wounded; when we were first attacked, It was near one o'Clock, & in this Confusion did we remain till near 5 in the Evening, our Men having then thrown away their 24 Rounds in the manner above mentioned, & scarce an officer left to head them. They then turned their backs, & left the Enemy in possession of every Thing. What officers were left, endeavoured to rally them at the first crossing of the River, but all to no purpose, terrified at the notion of having no Quarter & being scalped, they ran witht knowing where & most of them threw their Arms from them[.] The French & Indians not imagining our Pain & Consternation were so great, as they really were, pursued us no further than the first crossing otherwise 100 of them, might have cut the Remainder of us to Peices. We marched all night in the utmost horrour & distress, most of us wounded, without a bit of anything to eat & nothing to cover us. On Friday the 11th We arrived at Col: Dunbars Camp 56 Miles from the Place of Action. our Strength before the Engagement amounted to 1100 Men.

Killed & wounded 823.

Note: This account has never been in print, nor have I ever seen a reference to it. The manuscript, of which the preceding is a transcript, is Hardwicke 136, document no. 6, Manuscripts and Archives Division, New York Public Library, Astor, Lenox and Tilden Foundations. Dunbar's account follows one by British A in the Hardwicke Papers. The two reports are in different hands, probably neither that of the eyewitness himself. Both are rather rough. Ms. 6 is headed 'Extract of a Letter from Lieutenant Dunbar – Wills Creek 20th July 1755.'

ROBERT SOUTHEY
(1774–1843)
The Battle of Blenheim

I

It was a summer evening,
Old Kaspar's work was done,
And he before his cottage door
Was sitting in the sun,
And by him sported on the green
His little grandchild Wilhelmine.

II

She saw her brother Peterkin
Roll something large and round,
Which he beside the rivulet
In playing there had found;
He came to ask what he had found,
That was so large, and smooth, and round.

III

Old Kaspar took it from the boy,
Who stood expectant by;
And then the old man shook his head,
And, with a natural sigh,
''Tis some poor fellow's skull,' said he,
'Who fell in the great victory.

IV

'I find them in the garden,
For there's many here about;
And often when I go to plough,
The ploughshare turns them out!
For many thousand men,' said he,
'Were slain in that great victory.'

V

'Now tell us what 'twas all about,'
Young Peterkin, he cries;
And little Wilhelmine looks up
With wonder-waiting eyes;
'Now tell us all about the war,
And what they fought each other for.'

VI

'It was the English,' Kaspar cried,
'Who put the French to rout;
But what they fought each other for,
I could not well make out;
But everybody said,' quoth he,
'That 'twas a famous victory.

VII

'My father lived at Blenheim then,
Yon little stream hard by;
They burnt his dwelling to the ground,
And he was forced to fly;
So with his wife and child he fled,
Nor had he where to rest his head.

VIII

'With fire and sword the country round
Was wasted far and wide,
And many a childing mother then,
And new-born baby died;
But things like that, you know, must be
At every famous victory.

IX

'They say it was a shocking sight
After the field was won;
For many thousand bodies here
Lay rotting in the sun;
But things like that, you know, must be
After a famous victory.

X

'Great praise the Duke of Marlbro' won,
And our good Prince Eugene.'
'Why 'twas a very wicked thing!'
Said little Wilhelmine.
'Nay . . . nay . . . my little girl,' quoth he,
'It was a famous victory.

XI

'And everybody praised the Duke
Who this great fight did win.'
'But what good came of it at last?'
Quoth little Peterkin.
'Why that I cannot tell,' said he
'But 'twas a famous victory.'

ANNA MYERS
The Revolution Remembered (1)

The final defeat of the French by the British in North America in 1763 promised to bring lasting peace to the enormous region that lies between the Atlantic and the Mississippi, much of which had been fought over for a century and a half. The promise was not realized. Britain's attempt to make the English colonists pay for the cost of supporting the imperial garrison was seen as infringement of their rights. They had always been taxed, but by their own assemblies, not by London. The discontent over taxation swelled into a mood of general rebellion, which broke out in 1776. What Americans call 'the War of the Revolution' had begun.

It was a war that in many ways resembled the previous wars between the British and the French, in that many of the campaigns were fought in the difficult, wooded back country, where regular troops found themselves at a disadvantage. Both the British and French had enlisted Indian tribes as allies in this forest warfare, and the British did so again against the colonists after 1776. Indian fighting practices – ambush, raiding, massacre and the taking of hostages, who were often then adopted into tribal life – were repugnant to the colonists, but not unfamiliar. Those living at the edge of settlement along the line of the Appalachian mountains had endured Indian raids for several generations and were prepared to return savagery in kind. This was particularly the case in the country along the Mohawk river, a strategic corridor running due west from its junction with the mighty Hudson river at Albany towards Fort Oswego on Lake Ontario. Forts were key positions in the wilderness war, affording bases to British troops but also points of control over their Indian allies, where they could be paid, armed and given orders.

Anna Oosterhout Myers's account of her capture by the Indians of her native Mohawk Valley graphically conveys both the dangerous climate of frontier life before the War of the Revolution, and the deliberate rekindling of insecurity by the British in their effort to unhinge the rebellious colonists' hold on upper New York territory and their connections with the seat of revolution in Boston. The Mohawk river led directly to the Hudson from the British stronghold of Canada; the Hudson was, in George Washington's words, 'the key to the continent'. Anna's narrative, written long after the war was finished to justify her claim to a pension from the United States government, is not therefore simply a story of a personal ordeal. It also exemplifies the bitter, piecemeal but deadly serious campaigning for advantage in a key strategic zone of the struggle for mastery between the British and the Americans in the years 1776–82.

Note: The original settlers in New York colony were Dutch, hence Anna's use of that language. 'Tories' were American colonists who remained loyal to King George III and fought, often with great bitterness, against their revolutionary neighbours. The population of the colonies divided into a third revolutionaries, a third loyalists and a third who attempted, increasingly unsuccessfully, to remain neutral.

*   *   *

To Anna Oosterhout Myers (b. *ca.* 1747), as to many women of the frontier, Indian attacks were a normal part of life. When she was a child, her parents, four sisters, and a brother were murdered by Indians. She and another brother were captured, and she lived with the Indians long enough to forget most of her native Dutch.

The Revolution brought the Indians back to the Mohawk Valley in full force, and Anna Myers's deposition, forwarded in 1840 to support a claim for her husband's pension, is very much a personal narrative of her own experiences.

The Revolution on the frontiers of the upper Susquehanna, Delaware, and Mohawk rivers inspired a considerable body of romantic historical literature. Works such as William Leete Stone's *Life of Joseph Brant* (New York: A. V. Blake, 1838) and Jeptha R. Simms's *History of Schoharie County, and Border Wars of New York* (Albany: Munsell and Tanner, 1845) are classic narratives of Indian–white conflict that remain primary sources of information. Yet for no theater of the war is it harder to get concrete information on names, dates and places.

The standard sources do not confirm an attack at Canajoharie on 17 April 1778, when Anna Myers's account seems to suggest her settlement was overrun, but they are vague on many aspects of their subject. If she were ninety-three in 1840, her capture by Indians as a child of three would have occurred in 1750, several years before the outbreak of the French and Indian War. On the other hand, by the same reckoning of her age, her last child would have been born when she was fifty-five, improbable from a medical standpoint. It seems likely that she was eighty-six or eighty-seven at the time of her application. Even if her memory was faulty on dates, the narrative has an unquestioned air of credibility as to the reality of the events themselves.

Anna's husband and son eventually made it back home, and she went on to bear a total of twelve children. She lived at Minden, Sullivan, Hastings, and Mexico, New York. She was granted a pension for her husband's one year of military service. One could make a good case that she had earned it in her own right.

Says she is now, as near as she can recollect, about ninety-three years old; that she has no record of her age and therefore cannot state the precise time when she was born. Her maiden name was Anna Oosterhout. She was born in Canajoharie, in the present county of Montgomery, in the state of New York. During the French War, and when she was about three years old as near as she now recollects, she was taken prisoner by the French and Indians and carried to Canada. She well recollects the transaction. The house in which her father and his family resided was attacked and surrounded by the Indians, and her father and mother, four sisters, and one brother were killed by them, and she and a brother by the name of John, then about fifteen years of age, was taken prisoner. She understood and believes that the reason why she and her brother were not killed was that one of the Indians belonging to the party had lost children of about the same age, and wanted them to adopt. One other brother escaped, whose name was Frederick. At the time, he was sick with the whooping cough, and when the Indians saw him cough, they were frightened of it and let him alone. The Indians took her and her brother to their camp, but where or which way or how far they traveled she cannot state, but supposes and believes they went to Canada, as she recollects they called the place 'Canda'. She was with the Indians about three years, when she learned to speak the Indian language, and when she returned to the Mohawk, she had almost entirely forgotten her native language, the Dutch. Afterwards she was sent to Albany, where she was met by an uncle of hers who had come there to see if any of his brother's children were alive. She was taken by her uncle to his residence at Canajoharie.

She then went to live with her grandmother, Mrs Katharine Hess, with whome she resided until she was about fifteen years old. She was married to Henry Moyer or Myer [Myers] about the fifteenth day of May, 1770. She was married at the house of her father-in-law in Canajoharie, where she had been residing for several weeks previous. She was married by the Reverend Mr Ehle, a clergyman of the Low Dutch church. David Hess was present at the time of said marriage, as she well recollects he being a fiddler and played for the company to dance the evening of the marriage. She believes the said David is now living, and she knows of no other person who was present at said marriage. She knows of no record of said marriage. A record was made of said marriage in the family Bible, but the same was destroyed afterwards as will appear from what appears afterwards. She has had by said Henry Moyer twelve children, the oldest of whom is about sixty-eight years and the youngest about thirty-eight years. There are eight only of her said children now living. The said Henry Moyer was several years older than this deponent and died on the nineteenth January, 1830.

Soon after the Revolutionary War commenced, the valley of the

Mohawk became the scene of many important operations and bloody transactions. He was frequently called out for the purpose of defending the frontier from the incursions of the Tories and Indians and was on guard at the fort nearly the whole time. For about a year before the Battle of Oriskany, the said Henry held the office of ensign or lieutenant in the militia in a company commanded by Captain Diefendorf. As soon as it was announced, in the spring and summer of 1777, that Colonel St Leger was raising an army of Tories and Indians at Oswego for the purpose of invading the valley of the Mohawk, the whole country was in a state of excitement. General Herkimer issued a proclamation for every able-bodied man to turn out, leaving the old men and those who were not able to bear arms to guard the forts and other places where the women and children were assembled. The company commanded by the said Diefendorf turned out under General Herkimer and proceeded with him towards Oriskany. The said Henry was at that time an ensign or lieutenant in the company of said Diefendorf and went with the said Diefendorf as far as German Town, then called, about eight miles below Utica. The said Henry was there taken lame in consequence of having cut his foot, which had previously healed up, but in consequence of traveling it had broken out, and his foot had swelled to such a degree that it had cracked open when he returned. Said Diefendorf was killed in the Battle of Oriskany and was the brother-in-law of said Henry, having married his sister. It was said at the time that said Diefendorf was killed by an Indian who was in a tree. During the summer of 1777 the said Henry was absent most of the time in the service. After the Tories and Indians had left Fort Schuyler, in August or September, the said Henry returned to his home.

After the return of said Henry, as aforesaid, he was engaged for the greater part of that time and until the seventeenth day of April following in assisting about the erection of a fort in the present town of Minden, in the county of Montgomery and state of New York, about six miles east of Little Falls, which was called Fort Willett. Said fort was nearly completed on said seventeenth of April. It was intended for the people living near said fort to remove therein on the next Monday.

On Sunday, which was on the seventeenth day of April aforesaid, about sunrise in the morning, and while some of the children of this deponent were sent a few rods from the house to feed some calves, this deponent discovered the horses then owned by the said Henry run past the door of the house greatly frightened, and at the same time she heard her children scream. She went to the door to see what was the matter and there saw several Indians who had taken the two children who had been sent out as aforesaid. One of the Indians was near the door when she went out, and he yelled and whooped and seized her by the arm. The Indians took her and her four children about fifty rods from the house and stopped. Soon after they stopped, they were met by another party of Indians who had been up to a neighbor's by the name of Christian Durt, who had taken the said Durt, his wife, and one child, and the said Henry Moyer. A few minutes before she had been taken by the Indians, as aforesaid, her husband, Henry Moyer, had left the house and gone to the said Durt's to see about moving into the fort they had been building, as aforesaid, and while there, was taken prisoner with the said Durt and his family.

She was discharged by the Indians soon after the parties met, as aforesaid, with a sucking child then about two years old. Her husband, the said Henry, and three of her children were then taken away by the Indians, and where they went she does not know except from information. After she was discharged, as aforesaid, she returned to her house, which she found rifled of such articles as the Indians could carry and set on fire. The Indians had put brands of fire between one or two beds, which were on fire when she returned. She succeeded in getting the beds out of the house and extinguished the fire and prevented the building from being entirely consumed. About two hours after the Indians left, two of her children returned, who were daughters, leaving the said Henry [and their son], then about three years old, prisoners with the Indians. When her daughters returned, they informed this deponent that the Indians discharged them, and that their father also wanted the Indians to discharge the boy Henry, but they refused to do so and told the said Henry, her husband, that if he attempted to run away, they would kill his boy.

The wife of said Durt was also discharged by the said Indians, and her husband and child, a boy about seven years old, were carried off by the Indians. Alarm was soon made, and she on the same day went to a house called Fort House where the people had assembled and where she remained about a week, when she went into Fort Willett, where she remained for two or three years, until it was understood that it would be safe for the people to go onto their farms.

The said Henry, her husband, returned in the fall of 1779, having been absent more than a year and a half. When he returned, the said Henry informed this deponent, which she believes true, that the Indians took him to Niagara, where he was forced to run the gauntlet. While there, he was struck by an Indian with a tomahawk over the left eye, which produced a wen of considerable size and which remained there until his death. He also, at the same place, received a cut on the right side of the head which left a scar about three inches long. From Niagara, they went to Oswego. While there, he was set to chopping wood in company with a man by the name of Stimet near the lakeshore. While a party of the British were endeavoring to get a boat ashore for the purpose of receiving the wood, the same was capsized, when he and the said Stimet escaped and went up the Oswego River, which was then a wilderness. They went to Three River Point, about twenty-five miles south of Oswego, where they discovered that they were pursued. The party pursuing encamped overnight, and then he and the said Stimet crossed the river from the west to the east side and escaped. They were five weeks in the woods and finally were found by a party of friendly Indians about six miles from Schoharie in the present county of the same name. He remained there several days until he got recruited and had recovered his strength and then returned to Fort Willett, where this deponent was. This deponent's son remained a prisoner with the Indians until peace was declared, when he returned home.

During harvest the year before the said Henry was taken prisoner, the people in the neighborhood where she resided lived in a house called Fort Walradt. The fort was burned by the Tories and Indians after the people had escaped. This fort was situated about two miles from the Mohawk River, and an alarm had been made that the enemy were in the neighborhood, when the people left Fort Walradt and went to the river for greater safety. All the furniture, clothing, and all the household stuff of the said Henry was then destroyed, and also the Bible in which her marriage with the said Henry was recorded. And this deponent further says that she is now the widow of the said Henry Moyer, never having been married to any other person.

JOHN SCOTT OF AMWELL
(1730–1793)
The Drum

I hate that drum's discordant sound,
Parading round, and round, and round:
To thoughtless youth it pleasure yields,
And lures from cities and from fields,
To sell their liberty for charms
Of tawdry lace, and glittering arms;
And when Ambition's voice commands,
To march, and fight, and fall, in foreign lands.

I hate that drum's discordant sound,
Parading round, and round, and round:
To me it talks of ravaged plains,
And burning towns, and ruined swains,
And mangled limbs, and dying groans,
And widows' tears, and orphans' moans;
And all that Misery's hand bestows,
To fill the catalogue of human woes.

JACOB ZIMMERMAN
The Revolution Remembered (2)

Jacob Zimmerman, like Anna Myers, came from the frontier of settlement in the Mohawk Valley. He may, like her, have belonged to the original Dutch community or have been a German immigrant of the first or a later generation. By 1776 there were already sizeable communities of German immigrants in Pennsylvania and New York colonies, many of whom had left Germany for religious reasons.

Unlike Anna, who was claiming her husband's right to a pension, which descended to her, Jacob claims in his own right. He was a soldier in, presumably, the militia of New York Colony, an organization that predated the revolution; it was, indeed, the constitutional colonial defence force. The units of the militia on the frontier had a long experience of fighting Indian raiders. They initiated punitive operations, often on their own account, and had played a major role in the wars between the French and British. George Washington had begun his military life as an officer of the militia of Virginia, under British command. It was as a Virginia militia officer that he had marched with Braddock to the Monongahela.

Jacob's ordeal has a wider than American significance. The European powers enlisted local warlike peoples all over their empires, in India, Africa, Arabia and South-East Asia, as well as the Americas, sometimes to fight other locals, sometimes to fight Europeans in the struggle for imperial control. The European officers who managed or directly commanded such local, often called 'tribal', units attempted to impose European discipline and military ethics upon them, although with very varied success. They were often compelled to enter into a complicity with tribal military practices which would have been rejected with repugnance in the 'civilized' warfare of Europe itself. Native Americans – 'Redskins' – proved particularly difficult to discipline and French and British alike came to acquiesce in such habits of their allies as kidnapping, ritual torture and mutilation of the dead, even the wounded. After the disaster of the Monongahela, for example, French officers engaged in polite conversation with the British officers they had captured, while their Indian allies put the British private soldiers to slow and painful death.

Jacob's account of the forest skirmish in which he was captured may therefore be taken as highly authentic. The Tories (white loyalists) obviously made no effort to prevent their Indian confederates scalping their victims, while the British officers to whom he was eventually brought clearly accepted that, as a prisoner, he had become the Indians' property. He was lucky that one of them, Colonel Campbell, purchased his life for cash in Montreal. He was also lucky that the Indians spared his life after they found him wounded; had his wound been worse, they would undoubtedly have killed him as an impediment to their escape.

*   *   *

Jacob Zimmerman (*ca.* 1757–1835), a lifelong resident of Oppenheim, Montgomery County, New York, volunteered in the militia and served seventeen tours of duty on the New York frontier between 1776 and 1781. His narrative of service is long and repetitious, and only the last tour is published here.

While on a five-man scouting party traveling between Fort Zimmerman and Fort Walradt, 9 August 1781, he was wounded and captured. Although carrying a ball in his neck, he was taken by rapid forced marches to Swagotchie, now Ogdensburg, New York, and on to Montreal, where he was sold to a British officer. In spite of the excruciating pain and the privations of captivity, he provides an objective account of his Indian and British captors, free of the bitterness usually present in such narratives.

Zimmerman was returned by way of Boston in 1782. He applied successfully for a pension in 1833.

*

On the ninth day of August, 1781, he still belonged as a private in the company whereof Christian House was captain, in the regiment whereof Jacob Klock was the colonel. On said day he still resided in the said town of Oppenheim and was in Fort Zimmerman guarding same. He and about five others of said company, to wit, his lieutenant, John Zimmerman, his cousin Jacob Zimmerman, Adam Zimmerman, Peter Hellegas, and himself went from Fort Zimmerman on their way to Fort Walradt. He understood it was done by the order of Col. Marinus Willett, then stationed at Fort Plain in the town of Minden. The orders were that six men more from Fort Walradt were to go on a scouting party to see whether the traces of any Indians could be discovered in the neighborhood. On said ninth day of August, 1781, according to his recollection, they so started to go to Fort Walradt.

After going about a quarter of a mile or so from Fort Zimmerman on their way to Fort Walradt, they were fired upon by a large party of Indians and Tories who were concealed in the brushes, by which fire John Zimmerman, the lieutenant of said House's company, and said [cousin] Jacob Zimmerman were wounded and killed and scalped by the Indians. [Two others] made their escape. He [i.e. the Jacob Zimmerman who is the subject of this extract] was badly wounded in his neck and throat. The Indians did not discover his wound at first. They took him a prisoner together with Peter Hellegas, and he saw the Indians tomahawk and scalp his said lieutenant and said (cousin) Jacob Zimmerman.

The ball struck in his neck or throat. After going but a short way with the Indians, they discovered his clothes bloody and then saw his wound. They halted and ordered him to spit, to see, he supposed, whether he spit any blood and was dangerously wounded, and if so, to kill him also. As directed he did spit but not any blood, when they started off again on a hard trot, and he was obliged to keep up with them. In consequence, his wound gave him a great deal of pain. He several times began to feel faint and thought he should fall and be unable to proceed. His clothes were bloody. The Indians halted several times and made him spit, but as he did not spit any blood he was told he must go along. He suffered a great deal on the way.

Peter Hellegas, who was taken a prisoner also, would dip up water with his hands on their way to give him to drink, as he could not stoop to drink at the brooks by reason of his neck being swelled and stiff. They traveled about a week through the woods until they got to a place commonly called Swagotchie, where was a British fort. On the way, he lived chiefly on roasted cornmeal with which the Indians mixed water, almost the only food he could swallow. When they came to a stream of water, he was not suffered by the Indians to wet his feet, but they would take him on their backs across the streams. The ball still remained in his neck. His right hand he could hardly raise to his mouth by reason of the swelling of his wound, and he suffered more than he can express. The Indians treated him well enough, as much so as he could expect. Some leaves the Indians found and applied to his wound, which eased the pain some.

When he arrived at the fort at Swagotchie, he saw some Tories he had been acquainted with before the war. They examined him as to the state of affairs at home and whether the people had anything left to eat. He told them that Colonel Willett commanded at Fort Plain and was an active and good officer, that the Indians had destroyed much of their grain, etc., but that those whose property was spared would give to those that wanted, and thus they got along well enough. Major Ross it was said then commanded the fort at Swagotchie. He told some of the Tories he knew he wished the ball to be cut out of his neck. They told him that unless the Indians consented it could not be done. The Indians, however, consented. He was taken to the room occupied by a surgeon. He was placed on a chair with his head held back over the chair by an Indian. The surgeon cut or extracted the ball, and who told him that a few days more he would have died of his wound if the ball had remained.

From Swagotchie he was taken to Montreal, where he saw many of his fellow soldiers or countrymen prisoners of war. Colonel Campbell at Montreal purchased him of the Indians. He saw said Campbell pay the Indians some money on said purchase. A Captain Jones at Montreal he became acquainted with, who was a captain in the British service and who, he thinks, had to see to the prisoners;

to said Jones he had told the manner of his being taken a prisoner and his sufferings, etc. The Tories at Montreal wished him to enlist. He refused, telling them in substance he would rather perish on the spot than enlist among them. Captain Jones had previously informed him that the Tories dare not hurt him and he could freely express his mind to them. Captain Jones and his lady were kind to him. He would go often and see Captain Jones and ask him for a little tobacco, and he always got some, but once was refused when his lady told him, 'Oh! Do give him some coppers,' which the captain did, adding that he did not want to be troubled so much for tobacco and told him to go and buy some with the coppers.

From Montreal he was taken to Quebec, thence to Boston, and from Boston he traveled home on foot, to wit, to his own town of Oppenheim where he resided when he was taken a prisoner and has ever since his return from imprisonment. He returned home from his imprisonment about ten days before Christmas in the year 1782, that is, on the fifteenth or sixteenth day of December, 1782, according to the best of his recollection. He believes he has given correctly the time he was taken a prisoner and when he returned home but has stated same only from the best of his recollections, having no memorandum thereof. He was a prisoner as aforesaid from the time he was taken, as he considers, to the time he returned home, to wit, for one year, four months, and six days. A special law was passed by Congress allowing him a pension on account of his said wound as under which law he has received his pension up to the fourth day of March, 1833. He believes from information that Ogdensburg is now situated on or near the place called in that war Swagotchie or Oswegatchie. On reflection, he thinks that Captain Robenson commanded instead of Ross the fort.

DAVID CROCKETT
Davy Crockett

Davy Crockett (1786–1836), a Tennessee backwoodsman, hunter and Indian fighter, who became a popular Congressman and died fighting at the defence of the Alamo against the Mexicans, is to Americans the prototype of the frontiersman. Afraid of nothing, least of all the law of the United States, he represents the aggressive free spirit of those early citizens who fought their country's enemies – British, Indian, Mexican – to make it the dominant North American power.

His earliest military experience was during the Anglo-American War of 1812 (1812–15), when he joined General (future President) Andrew Jackson's campaign against the Creek Indians of Alabama. The Creeks had allied with the British and on 20 August 1813 surprised the American garrison of Fort Mims, now in Alabama, and massacred over two hundred soldiers and civilians sheltering inside the fort. The punitive campaign led by Jackson that followed exemplifies the savagery of the war white Americans were prepared to wage against native 'Redskin' Americans in the struggle to secure the territory of the United States. Crockett deplores the military practices of the Creeks, but is clearly prepared to overlook the cruelties of the Chickasaws and Choctaws who were fighting on his side if it furthered the American cause.

*　　*　　*

Crockett has joined the militia in the fight against the Creek Indians following their attack on Fort Mims, Tennessee.

When we marched from Fort Montgomery, we went some distance back towards Pensacola; then we turned to the left, and passed through a poor piny country, till we reached the Scamby river, near which we encamped. We had about one thousand men, and as a part of that number, one hundred and eighty-six Chickasaw and Choctaw Indians with us. That evening a boat landed from Pensacola, bringing many articles that were both good and necessary; such as sugar and coffee, and liquors of all kinds. The same evening, the Indians we had along proposed to cross the river, and the officers thinking it might be well for them to do so, consented; and Major Russell went with them, taking sixteen white men, of which number I was one. We camped on the opposite bank that night, and early in the morning we set out. We had not gone far before we came to a place where the whole country was covered with water, and looked like a sea. We didn't stop for this, tho', but just put in like so many spaniels, and waded on, sometimes up to our armpits, until we reached the pine hills, which made our distance through the water about a mile and a half. Here we struck up a fire to warm ourselves, for it was cold, and we were chilled through by being so long in the water. We again moved on, keeping our spies out; two to our left near the bank of the river, two straight before us, and two others on our right. We had gone in this way about six miles up the river, when our spies on the left came to us leaping the brush like so many old bucks, and informed us that they had discovered a camp of Creek Indians, and that we must kill them. Here we paused for a few minutes, and the prophets pow-wowed over their men awhile, and then got out their paint, and painted them, all according to their custom when going into battle. They then brought their paint to old Major Russell, and said to him, that as he was an officer, he must be painted too. He agreed, and they painted him just as they had done themselves. We let the Indians understand that we white men would first fire on the camp, and then fall back, so as to give the Indians a chance to rush in and scalp them. The Chickasaws marched on our left hand, and the Choctaws on our right, and we moved on till we got in hearing of the camp, where the Indians were employed in beating up what they called chainy briar root. On this they mostly subsisted. On a nearer approach we found they were on an island, and that we could not get to them. While we were chatting about this matter, we heard some guns fired, and in a very short time after a keen whoop, which satisfied us, that where ever it was, there was war on a small scale. With that we all broke, like quarter horses, for the firing; and when we got there we found it was our two front spies, who related to us the following story: As they were moving on, they had met with two Creeks who were out hunting their horses; as they approached each other, there was a large cluster of green bay bushes exactly between them, so that they were within a few feet of meeting before either was discovered. Our spies walked up to them, and speaking in the Shawnee tongue, informed them that General Jackson was at Pensacola, and they were making their escape, and wanted to know where they could get something to eat. The Creeks told them that nine miles up the Conaker, the river they were then on, there was a large camp of Creeks, and they had cattle and plenty to eat; and further that their own camp was on an island about a mile off, and just below the mouth of the Conaker. They held their conversation and struck up a fire, and smoked together, and shook hands, and parted. One of the Creeks had a gun, the other had none; and as soon as they had parted, our Choctaws turned round and shot down the one that had the gun, and the other attempted to run off. They snapped several times at him, but the gun still missing fire, they took after him, and overtaking him, one of them struck him over the head with his gun, and followed up his blows till he killed him.

The gun was broken in the combat, and they then fired off the gun of the Creek they had killed, and raised the war-whoop. When we reached them, they had cut off the heads of both the Indians; and each of those Indians with us would walk up to one of the heads, and taking his war club would strike on it. This was done by every one of them; and when they had got done, I took one of their clubs, and walked up as they had done, and struck it on the head also. At this they all gathered round me, and patting me on the shoulder, would call me 'Warrior – warrior.'

They scalped the heads, and then we moved on a short distance to where we found a trace leading in towards the river. We took this trace and pursued it, till we came to where a Spaniard had been killed and scalped, together with a woman, who we supposed to be his wife, and also four children. I began to feel mighty ticklish along about this time, for I knowed if there was no danger then, there had been; and I felt exactly like there still was. We, however, went on till we struck the river, and then continued down it till we came opposite to the Indian camp, where we found they were still beating their roots.

It was now late in the evening, and they were in a thick cane brake. We had some few friendly Creeks with us, who said they could decoy them. So we all hid behind trees and logs, while the attempt was made. The Indians would not agree that we should fire, but pick'd out some of their best gunners, and placed them near the river.

Our Creeks went down to the river's side, and hailed the camp in the Creek language. We heard an answer, and an Indian man started down towards the river, but didn't come in sight. He went back and again commenced beating his roots, and sent a squaw. She came down, and talked with our Creeks until dark came on. They told her they wanted her to bring them a canoe. To which she replied, that their canoe was on our side; that two of their men had gone out to hunt their horses and hadn't yet returned. They were the same two we had killed. The canoe was found, and forty of our picked Indian warriors were crossed over to take the camp. There was at last only one man in it, and he escaped; and they took two squaws, and ten children, but killed none of them, of course.

We had run nearly out of provisions, and Major Russell had determined to go up the Conaker to the camp we had heard of from the Indians we had killed. I was one that he selected to go down the river that night for provisions, with the canoe, to where we had left our regiment. I took with me a man by the name of John Guess, and one of the friendly Creeks, and cut out. It was very dark, and the river was so full that it overflowed the banks and the adjacent low bottoms. This rendered it very difficult to keep the channel, and particularly as the river was very crooked. At about ten o'clock at night we reached the camp, and were to return by morning to Major Russell, with provisions for his trip up the river; but on informing Colonel Blue of this arrangement, he vetoed it . . . and said, if Major

Russell didn't come back the next day, it would be bad times for him. I found we were not to go up the Conaker to the Indian camp, and a man of my company offered to go up in my place to inform Major Russell. I let him go; and they reached the major, as I was told about sunrise in the morning, who immediately returned with those who were with him to the regiment, and joined us where we crossed the river, as hereafter stated.

The next morning we all fixed up, and marched down the Scamby to a place called Miller's Landing, where we swam our horses across, and sent on two companies down on the side of the bay opposite to Pensacola, where the Indians had fled when the main army first marched to that place. One was the company of Captain William Russell, a son of the old major, and the other was commanded by a Captain Trimble. They went on, and had a little skirmish with the Indians. They killed some, and took all the balance prisoners, though I don't remember the numbers. We again met those companies in a day or two, and sent the prisoners they had taken on to Fort Montgomery, in charge of some of our Indians.

I did hear that after they left us, the Indians killed and scalped all the prisoners and I never heard the report contradicted. I cannot positively say it was true but I think it entirely probable for it is very much like the Indian character.

JOHN D. HUNTER
Captivity Among the Indians

The nomadic Indians of North America must be counted among the most warlike peoples the world has known. Others – the Mongols, the Pathans, the Turks, the Zulus – terrified their enemies by their contempt for human life, including the preservation of their own in the heat of battle. Native Americans brought to warriordom, however, an element of fearlessness found in scarcely any other culture, a sense of obligation to participate in the most horrific cruelty that could be inflicted on an individual if he fell into enemy hands. The European code of military honour demands that the soldier risk his all in combat; the distinction he thereby wins is held to entitle him to a dignified immunity from further danger should he become a captive. Native Americans thought otherwise. The warrior's ordeal had only truly begun when he became a defenceless prisoner. It was then, under torture and promise of unavoidable death, that he began to show his quality. The ritual torment of captives was as central to native American warfare as the preliminary war dance. Both were religious in essence.

The intensification of inter-tribal warfare owed much, nevertheless, to European influence. Before the coming of the Spanish from the south and the English from the north, the tribes of the Great Plains were too immobile and poorly armed for their disputes to lead to bloodletting of any bitterness. The acquisition of the horse from the Spanish in the sixteenth century and of firearms from the English in the eighteenth dramatically enhanced their ability to travel rapidly as raiders and to do damage when they met their enemies. When the separate processes of diffusion of the horse and of gunpowder weapons overlapped, as they did west of the Mississippi at the beginning of the nineteenth century, a new and vicious style of Indian warmaking was born. Mounted Indians with guns were to present the white American immigrants moving westward across the Mississippi–Missouri Line after 1849 with the most important impediment to settlement. Already, however, the horse and the gun had transformed the character of warfare between Plains Indians themselves. John Hunter, a white American taken captive by Plains Indians, who spared his life as was occasionally their custom, and who lived a tribal life for several years, describes the horse-and-gun culture of his captors.

*   *   *

The mode of life peculiar to the Indians exposes them to the optional encroachment of all their hostile neighbours. For their security they are therefore indebted to personal bravery, and skill in attack and defence; because, in their active warlike operations, they obey only general instructions; each warrior accommodating his manoeuvres according to his own judgment on the exigency of the occasion. Hence, the cultivation of martial habits and taste becomes essential, and constitutes the chief employment of every individual in their respective communities, first, of the squaws and old men, in relation to precept, and then of the warriors, in respect to example.

Under such guidance, the love of war becomes almost a natural propensity. Besides, they are taught to believe that their happiness here and hereafter is made to depend on their warlike achievements; and daily example confirms it as a fact, so far as the indulgence of their affections is concerned; for the females, both young and old, affect to despise the Indian who openly becomes the lover, without the authority of having acquired distinction either in the chase, or in fighting against the enemies of his country.

It is not, therefore, extraordinary that they should love war, since so many and important results are believed to depend on their success in it: their happiness, their standing in society, and their sexual relations, make it necessary that they should excel, or at least strive to, in whatever is connected with their mode of existence. Hence, they court opportunities for self-distinction, and, in fact, when wanting, often make them, in opposition to justice, and the welfare of their nation; and the indulgence of this disposition is one of the principal causes of the frequency of war among the Indian nations.

They regard their hunting grounds as their birthright; defend them with the most determined bravery; and never yield them till forced by superior numbers, and the adverse fate of war. They are exceedingly tenacious of their rights, and chastise the slightest infringement. Hence, they are almost constantly engaged in warfare with some of their neighbours.

Their instruments of war were formerly the scalping-knife and tomahawk, formed from flinty rocks, the bow and arrow, the war-club, and javelin or spear; and, among some tribes, shields made of several folds of buffalo skin. Latterly those have been pretty generally superseded by the rifle, and steel tomahawk and scalping-knife, procured from the traders.

When a sufficient cause for war is thought to exist, it becomes the subject of private conversation, till the opinions of the warriors are pretty well understood; a council is then convened, and it undergoes a thorough discussion. If determined on conditionally, the offending tribe is made acquainted with all the circumstances; otherwise, they generally keep the affair secret, at least so far as respects the subject of their hostility. On some occasions, when the chiefs from prudential motives think it advisable not to go to war, and omit to convene a council to try the question, the discontent of the warriors reminds them of their duty. They discover it by planting painted posts, blazing trees, ornamenting their persons with black feathers, and omitting to paint, or painting their faces after the manner practised in war. These symptoms are discoverable among the young and uninfluential warriors; but they nevertheless produce the intended effect and lead to a formal expression of the public feelings.

On adjournment of the council, the warriors repair to their respective homes, and, having painted their necks red, and their faces in red and black stripes, they reassemble at some place previously fixed on, and discover their hostile intentions in the dances and songs that follow. They next prepare their arms, and provide the munitions for war; and then follow the ceremonials of fasts, ablutions, anointings, and prayers to the Great Spirit, to crown their undertaking with success. They take drastic cathartics, bathe repeatedly, and finally anoint themselves with bears' grease, in which yellow root has been steeped. They abstain from sexual intercourse, eat sparingly from their military provisions, and take freely of the Kutche-nau, a plant which operates on the human system something like opium, without producing the same comatose effects. They then perform the war dance, which is not less appropriate to this occasion than are all their festive ones to the events for which they have been adapted. Whole days are sometimes spent in making preparations for it. Robes, stumps, posts, &c. are painted red or black; every movement and appearance bespeaks the interest and solemnity that are diffused through the tribe.

The warriors, arrayed in their military habiliments, at a proper signal, assemble and commence the dance. It consists in imitating all the feats of real warfare, accompanied with the alternate shouts of victory, and yells of defeat. In short, they perform every thing which is calculated to inspire confidence in themselves, and to infuse terror into their enemies. They are celebrated only at the dawn of a campaign. After this dance, they commence their march to the cadence of the shouts, songs, and prayers of the old men, women, and children, who usually attend them a short distance on their way.

Their equipments and stores amount merely to indispensables, which consist of their arms, buffalo suet, bears' oil, parched corn, anise and wild liquorice roots – and pipes and tobacco.

Their progress differs according to the make of the country, the prevalence of woods, or hiding-places, &c., through which they have to pass. It sometimes amounts to fifty or sixty miles in a day; but usually to about thirty or forty. This difference arises in general from the circumstance whether they are the pursuers or pursued. They use great precaution in travelling so as not to leave traces for their enemies to follow them. They march by families, or small parties separated from each other, within hearing distance, in single file; and step high and light.

They make various kinds of whoops, by which they communicate intelligence one to another, to any distance within hearing; such as those of war, which are to encourage their own adherents, and intimidate their foes; those of alarm, which advise secrecy or flight, as the exigency may require; those of the chase, &c. They imitate the barking of the fox, the cry of the hawk, or the howl of the wolf; at short intervals of time, so as to maintain their regular distances, and give each other notice in case of danger. These imitations are varied, and accommodated to circumstances previously agreed on, and are as well understood as the telegraphic signals practised among civilized nations.

When arrived within the neighbourhood of their enemies, a

whispering council is held, which is constituted of the principal and subordinate chiefs, and their deliberations are guarded by sentinels, secreted at convenient distances, to prevent a surprise. They then separate and remain hidden, till intelligence from their spies authorizes an attack.

Their modes of fighting vary according to circumstances.

They generally aim at surprising their enemies, and, with such views, secrete themselves and wait patiently, for many days together, for an opportunity. During such times they neither visit nor converse with each other, but lie the whole time, without varying their position more than they can possibly help.

They are implacable in their enmities, and will undergo privations that threaten their own existence, and even rush on certain death, to obtain revenge; but they are grateful for benefits received, and ardent and unchangeable in their friendship. When battle rages, and death is in every aim, the Indian, at the risk of his own life, will save his friend, though arrayed against him in the combat.

Shin-ga-was-sa, while young, visited the Kansas during a hunting excursion. The wife of a distinguished warrior paid him some attentions without the approbation of her husband, which resulted in her repudiation, and threatened the existence of her gallant. Pa-ton-seeh, a young Kansas, secretly interfered, and Shin-ga-was-sa made his escape, without coming in collision with his justly irritated foe. Many years afterwards, the Grand Osages and Kansas were involved in war: a battle followed, in which an Osage had shot down Pa-ton-seeh, and was in the very act of taking his scalp, when Shin-ga-was-sa arrested his hand, and preserved his friend.

In another instance, a Pawnee, who had rendered himself an object of public resentment to the Kansas, and was about to expiate his offences by suffering torture, was, to the astonishment of the whole tribe, preserved by the daring intrepidity of his friend. The circumstance was as follows: The Pawnee had on some former occasion laid his preserver under particular obligations, by an act of which I am now ignorant. In return for it, Sha-won-ga-seeh, the moment he knew of the captivity of his friend, intrigued with the young warriors, who, with some of his friends, interrupted the ceremonials that had been authorized by a national council; cut the bonds of the prisoner: mounted him on a fleet horse, and commanded him to fly for his life.

This daring Kansas had previously so disposed of their horses, that pursuit was out of the question; and the boldness of the measure so completely paralyzed the volition of the Indians, that a single effort was not made to arrest its success. The excitement produced by this affair at first threatened tragic consequences: but Sha-won-ga-seeh's friends rallied to his defence; an explanation ensued, and he finally was much commended for an act that might have cost him his life, without the propitiation demanded for murder on all other occasions.

I could relate many circumstances of a similar nature, which would place this trait in their character beyond all doubt; but the limits prescribed to my work will not authorize it.

In taking a scalp, they seize the tuft of hair left for the purpose on the crown of the head in the left hand, and, raising the head a little from the ground, with one cut of the scalping-knife, which is held in their right hand, they separate the skin from the skull.

During an engagement quarters are very seldom asked or given; but should a combatant throw down his arms, his life is spared, and he is placed in charge of those who are entrusted with the wounded. When it is over, the prisoners are all assembled, and marched to the villages of the captors, either slow or fast, according as they apprehend danger from pursuit: should this, however, be pressing, they destroy all, sparing neither the aged, women, nor children.

When arrived within hearing distance of their homes, the warriors set up the shout of victory, and after a short pause utter as many distinct whoops as they have taken prisoners and scalps. At this signal all the inhabitants tumultuously proceed to meet them, and, after the first greetings and salutations are over, commence an attack, with clubs, switches, and missiles, on the captive warriors. The women are exceedingly barbarous on such occasions, particularly if they have lost their husbands, or any near relatives, in the preceding fight.

Every village has a post planted near the council lodge, which is uniformly painted red, on the breaking out of a war. It is the prisoner's place of refuge. On arriving within a short distance of it, the women and children, armed as above, and sometimes even with firebrands, place themselves in two ranks, between which the warriors, one by one, are forced to pass: it is in general a flight for life; though some, who are sensible of the fate that awaits them, should they survive, move slowly, and perish by the way. Those who reach it are afterwards treated kindly, and permitted to enjoy uninterrupted repose, under the charge of relief guards, until a general council finally determines their fate. The women and children are at once adopted into the respective families of the captors, or some of their friends.

Such warriors as are exempted from their vengeance, generally marry among them, and constitute members of their community. They, however, have it in their power to return to their relatives and nation whenever a peace has been concluded; but, as such conduct would be esteemed ungrateful, instances of the kind very seldom occur. Those who are condemned to death, suffer with great magnanimity the most cruel tortures which revenge can invent. They are generally bound hand and foot, sometimes together, and at others to separate posts or trees, and burned with small pieces of touchwood; pierced with goads, and whipped with briars or spinous shrubs, at different intervals, so as to protract the periods of their tortures.

These victims to a mistaken policy, during their sufferings, recount, in an audible and manly voice, and generally with vehement eloquence, all their valorous deeds of former times, and particularly those which they have performed against their persecutors. They contrast the bravery of their own people with the squaw-like conduct of their enemies: they say that they have done their duty; that the fortune of war happened to be against them; and that they are only hastened into more delightful hunting grounds than those they possess here, by squaws who are incapable of appreciating the merits of brave warriors.

They speak of their own deaths as a matter of no consequence; their nation will not miss them; they have many fearless warriors, who will not fail to revenge their wrongs.

As they grow feeble from suffering, they sing their death songs, and finally expire, without discovering the slightest indication of the pains they endure. Indeed nothing can exceed the indifference with which the Indians apparently suffer the tortures and protracted deaths, inflicted on them by their relentless and unfeeling foes.

In these executions the prisoners often make use of the most provoking language, with a view, no doubt, to shorten the period of their tortures; and they generally succeed; for the outraged party, unable to resist the desire of revenge, despatch them at once with the tomahawk, or some other deadly weapon.

I have known an instance, and others have occurred, in which a female had the temerity to risk the public resentment, by interfering in behalf of the captive. It was at the Kansas village. The subject was a young Maha, who had rendered himself particularly odious, from having taken the scalp of one of their distinguished warriors. He had been bound, and his tormentors had just commenced their dances, and fiend-like yells as the prelude to his destruction, when Shu-ja-he-min-keh, a beautiful girl of eighteen, and daughter of one of their chiefs, abandoned her countrywomen, and, as it were, her country, clasped the destined victim in her arms, implored his life, and would not be separated till her prayers were granted.

Attempts of this kind are not, however, always successful, the Indians being governed somewhat by the number of those condemned, and by the respective standings, and character of the supplicants.

The sufferers, in these instances, believe that to die courageously will entitle them to the particular favour and protection of the Great Spirit, and introduce them into the councils and society of the brave and good, in the delightful regions of perpetual spring and plenty, where, under a cloudless sky, they are destined to enjoy with heightened zest the consciousness of this life unalloyed by its anxieties, pains, and afflictions.

With the Indians, the passion of revenge ceases with its object; and these tragic scenes close with the burial of their victims, which are universally respectful, and attended with very nearly the same exterior ceremonials that are observed in the interment of their own dead; especially if their conduct at the closing scene had been brave and consistent.

In their campaigns, the Indians are always accompanied by some who officiate, when necessary, in the character of surgeons and physicians, but who ordinarily perform the warrior's duty. They do not, however, attend to the wounded till the battle is over, unless they should be in imminent danger, or it should prove of long duration, and the number of sufferers or prisoners becomes considerable. In such cases they become non-combatants, and perform the two-fold duty of surgeons and guards. I shall omit the description of their surgical operations for another occasion. The wounded are borne off on litters to some place of safety: in cases of retreat they are sometimes abandoned; but, in general, they are kept in the advance, and defended with the most obstinate bravery and resolution. They observe the same pertinacious courage in regard to their dead; though, when obliged to abandon them, they do not, if they can possibly avoid it, permit their scalps to fall into the possession of their enemies, and always return and collect their bones, as soon as they can do it with safety. When at a great distance from home, they inter their dead temporarily, but always return, when the proper period has arrived, for their skeletons, and pay them the same honours as though they were enveloped in their muscular integuments.

Nothing can exceed the joyous exultations of the old men, women and children, who have not lost relations, on the return of the warriors from successful warfare; while with those who have, the expression of grief is equally extravagant.

The afflicted associate themselves on the occasion, apart from the festive circles, and the duration of their grief is generally in the inverse ratio of this violence: it does not last long, and they soon join in the rejoicings, which are continued for several days. They are consummated by the scalp dance, in which the squaws bear the trophies, such as scalps, arms and apparel, won by their husbands from the enemy, by songs, the torture of their enemies, and finally by feasts. In the performance of the scalp dance, the squaw usually attaches all the scalps that are in her family to a pole; which she bears on the occasion. As they dance round the council lodge or fire, they alternately sing and recount the exploits that were achieved on their acquisition. The one who sings is for the time the principal, and all the others obsequiously follow her. The men and children join in the whoops and rejoicings. During these festivities, marks of favour are lavished, particularly by the squaws, on all such as have distinguished themselves. The most worthy are seated by the old men and chiefs; the women dance round them, decorate their persons with dresses ornamented with feathers, and porcupine quills stained of various colours; and crown them with wreaths of oak leaves, fantastically interwoven with flowers, beads, and shells.

The reception of the warriors from an unsuccessful expedition is different in the extreme, from the reverse of the circumstance. The mournings are general, and last for several days. The men are morose and gloomy, and only break silence in their prayers to the Great Spirit for support in the revenge they may dilate, or in imprecations denounced against their enemies. After the mournings are at an end, the women appear apprehensive and reserved, and do not generally renew their caresses for some time, unless invited to by the occurrence of more fortuitous events.

THOMAS CAMPBELL
(1777–1844)
Hohenlinden

On Linden, when the sun was low,
All bloodless lay the untrodden snow,
And dark as winter was the flow
Of Iser, rolling rapidly.

But Linden saw another sight,
When the drum beat, at dead of night,
Commanding fires of death to light
The darkness of her scenery.

By torch and trumpet fast arrayed,
Each horseman drew his battle blade,
And furious every charger neighed
To join the dreadful revelry.

Then shook the hills, with thunder riven;
Then rushed the steed, to battle driven;
And, louder than the bolts of heaven,
Far flashed the red artillery.

But redder yet that light shall glow,
On Linden's hills of stainèd snow;
And bloodier yet, the torrent flow
Of Iser, rolling rapidly.

'Tis morn; but scarce yon level sun
Can pierce the war-clouds, rolling dun,
Where furious Frank, and fiery Hun,
Shout in their sulphurous canopy.

The combat deepens. On, ye brave,
Who rush to glory, or the grave!
Wave, Munich, all thy banners wave,
And charge with all thy chivalry!

Few, few shall part, where many meet!
The snow shall be their winding sheet,
And every turf, beneath their feet,
Shall be a soldier's sepulchre.

Note: On 3 December 1800, the French General Jean Victor Moreau crossed the Rhine with his troops, decisively defeating the Austrian and allied forces in the Bavarian village of Hohenlinden. The battle marked a turning point in the French Revolutionary Wars.

PART II

This second series of extracts concerns the warfare of regular armies in the age of established European states. Such armies shared a common military culture and utilized a closely similar technology. Victory or defeat in the battles they fought was accordingly the outcome of superior or inferior generalship, or logistics. The European impulse to empire nevertheless brought such armies frequently into conflict with military cultures that were not their own, particularly in India and Africa. Western military technology did not necessarily prevail in the face of peoples animated by a primordial warrior ethos.

SERGEANT WILLIAM LAWRENCE
Fugitive and Recruit 1804–1806

The eighteenth-century wars of empire between Britain and France had been fought not only by local settler populations and their native confederates, but by regular units recruited from the home populations and formed for the most part from agricultural labourers. William Lawrence, a Dorset boy who joined the British army in 1804, at the beginning of the War of the Third Coalition against Napoleon, the decisive passage of the Napoleonic Wars, tells a familiar story. Apprenticed to a hard taskmaster, he ran away and decided to join the army. It was a headstrong choice. While France, since the Revolution, had conscripted its young men as soldiers and made military service a civic duty, soldiering in Britain remained a despised calling. The common soldier effectively surrendered his legal rights, lost his freedom to marry, became subject to corporal punishment at his superiors' will and was shunned by decent society. Respectable families were shamed by having a son a soldier. It is not surprising, therefore, that Lawrence's parents tried so hard to win him back from the clutches of the recruiting sergeant. His determination to enlist all the same may be explained in a number of ways. An apprenticeship as a labourer, on skimpy wages, was a form of service little different from that of the military. The boredom of rural life in a remote West Country region was burdensome to a young man of spirit. The labourer's smock was a less glamorous uniform than the soldier's red coat. The recruiter's guineas were more money than he had ever seen or could dream of possessing. Many other young Englishmen from the agricultural counties yielded, as William Lawrence did, to the enticements of the recruiting sergeant during the Napoleonic Wars. Few were, like him, literate and so able to recount their experiences. His literacy probably helped him to become a sergeant, a rank that lifted him back into respectability.

Note: The 40th Foot was the 2nd Somerset Regiment, from the county that Dorset borders.

*   *   *

Dorchester was only eight miles from my parents' house, but I never seriously thought of going to them. Unable to make up my mind what to do, or where to go, I ambled through the town watching the preparations for the fair, which was to take place the next day. I wandered into the stable-yard of one of the principal inns and was brought to my senses when a voice sang out: 'Hey you! What do you want?'

It was the ostler. I told him I was hungry but had no money, and was in search of employment. He said if I brushed about a bit and helped him rub down the horses, he would find me plenty to eat. I did so and, sure enough, he brought me a lump of bread and beef, enough for two or three meals. I ate as much as I wanted. Afterwards I felt tired. I made up a bed with some straw and, putting the remainder of my meal into my handkerchief to serve as a pillow, I lay down. The ostler had given me a rug and this I pulled over me. I slept soundly all night.

In the morning I did some more work in the stable, then walked out into the street with my new friend. We saw some soldiers and I said I wanted to be a soldier too. The ostler knew where he could enlist me and took me straight to the rendezvous which was a public-house. Inside was a sergeant of artillery, who gave him two guineas for bringing me and myself five for coming. My measurements were taken – which caused a lot of amusement – and I was put into an old soldier's coat. With three or four yards of ribbon hanging from my cap, I paraded around town with the other recruits, entering almost every public-house, treating someone or other.

In the very first inn sat a Briantspuddle farmer, a man I knew well. He exclaimed in surprise at seeing me. I begged him not to tell my father and mother where I was, and how he had seen me, and hurried out. Then later in the day I encountered my father's next-door neighbour. He recognized me immediately. I offered him the price of a gallon of ale not to say anything to my parents. He took the money and promised he wouldn't. How I spent the rest of the night can better be imagined than described, but the next morning, I had to be sworn in at the Town Hall. I was on my way there with an officer when who should meet us but my father and mother. As soon as the neighbour had got home, he had gone and told them what I was up to. They told the officer I was an apprentice, and he gave me up to them without any trouble, but he asked me what had become of the bounty of five guineas. Discovering that I had only seventeen shillings and sixpence left, he kindly relieved me of even that. My parents marched me off home, and my father went to see a magistrate to find out what he should do about me. The magistrate advised him to take me back to Dorchester to be tried at the next sitting. This my father did and I was severely reprimanded by the bench. They gave me the choice of serving my time as an apprentice or going to prison. Of course, I chose the former, so they gave me a letter to give to Henry Bush [his employer].

When I got downstairs, the officer was there. He said that if my master was unwilling to take me back, he would enlist me again. He asked if I had any money. I didn't, so he gave me a shilling and wished me well.

My father sent me off from Dorchester immediately, giving me strict orders to get back to Studland as quickly as I could. I received no blessing, or anything else so, with a heavy heart, I set off. I hadn't gone far when I was overtaken by a dairy cart. The dairy-man offered me a lift and I accepted. He asked where I was going. I told him some of my story and showed him the letter, getting him to open it so that I could find out what was inside. He said my master would not be able to hurt me, that it was safe to go back to Studland. That was cheering, but I didn't intend to go back anyway.

I rode with the man as far as he went, then continued on foot to a village called Winfrith. Being hungry I went into a public-house and ordered some bread and cheese. A soldier was there and the sight of him revived my spirit, and my longing to be like him. I got into conversation with him and discovered that he was on furlough, bound for Bridport. I said I wanted to be a soldier too. Straight away he said that he could enlist me in the 40th Regiment of Foot which gave 16 guineas bounty. It sounded a great deal of money. I thought that if I got hold of it I would not want for money for a long time, so I accepted his proposal without hesitation.

We headed for Bridport but, afraid of finding myself in Dorchester again, I tried to persuade the soldier to go around it. He wouldn't, but we slipped through at night, safely reaching Winterborne, where we put up.

Next morning we got the coach to Bridport and when we arrived, the coachman surprised me by remarking that it was only yesterday that my father had got me out of the artillery! He meant well but, of course, the soldier then asked me if I was an apprentice and I had no choice but to admit I was. He promptly made me get down. He took me across some fields to his home and there kept me quietly for three days.

As the barracks of the 40th Regiment were in Taunton, Somersetshire, it was there we thought it best to go. We went to see the colonel and the soldier told him that I was a recruit. The colonel asked me what trade I was in.

'I'm a labourer,' I replied.

'Labourers make the best soldiers,' he said and offered me a bounty of 2½ guineas, which was considerably less than the sixteen we had been expecting so we decided to try the Marines. Their recruiting sergeant promised us 16 guineas bounty when I arrived at their Plymouth headquarters but this did not suit my conductor because, after paying the coach expenses, there would have been nothing left over for him. He asked me what I intended to do, advising me to go back to my master, and forget about the expense he had gone to for me. But I had destroyed the letter so I told him I preferred the 40th Regiment. We went back to the colonel, he gave my companion 2 guineas, and I was sent into barracks.

Next day I received my clothes, and about a week later was sworn in before a magistrate, receiving my bounty at the same time. I was very mistaken about the money lasting.

Shortly afterwards orders came for the regiment to march to Winchester. There we remained for about a month. I had begun to drill twice a day. I soon learnt the foot drill and was then put on musketry drill.

After Winchester, we moved to Portsmouth. We were there a week before being ordered into barracks at Bexhill in Sussex. Our 1st Battalion was there and, in order to make it 1,000 strong, a number of men were drafted into it from our Battalion – the 2nd. I was one of them. Soon orders came for us to go to Portsmouth; we were about to embark on foreign service.

SERGEANT WILLIAM LAWRENCE
Badajoz March–April 1812

By 1812, Lawrence was a seasoned soldier who had spent three years campaigning in Portugal and Spain under Wellington. The campaign – the Peninsular War – had been launched by Britain both to liberate Spain from French occupation and to bring diversionary pressure to bear upon Napoleon, who was meanwhile conducting a series of offensives against Britain's Continental allies, particularly Austria but by 1812 also Russia.

Wellington's firm base was in Portugal, where he fortified the countryside north of Lisbon, to secure an impregnable position. From these 'Lines of Torres Vedras' he then launched a series of probes from Portugal into Spain through the passes in the mountains separating the two countries. One of the most important passes was guarded by the walled city of Badajoz, which Wellington captured in 1811. Forced to abandon it when the French reappeared in strength, he returned to the city in the spring of 1812 with the intention of taking it and breaking through into the Spanish heartland.

Wellington was weak in artillery and so lacked the means to

batter Badajoz into submission by bombardment. As a result, he was forced to use his infantry to storm its walls in an old-fashioned escalade – that is, with ladders. It was a dangerous venture and bound to be very costly of the lives of his soldiers.

Lawrence's account of the attack provides a vivid description of one of the most vicious operations of war, the hand-to-hand struggle for possession of a stronghold. He volunteered for the 'forlorn hope', the advance party detailed to seize a foothold which the followers could then exploit. Lawrence makes no bones about his motive, which was loot. The protocol of siege warfare allowed successful besiegers to plunder a city which had refused to surrender when called upon to do so. In this case, the unfortunate inhabitants were caught between two fires, the determination of the French occupiers to resist and the determination of Wellington to conquer at all costs. The French commander, nevertheless, gave those inhabitants who chose to leave permission to do so. Those who stayed would have been well advised to go. Perhaps they calculated that the assault might fail – as it nearly did. Perhaps they hoped that they would be able to deter the British soldiers from despoiling the city even if it succeeded. If so, they were wrong. Lawrence's comrades who survived the extreme peril of the assault took the normal compensation. It was paid in plunder, female virtue and even the lives of the citizens. The common soldier of Wellington's army was a rough sort at best, and at worst, when allowed to drink his fill, became a savage brute. Wellington himself preserved no illusions about 'that article' who was the instrument of his victories.

*   *   *

Wellington had taken Ciudad Rodrigo, now he had to take Badajoz. The French army of Marshal Soult was tied down besieging Cadiz, and that of Marshal Auguste Marmont (who had superseded Massena) did not have the resources to tackle Wellington's army alone, despite being urged by Napoleon to do so. It was an opportune moment for Wellington to renew his efforts at Badajoz so he sent his army south to invest it again.

Our stay at Rodrigo was short and we – the 4th Division – along with the 3rd and Light Divisions under Marshal Beresford and General Picton, were ordered south to invest Badajoz, another long and tedious march of over 150 miles.

We arrived at Badajoz at the beginning of March and immediately began work in the trenches, throwing up breastworks and batteries. Heavy rains set in but our troops persevered. A cannonade was kept up from the town and fortunately did little damage, but on 19 March the garrison came out and attacked us. They were driven back but only after we had lost 100 men either killed or wounded. They lost more.

I myself killed a French sergeant. I was in the trenches and he came on the top. Like me, he had exhausted his fire and so made a thrust at me with his bayonet. He overbalanced and fell, and I pinioned him to the ground with mine. The poor fellow expired. I was sorry afterwards and wished I had tried to take him prisoner, but with the fighting going on all around there had been no time to think, and he had been a powerful-looking man. Tall and stout, with a moustache and beard which almost covered his face, he had been as fine a soldier as I had ever seen in the French army. If I had allowed him to gain his feet, I might have suffered for it, so perhaps what I did was for the best? At such times it is a matter of kill or be killed.

In case of another attack, a large number of men afterwards formed a covering party for the 800 of us who were busy in the trenches every night. And still it rained. The trenches were so muddy that our shoes were covered. It poured down so fast that, in places, bailers had to be employed. During the day we were employed finishing off what we had done during the night, for little else could be achieved owing to the enemy's fire.

After a few days we were within musket shot of a fine fort, situated a little distance from the town. Garrisoned by 400–500 of the enemy, they annoyed us during our operations. One night, I was working in the trenches when, just as the guard was about to be relieved, a shell from the town fell amongst us and exploded, killing and wounding about 30 men. The next morning a terrible scene presented itself to us, for the remains of our mangled comrades strewed the ground in all directions. I never saw a worse sight of its kind. Some of them had their arms and legs – and what was worse, their heads – completely severed from their bodies. Working near me at the time was Pig Harding. Like me, he had become hardened to the worst sights on the Peninsula.

'Lawrence,' he said, 'if anyone is in want of an arm or a leg, he's got a good choice here.'

The fort was very troublesome and had to be dealt with. Suspecting it had been mined, engineers were sent for. In the dead of night, between the fort and the town, they searched for a [powder] train. Finding that the earth had been disturbed, they dug down, found the train and cut it off. On the next night, the 87th and 88th Regiments were ordered up to storm the fort. After a brisk action, they succeeded in gaining it, but most of the garrison escaped into the town.

Next morning, with the rest, I entered the fort. We saw wounded Frenchmen and relieved their pain a little by giving them some of our rum and water before conveying them to the rear. Most of their wounds – they looked like bayonet wounds – were bad, but not mortal.

Having taken this fort, we were able to carry on our works much nearer the town and, by the beginning of April, two batteries were formed within 300–400 yards of it. Within five days our twenty-four pounders had made three practicable breaches in the walls.

Lord Wellington asked the town to surrender. The answer was no, so he asked that the inhabitants be allowed to leave, as he intended to take it by assault. Thousands left and he ordered that, on the night of the 6th, the town should be attacked. From each of the 3rd, 4th and Light Divisions, a storming-party was selected and assigned to one of the breaches. I joined our forlorn hope. With me was Pig Harding and another comrade, George Bowden. All three of us had been quartered at Badajoz after the battle of Talavera so we knew where the shops were located. Having heard a report that, if we succeeded in taking the place, three hours' plunder would be allowed, we arranged to meet at a silversmith's shop. Pig even provided himself with a piece of wax candle in case we needed to light our way.

Those in the forlorn hope were supplied with ladders and grass bags to carry. We ate our rations and, at about half-past eight, fell in to await the signal for all to advance. Our men were particularly silent. At last the deadly signal was given, and we rushed towards the breach.

I was one of the ladder party. At the breach, a French sentry on the wall cried out three times, 'Who comes there?' No answer being given a shower of shot, canister and grape, together with fire-balls, was hurled amongst us. I lost sight of Bowden, poor Pig received his death wound and I received two small slug shots in my left knee and a musket shot in my side. Despite my wounds, I stuck to my ladder and got into the entrenchment. By now, many had already fallen, but on the cry of 'Come on, my lads!' from our commanders, we hastened to the breach. There, to our dismay, we encountered a cheval de frise [spiked tree trunk] and a deep entrenchment, from behind which the garrison opened a deadly fire on us. The cheval de frise was a fearful obstacle and although attempts were made to remove it – my left hand was dreadfully cut by one of the blades – we had no success. We were forced to retire for a time and remained in the breach weary with our efforts to pass it.

My wounds were bleeding and I began to feel weak. My comrades persuaded me to go to the rear, but it was difficult because, when I arrived at the ladders, they were filled with the dead and wounded. Some were hanging where they had fallen, with their feet caught in the rungs, and all around I could hear the implorings of the wounded. I hove down three lots of ladders, and on coming to the fourth, I found it completely smothered with dead bodies. I drew myself up over them as best I could and arrived at the top. There I almost wished myself back again, for what greeted me was an even worse sight – nothing but dead lying all around, and the cries of the wounded mingled with the incessant firing from the fort.

I was so weak I could hardly walk; on my hands and knees I crawled out of reach of the enemy's musketry. I hadn't gone far when I encountered Lord Wellington and his staff. He wanted to know the extent of my wounds and what regiment I belonged to.

'The Fortieth,' I said, and told him I had been one of the forlorn hope. He enquired whether any of our troops had got into the town. I told him no, and that I did not think they ever would, because of the cheval de frise, the deep entrenchment, and the constant and murderous fire from the enemy behind them. One of his staff bound up my leg with a silk handkerchief and, pointing to a hill, told me that behind it I would find a doctor to dress my wounds. And so I did – my own regimental doctor. I was lucky – the musket shot in my side would have been fatal had it not penetrated my canteen first, making one hole going in and one coming out.

After I arrived, Lieutenant Elland was brought in by a man called Charles Filer, who had found him at the breach, lying wounded with a ball in the thigh. The lieutenant had asked to be conveyed from the breach so Filer had raised him onto his shoulders. The night was so dark, and the clamour of cannon and musketry and the cries of the wounded so noisy, that Filer did not notice when a cannon-ball took the lieutenant's head off, so he was astonished when the surgeon asked why had he brought in a headless trunk? Filer declared that the lieutenant had had a head on when he had found him, for how else could he have asked to be taken from the breach?

Poor Filer. He was hardly composed – the exposure of his person at the breach and the effort of carrying what proved to be a lifeless burden for nearly half a mile, would have unnerved a harder temperament than his. Of course, the story spread through the camp and caused a lot of amusement at his expense. 'Who took a headless man to the doctor, then?' was one of the comments.

Lord Wellington realized it was useless to face the breach with the chevaux de frise so, as more success had been achieved in the other breaches, he withdrew the men from ours to reinforce theirs. He ordered the castle to be attacked. For this troops were supplied with long ladders which they raised against the castle walls, but the enemy showered down on them such a mass of heavy substances – trees and large stones, and deadly bursting shells – that the ladders were broken and the men tumbled down, crushing some of their comrades underneath. There was a long delay while more ladders were procured. As soon as they arrived they were quickly hoisted. This time, the precaution was taken to fix them farther apart so that if more beams were rolled over, they would not make such a deadly sweep. This second attempt was more successful. The ramparts were gained and the French driven back. A footing was soon established for others, who succeeded in turning round some guns and firing them along the ramparts, sweeping the enemy from them. The garrison was forced back into the town. The ramparts were scoured, the breaches cleared, the chevaux de frise pulled down, and the main body of the English entered the town. In the streets there was still some opposition, but that was soon cleared away. The French escaped to Fort San Cristoval.

Our troops found the city illuminated to welcome them, but it counted for nothing and they began to engage in the plunder, waste, destruction of property, drunkenness, and debauchery, that usually follow a capture by assault. When the town was taken I was in camp at least a mile off but, after the sound of guns and muskets had ceased, I could distinctly hear the clamour of the rabble. The next morning, with the help of a sergeant's pike chopped up to form a stick, I hobbled into the town. There I found a pretty state of affairs!

Pipes of wine had been rolled into the streets, tapped by driving the heads in, and then left for anyone to drink. To try to keep order, officers had poured away all they could. The streets therefore were running with all sorts of liquors and some men, already very drunk, lay down and drank out of the gutters.

Throughout the city, doors had been blown open by placing muskets at the keyhole to remove the locks. I saw some of our men launch a naked priest into the street and flog him down it – they had a grudge against him for the way they had been treated at a convent when they were in the town previously. I met one of my company who was wounded in the arm but said he had something which compensated him a little. He showed me a bag of about 100 dollars and said I should not want whilst he had it.

Although some of our soldiers engaged in debauchery, others did everything in their power to stop it. That morning I met many who said how sorry they were that soldiers should go to such excess, ransacking respectable houses with no regard to the entreaties of the inhabitants who had remained, and destroying what could not be taken. Men were threatened if they did not produce their money. Women too. No doubt, some murders were committed. Two or three officers were killed trying to keep order and I understand that some men in the 5th division, having arrived after most places had been plundered, stole from their drunken comrades, and even killed some of them.

Not till the drunken rabble cropped into a sound slumber – or had died of their excesses – did the unhappy city became composed. In the morning, fresh troops were placed on guard and several gallows erected, but not much used. Lord Wellington punished the offenders by stopping their grog, but such scenes were not unusual after a place had been fought for so hard.

The garrison that surrendered numbered about 5,000. 1,200 had been slain in the assault, and the rest made prisoners. Nearly 150 guns, 80,000 shots, and a great quantity of muskets and ammunition were taken. Our loss was severe. Nearly 5,000 of our men – including 300–400 of officers – had been killed or wounded. When you think of what our troops had to contend with, it was a wonder they entered the town at all that night. The storming of Badajoz was one of the worst engagements of the whole Peninsular War.

When everything was over, I remembered Pig Harding, George Bowden, the meeting we had planned, and how it had all come to nothing. Poor Pig had received seven shots in his body, and both George Bowden's thighs had been blown off. They must have died instantly. We missed Pig Harding more than anyone. He had been a thoroughbred Irishman whose jokes had helped to pass the time pleasantly, and whose roguish tricks had supplied us with many an extra piece of tommy [bread].

I resolved never to make any more arrangements under such fearful circumstances.

> *Note*: The taking of Badajoz was indeed one of the worst engagements of the Peninsular War. Wellington thought the assault a 'terrible business', having known before he ordered it what the human cost would be. On the day after the storming, when he stood on the city ramparts and looked out over the

carnage, he wept. That same day he wrote to the War Minister: 'The capture of Badajoz affords as strong an instance of the gallantry of our troops as has ever been displayed, but I greatly hope that I shall never again be the instrument of putting them to such a test.'

A GENTLEMAN VOLUNTEER
The Battle of Vitorio

By midsummer 1813, Wellington had captured most of the Spanish heartland, in the campaign that had opened with the capture of Badajoz the year before, and was advancing towards the Pyrenees, the mountain barrier that forms the frontier between Spain and France.

A major French army, commanded by Napoleon's brother, Joseph, whom the Emperor had appointed King of Spain, still lay in his way. On 21 June 1813, Wellington found Joseph near the city of Vitorio. The French had 65,000 men, Wellington 80,000, including Spanish and Portuguese regiments as well as British.

The 'gentleman volunteer' who reports on the Battle of Vitorio that followed was an acute observer both of the events of a gunpowder engagement, and of the everyday routines of army life in an age before railways or motor transport supplied soldiers with their necessities. As always in those days, the first thought of the rank and file, even in the immediate aftermath of combat, was how to provide for their own sustenance – and for that of the women who, often against army orders, accompanied them on the line of march. When hungry soldiers were presented with the windfall of sheep killed in crossfire, therefore, they set to with a will. The meal improvised after the victory of Vitorio fuelled the subsequent advance to the passes of the Pyrenees, and thence to the battles inside France that would lead to Napoleon's abdication and exile to Elba in 1814.

*　　*　　*

Camp on the banks of the river Aragon, near Caseda,
2 lea[gues] S. of Sanguessa.
June 29th 1813.

DEAR BROTHERS,

This is the first hour I have had to write since my last of the 20th inst. We halt today. One reason is [that] if we did march we should leave half behind us – they are completely knocked up. Every hour since my last is so full of incident that I shall give you it as much at length as I can.

On the 21st our own division marched. I was in orders for the charge of the baggage which was to march at 7 o'clock. We did so & went round that mountain from which we had heard firing the night before. After marching ½ a league the baggage halted for the reserve artillery & Household troops to pass. We then mov'd up the hill & I there learned that a general engagement was begun. I gave the sergeant charge of the baggage & immediately galloped on & at the top of the hill saw the whole of the French army in a plain with Vitoria in their rear & a smart skirmish on my right. Ours was the first brigade.

On arriving I found the 43rd in front & came up to them as it was going round a hill to a wood on the top, the river running just below. We remained there about an hour & then marched round by No. 2 bridge which was *not* defended, to a hill overlooking the whole plain with a wall in front breast high & we formed behind it. They gave us a salute with one or two guns from bridge No. 3 which did no damage. This was to begin with.

I have called this spot a plain because it is so in comparison with this mountainous country surrounding. It is an oval form but it is crowded with little hills 3 or 400 yards asunder, very disadvantageous for a retiring army. The hill in front was the highest. Behind was a line of artillery and infantry well formed to defend the bridge.

The 2nd Division began the battle on our right. In the meantime our main force, viz: 1st, 7th, 3rd & 4th divisions moved round to their right. We had between 70 & 80,000 men: French 60,000. We all expected they would have defended the bridges obstinately and their positions more so, but they most shamefully and very unlike Frenchmen gave up both with scarcely a shadow of defence.

About 10 o'clock the 1st 95th moved along the river side, the 2nd brigade moving round to attack bridge No. 3. The river is fordable in some parts. The bridge was lined with skirmishers, 2 or 3 troops of cavalry & 2 guns. The 2nd brigade sent skirmishers below the bridge. They opened a fire just as Graham's columns came in sight. At a distance of ¼ mile the 95th was running down to flank them [the French]. They had a large body in squares. All now began to return to them as fast as they could run. We then moved down the hill in open column of companies and, formed line and, with a few skirmishers in front, were ordered to attack the hill. A sergeant was in the centre with [one] Colour and myself with the other. We were to make for the centre of the middle & highest hill, the 17th Portuguese supporting us. When we were ½ way up the hill, they [the French] disappeared without firing a shot. During this time the 2nd brigade were moving round the hill and our columns were crossing the river. Their [French] line moved off before them. We moved round the hill & heard a heavy fire on our left & very near. We moved on to a hill 300 yards farther & there the cannon balls began to hiss over our heads. We mov'd on to a hill 100 yards from the village, a very heavy fire continuing on our left. We formed line about 20 yards from the bank (2 yds high). Here we had a very strong fire from a battery of theirs, of balls & shells, while the 95th & some other troops were attacking the village which they defended well.

The first ball that came was a spent one. It struck the ground about 50 yards from us & was coming straight for me but it rebounded about 10 yards [from me] & went to my left, just over the heads of the men & struck our old colonel [Daniel Hearn] on the arm. He called out but was not much hurt as it came about as swift as a swallow flies. Finding the fire heavy, we moved under a bank & lay down. At that moment a shell came gently hopping direct for me but it was polite enough to halt on top of the bank about 6 yards from us. We lay down & in about 1 minute it burst doing no harm. In another minute a ball struck the close column of the 17th Portuguese not a yard from the place the 43rd colours had just left & about 16 yards from us. It killed a sergeant & took off the leg of each of the ensigns with the colours. This was about 2 o'clock.

We halted there a ¼ of an hour until the village was taken by the 95th, who captured their cannon. We then moved in open column to our right, the battery firing as fast as possible all the time. During this time some Spanish troops were skirmishing with a flock (for they were all scattered) of French on the side of the mountain and a body of French between us & them, which the 4th Division was chasing, kept retiring and taking up good positions & most cowardly abandoning them till we came to the village. This is to be said for them that their principal force was on our left & was retiring hotly engaged. The 3rd Division had most of it (the attack) & were nearly a mile before them so that they were afraid of being cut off. At the same time they had to support their skirmishers on the mountain as they were more behind still. We were about even [i.e. level] with them & against a village, on the right of which there was a wood. We moved to our right hoping to cut them off but they ran too fast for us.

Just after we came out of the wood we found a little valley & they had regularly taken up an excellent position. We formed line on the hill while the one opposite us was taken. The 45th was in line in the hollow ready to charge up the hill, more in the rear, on which were 18 or 20 pieces of cannon playing upon our regiment. They were ½ a mile distant. We kept driving in their skirmishers to a line formed on a hill to our right, No. 7. After this a Portuguese regiment moved round their left &, appearing on their flank, they all set off from that, leaving all their artillery.

(*Note.* The French had no idea of our attacking them today. If we had done so they would have had batteries all along the hill, & if they had been obliged to retire, would have returned to this position & suddenly opened upon us a most destructive fire. It was their intention to attack us, but Lord Wellington had a better head than any of their miserable generals who commanded them on that day.)

All this time Genl. Graham was pressing their main force on the[ir] right. It was near here that two of our officers [Major John Duffy and Lieutenant George Houlton] were wounded. They did not maintain their position more than ½ an hour. Several of our officers remarked, & I think it just, that cannon make more noise and alarm than they do mischief. Many shots were fired at us but we suffered little from them. A young soldier is much more alarmed at a nine pounder shot passing within 4 yds. of his head than he is of a bullet at a distance of as many inches, although one would settle him as effectively as the other. Artillery makes great havoc when in close column. The French are very correct in aiming their artillery. [At] ½ past 3, in passing the line the 45th occupied in the valley, I saw 10 or 12 killed or wounded in the space of as many yards owing to the fire kept up on them from the hill.

Upon their cannons ceasing to fire, our guns galloped after them as fast as they could move. They began to run faster than we could follow. We chased them by Vitoria in grand style, leaving them no time to save their immense baggage. They took up a position 1½ miles beyond Vitoria which they abandoned as soon as some 9 pounders from a hill close to Vitoria played upon them. As they went up the hill in the greatest disorder, scattering like a flock of sheep, we kept moving forward leaving Vitoria ½ a mile to our right. After this the Household troops, viz: the [Life] Guards & Blues, came galloping by us. I do not know what good they did, if any I am sure you would hear of it.

At 6 o'clock we came up to a village where were 8 or 10 wagons overturned with all kinds of valuable baggage attended by dragoons, Spaniards & stragglers plundering them. The smell of French brandy was very strong, I am sorry I cannot tell the taste. The soldiers were not allowed to touch a drop. If they attempted it the officers knocked it out of their hands. However, as they had to march over a great quantity it could not be entirely prevented. We saw the French no more this day. We continued marching till dark, 9 o'clock, passing more wagons overturned & baggage of every description, including flocks of sheep, goats, bullocks, asses, mules, horses &c.

At last we halted at the side of a narrow lane stopped up with wagons, our artillery in front of us firing as long as they could see. We here found a flock of sheep, mostly killed & tumbled one upon another into a deep ditch. In the same ditch were 2 or 3 pieces of cannon, overturned, horses & mules with them. You can hardly form any conception of the scene here – everyone busy & most employed in getting the sheep out of the ditch, while others were skinning them, the whole, as you may suppose, knocked up. In the morning I was fortunate in being on the baggage so [had] had an excellent breakfast & took the precaution to put some bread & meat in my pocket. Many of the officers had nothing all day. From bridge No. 1 to Vitoria is 2 leagues & we marched 1½ leagues beyond it. The division then formed upon a hill close by. The baggage, of course, could not get up.

I shall now conclude this with a short description of the scene presented on halting. The wood of the wagons supplied fuel & about every 2 yds. square was a fire & a circle round it. One will describe the whole – one making dough boys (flour & water mixed) swearing all the time at one for not producing a frying pan, at another for getting in his light; another giving a young soldier a thump for crossing between him & the fire while he plastered his blistered feet. The poor creature is turning round to beg his pardon, when he treads upon another, who threatens to upset him if he does not sit down. A woman who is undressing by his side (perhaps the wife of one of the party) raises her shrill voice & blasts him for not being quick. An old soldier sits smoking his pipe & frying the mutton or skimming the pot, while a dirty fist seizes the mutton, and another equally so lays hold of it & it is torn asunder by a knife with edge & back alike. The whole is shortly devoured & they lie down to sleep in their blankets. It was a cloudy but fine day. I had not the slightest touch of shot or shell.

Yours &c.
G. H.

HELEN ROEDER
Captain Roeder

In May 1812 Napoleon decided to declare war on Russia and march on Moscow. To defeat the Tsar would, he believed, complete his conquest of Europe and give him complete control of Continental Europe. On 24 June he crossed the river Niemen, which marked Russia's western frontier, at the head of 600,000 men. The majority were French but the army also contained numerous contingents from nationalities that had been incorporated into the French Empire – Dutch and Italians – or from states that had been forced into alliance with it. They included many regiments from the small German princely states. One of them was the Lifeguard Regiment of the Grand Duke of Hesse, which, with the Prince's Own Regiment, formed a brigade of four battalions. As household troops of high quality, they were attached to Napoleon's own Imperial Guard.

Franz Roeder was thirty-eight in 1812, a professional soldier who had fought at the battles of Friedland (1807), Wagram (1809) and Aspern-Essling in the same year. He was also an educated man of considerable literary talent. In 1812 he commanded a company in the 1st Battalion of the Lifeguards, at the head of which he rode into Russia in July. Their road lay almost due eastward towards Moscow. At first the weather was unbearably hot. Roeder was taken ill with fever and so missed the battle at Smolensk on 17 August. He recovered enough to take part in the subsequent advance towards Viazma in September, but again missed the Battle of Borodino and the entry into Moscow. The Lifeguards were left behind to protect the line of communication and to convoy supplies forward. Before they could start the latter task, however, they met the Grand Army coming back. Almost as soon as Napoleon entered Moscow on 14 September, to find the city in flames, he decided that he could not bring the Tsar's army to decisive battle before the onset of the Russian winter and that he must in consequence retreat the way he had come.

The retreat that followed turned into one of the greatest military disasters in history. In early November the snows began, slowing the army's withdrawal. It was also harried by bands of Cossacks hanging about the edges of the long columns dragging their way back to Poland. Stragglers were cut off, robbed and killed. Those still strong enough to defend themselves stumbled on but were often killed by the cold at night, many having to sleep in the open without shelter. Others starved to death, the army's supply system having broken down, leaving thousands without food. The worst ordeal came at the Beresina, the river blocking the Grand Army's escape out of Russia. The winter had filled the waterway with ice, which was not hard enough to bear the fugitives' weight. Many of those who tried to cross beside the only bridge drowned; others froze to death while waiting on the banks.

Franz Roeder, who suffered as much as any of the fugitives, survived, to make his way back to Hesse and to his wife and children. His journal of the campaign is not only one of the most important narratives of the Retreat from Moscow, but a dramatic record of the grandeurs and servitudes of the military life.

*   *   *

From the diary of an officer in the First Battalion of Hessian Lifeguards during the Russian campaign of 1812–13.

At three o'clock on the morning of [November] the 17th the Hessians prepared to march against the enemy. Captain Roeder's company had been reduced to seven sergeants and twenty-seven men. The total strength of the Lifeguards, who had numbered 660 combatants when they left Viazma some ten days earlier, was now twenty-six officers, 442 men, while the Prince's Own (Regiment) could only number twenty-three officers, 450 men. 'What a brigade of Guards of four battalions! And yet we are much stronger than the French!'

The Captain wrote a little letter of farewell to his family, and then they marched for about two hours back along the road to Smolensk. They took their stand in battalions to the left of the road, and there they remained until about eight o'clock. 'I found it terribly hard; cold and drowsiness.' At eight o'clock a Russian corps approached towards the town, and by nine the Hessians stood facing the enemy. From half past nine until half past twelve they were 'exposed to the fire of about ten cannon and two howitzers, and especially of a battery of about six pieces lying a little to the left, which fired at us unceasingly and with great violence, so that even in the great battles of Wagram and Aspern we had never had to stand up to such a cannonade of such long duration. I left my place for a moment to have a word with Captain Schwarzenau, and just before I returned to it a ball passed through prodigiously close to Lieutenant Succow, who had stepped in, killing outright the men who were standing in the second and third rank to the right. The first of these was my old cook, Heck, an honest fellow, who died a noble death. In all, from twenty-seven men including officers, for I had lost many more from weakness during the march, I lost one dead and three wounded. Yet another shot passing close to my eyes, tore through a gap in the rank without damage, but struck the hand off a drummer in the Fourth Company.

'The Prince's Own, which was close to the Russian cavalry, unfortunately had to form a square, and in a short time suffered a loss of ten officers and 119 men dead and wounded. All the wounded officers fell into the hands of the Russians.

'The First Corps, like ourselves, were stationed in a wood filled with Russian sharpshooters, but several times they had to form closed columns and attack at the double. The Russians did not yield, and nothing was done to circumvent or dislodge them *par force*, for this fight only aimed to hold back a little the corps which was stationed there under General Orlov, or to see whether they were supported by their main army or by a strong corps. We withdrew therefore at one o'clock, and, covered by a weak division, retreated with all speed as far as the frontier of Old Russia five hours beyond the little town of Lodsi.'

The Captain, 'fearfully weary and suffering from a total lack of any kind of food', tried to find the recruit who had been supposed to hold his horse on the neighbouring hillside, but the boy had taken himself off to a place of safety with the horse and such meagre supplies as might be in the saddle-bags. It was a Russian pony commandeered to replace the big horse that had collapsed a few days before. The starving Captain looked round him hopelessly; already the looters were at their work, stripping the corpses almost before they were dead. One of these men approached him with a bloodstained fur coat ripped by the cannon ball which had killed its wearer, a French subaltern in the Voltigeurs. The Captain gave a gratuity to the corpse-stripper; the coat was torn but at least it was warm, and this was neither the time nor the place for feelings of refinement. There was no need to freeze even if one did have to starve. He put his frozen hand into the pocket and found there a piece 'of the most excellent sugar'. So at least he had something to gnaw. Another soldier brought him a small bag of barley coffee, 'which had been found in the pocket of the fallen Heck, and since nobody wanted it I put it in my pocket'. So even in death his old cook provided him with a meal, for he managed to nourish himself that day upon six cups of barley coffee with the sugar added, 'ladled out so that the roasted grain could be eaten too'.

The bivouac was horrible. They spent the night without shelter and 'marched on, fasting, at five o'clock'. The Captain's feet were beginning to swell dangerously and his hands also were frost-bitten, for he had lost the gloves which Mina [his wife] had knitted for him. The recruit did not return, having allowed the saddle-bags to be pillaged by the French, so his small store of food was gone. 'I begged a piece of bread from Prince Wittgenstein, and then gave it to Amman because I thought that his need was the greater. Also I gave Captain Hoffmann my reserve flask of brandy. He promised me bread in return but gave me none.' He tried to take note of the country through which they were passing, but was no longer able to do so; all he could do was to stagger on somehow. 'We went by Kazani, where there was only one bridge, and this blocked by vehicles, so that the greater part of the infantry had to go through the water and ice, which terribly retarded our march. We laid stakes across it and passed over very slowly, and still most of us fell into the water. Here I ate some horseflesh grilled on the cinders and found it excellent. We went on for about another hour and a half beyond Kazani and bivouacked on the road by a great church. I lodged in a house with a number of officers of the Sixth Tirailleur Battalion. I have no batman.'

The next day he woke shivering and streaming, with his feet so swollen that he was unable to draw on his boots, so that he was constrained to borrow shoes from a soldier and they were too big for him. 'Before the march out I lost my blue handkerchief in the straw. I could not search for it in the room full of officers.' It is difficult for us in the twentieth century to understand how they could in such circumstances have continued to observe the punctilio which was considered proper to officers in the Guards. And yet, if they survived at all it may have been in some part due to the observance of a rigid code. They had no boots, sometimes no feet, but they knew how to die with dignity.

'At half past four this morning between our encampment and the town of Dubrovna we were harassed by Cossacks, but we are not being pursued as we should be.

'I was shivering from riding and from my indisposition, so I asked Prince Emil for a drop of the schnapps which he had offered me yesterday, and I also had to accept from him a slice of the Göttingen sausage which Mother had sent me and I had presented to him. It was excellent.

'We occupied an odious bivouac to the right of the road towards Orsza, where it was impossible to make even a decent fire. I had nothing to eat, but managed to purchase three platefuls of groats [crushed grain, usually oats] for three francs from one of my soldiers. Amman was still my guest and slept by my fire. Coffee I still had, but the sugar of the dead French officer was all the solid nourishment I had taken until then, and even that I had shared too liberally. My lad Dietrich with all my best effects has not yet put in an appearance. Musketeer Alt with the furs, fodder and cooking pots may well be utterly lost.

'*20th*. Very ill. After one and a quarter hours we reached Orsza, where I rode straight over the Dnieper Bridge 125 paces long. On the opposite side I found Colonel Follenius on a hill with a number of officers, who had mustered all those of our men who had gone ahead. Their number was equal to that of the regiments we had with us. Upon this hill we were informed that the army was to take three different routes via Minsk, Vilna and Vitebsk respectively. We were to take the first route with the Emperor.

'There was to have been a great seizure of flour and brandy here, and the men were each given a *schoppen* of brandy to empty the magazine, but those who were to have removed the stores immediately became so sozzled that the twenty sacks of flour could not be brought away. So in spite of the superfluity, the soldiers in general received nothing, for only very starving men could wrest some of it from the universal pillaging and bring it over the bridge. I bought a little schnapps extremely dear.

'My batman Dietrich has arrived safely with my best effects. I have just learned that my groom, Gottfried Köppinghof, died at the first night station after Smolensk, after I myself had left him quite cheerful and well provided and able to make the march on foot. I had thought that the hope of soon being back in his native land would have helped him to a complete recovery. The news came to me as a great shock and I was very sorry to hear it. Colonel Follenius has invited me to take a place in his chaise, so that I shall procure him night quarters and bring him through the French.

'Riding back over the Dnieper my horse slipped on the bridge and lay with both back legs over the side. I had to fling myself quickly over its head into the throng of wagons and horses. But being an intelligent pony, he knew so well how to balance himself and remained so quiet that it was possible to help him up.

'In the evening, after standing about in mud and darkness for a distribution at which there was nothing to distribute, we went on for about three quarters of an hour to a village to the right of the road, where we bivouacked.

'*21st*. While we were on the march today about twenty Cossacks approached and carried off a wagon and two horses under the noses of our brigade and the cavalry, which rode on instead of letting fly at them. Our Schützen [light infantry] and the brigade thereupon opened fire, but naturally they made off with all speed. We marched for about seven hours, crossed a river and bivouacked at Kochanovo. I reported sick.'

They were approaching the Beresina and the worst of their ordeal was yet to come. [A Russian officer, a major, who also left an account of the retreat] gives a strange picture of their plight between Krasnoi and the terrible crossing:

'The second period of the retreat began at Krasnoi and continued to the Beresina; a distance of about twenty-six leagues. At first things appeared to be more favourable for the French army, for, once across the Dnieper, they expected to link up with the corps of Victor and Oudinot and Dombrowski's division, which together were over 30,000 men strong. Also the pursuit had been somewhat retarded by the fight with the Ney Corps on the 18th. Thirdly the army had now entered the area of its magazines and was in a country which it could regard as its ally, and fourthly the weather had grown somewhat milder. All these ameliorations collapsed before the fact that Admiral Tschitschagov with the Army of the Danube had pressed on via Minsk to catch the French army at the Beresina, and Count Wittgenstein was approaching from Tschasnik with his corps reinforced by General Steinheil, in order to link up with the Army of the Danube. By the movements of these armies the French were placed in great peril, and the least they could expect was a repetition of Krasnoi. Napoleon, perfectly well aware of the danger of his position, hurried to the Beresina by swift marches. When he came through Orsza he found the deputies of the Province of Mohilev waiting to receive the Emperor's orders. The Emperor, usually so ready to avail himself of this kind of attention, sent them packing without seeing them. He had every reason for not wishing to exhibit his army, which had certainly lost some of its demeanour in the course of the march and was somewhat fantastically attired in priestly vestments and even women's gowns as a protection against the cold.

'As soon as Napoleon had taken on his reinforcements, he sent the Poles to the left against Borisov, which town had been occupied by Admiral Tschitschagov, and threw the Victor Corps to the right against Count Wittgenstein. Under cover of these detachments he reached the Beresina with the remainder of the army on the 25th, flung a bridge across it fifteen versts [a verst is approximately 1,000 metres, or two-thirds of a mile] above Borisov at Semlin, and crossed without losing time. Because of its horrors the crossing of the Beresina will live long in the memory of soldiers. For two days the crossing continued. Right from the beginning the troops surged over in disorder, for in the French army order had long been abandoned, and already many found a watery grave. Then, as the Russians forced back the corps of Victor and Dombrowski and everyone surged across the bridge in wild flight, terror and confusion reached their summit. Artillery and baggage, cavalry and infantry all wanted to get over first; the stronger threw the weaker into the water or struck him to the ground, whether he were officer or no. Many hundreds were crushed under the wheels of the cannon; many sought a little room to swim, and froze; many tried to cross the ice and were drowned. Everywhere there were cries for help, and help there was none. When at last the Russians began to fire on the bridge and both banks, the crossing was interrupted. A whole division of 7,500 men from the Victor Corps surrendered together with their general. Many thousands were drowned, as many more crushed and a mass of cannon and baggage was abandoned on the left bank. This was the end of the second period. To the Russians it brought over 20,000 prisoners, 200 cannon and immeasurable booty.'

The Captain did not travel long in the Colonel's chaise. On the 23rd, when the cold weather set in once more, he marched on foot for seven hours to Bobr. 'I did not think I should come through today. Asked the Colonel for his chaise, but it was already occupied. Rode. My horse, searching for water, broke the ice, stumbled into a water hole and I fell in up to the stomach.' The horse was drowned; the Captain, with dysentery, violent coughing and with frost-bitten feet, dragged himself out somehow. 'Now I saw that I must go on stoutly or perish. I pulled myself together with all the strength of my body and soul and covered seven or eight hours on swollen feet. Strecker is suffering from some sort of stroke. Hoffmann is feigning illness to get preferential treatment. He behaves with considerable animosity towards myself. Dietrich remains absent; consequent privation.

'*24th*. The Emperor has stopped on the way at Losznita, in a great church to the right of the road, perhaps because a violent cannonade can be heard, which means that a battle is in progress. He has also received despatches.

'Cadet Becker has died. All of us officers are lying together in a barn with Prince Emil. Extremely wretched and fearfully crowded. Today I should have remained lying had not Prince Emil sent back his own saddle-horse for me. I feel very ill.'

On the 25th they reached the first bend of the winding Beresina at a market town called Njemonica. The whole division could now only form one weak battalion; the Lifeguards had seventy-five men, the Prince's Own, twenty-five. The Captain was once more travelling in the Colonel's chaise, for the Colonel set great store upon bringing it through the ever-increasing throng of men and vehicles.

The French army must still have had some fight in them for 'the day before yesterday the Second Corps beat the Russian Lambert Division. Six cannon were taken and the Russians flung back over the Beresina.

'*26th*. This morning at nine a violent fight began to the left; the Second Corps with the Russians on the other side of the river. They must have been victorious because the noise died away in the direction of Borisov. The Third Corps of Guards in reserve were on the right of the road along which we travelled.

'At Borisov the long bridge crosses over lake and swamp and at the entrance of the town there are two marshy rivers. After this had been crossed the column changed its course from the direct road to Minsk, because this led close to a Russian entrenchment upon a hillside only a quarter of an hour away. According to my map I thought that the road must lead to Semlin, where we should be under necessity to repair and cross the great bridge leading over lake and swamp.'

After waiting his turn for four hours, the Captain, who was fortunate enough to be in the Colonel's chaise, managed to cross this bridge 'with indescribable difficulty, struggling through with the Colonel's excellent coachman, Jacob. The battle went on all round us to the left, reminding me vividly of my own first battle in the Schorlmberge terrain, with almost the same violent fusillades. It lasted until eight in the evening.'

He rejoined his company at the small town of Vesselevo; 'the bulletins say it was Studianka.' Although they had meat and flour from Borisov, they were unable to cook them, for there was only one iron cooking pot. 'My turn never came. My irritability with my servants increases as my strength fails. When we broke camp in the darkness my overcoat was stolen by one of my batmen, my jar of honey pilfered by another, and my coffee left behind. I am in no state to think or notice anything. Physically I am suffering extremely, especially from violent coughing. The rent in my fur coat has not yet been repaired, so that I cannot put it on or take it off without a long struggle, especially in the darkness with my swollen hands, although it is a great comfort.'

The next day, November 27th, they reached the long, fatal bridge over the Beresina. The Captain, desperately ill and faint with star-vation, hoped to be able to travel once more in the Colonel's chaise, but found it already occupied by Captain Schwarzenau. His nerves, already at breaking point, snapped in a violent rage and he stumbled off to mount his wretched pony, only to find it 'without a bridle and with one stirrup two spans too long'. They were early at the bridge, but already the press was terrible; what hope could there be for a sick man on a starving pony with no bridle and one stirrup? Imagine him, the gaunt, fainting figure in a torn fur coat and tall cockaded hat; the medals still on his chest, the sword with its *porte-épée* [sword knot] slapping the swollen leg in a torn blue stocking; the frost-bitten foot in a soldier's shoe two sizes too big for it groping for the dangling stirrup. And thus he was to cross the Beresina! Then, through the struggling mass of men and horses a big man came pushing his way and shouting:

'Cap'n Roeder! Cap'n Roeder, sir! Don't you worry, Cap'n. Leave it to me, sir. Just you lean on me, sir.'

It was Sergeant-Major Vogel. 'He led my horse by the mane and forced his way through, while I, like a poor sinner, clung to its neck.'

Somehow they got over. 'I do not know which way we came; I could not notice it. We have taken up our position half an hour beyond the long bridge, and here we are to stay the night. I feel very wretched, but fortunately I have some good hay in which to bed myself. My right breast gives me great pain with coughing.'

The next day the regiment took part in a battle, but the Captain 'could not even put in an appearance', for his horse was so weak from hunger and thirst that it was unable to climb the hill, and he himself 'for sheer misery' was hardly conscious of what was going on around him. That night he wrote: 'I am bivouacking in the open air with the brigade flag (under which we have no more protection), suffering prodigiously by a strange night fire (or mostly no fire at all). I am trying to sleep huddled in the most wretched camp. A terrible night. Violent cold and cutting wind.

'*29th*. My horse stolen, I thought it would have perished. Now I had to march and, supported by my Sergeant-Major, to cross another long bridge. I think it was Zembin. I could hardly go for the pain in my right breast. Found a wretched pony by the roadside, which had been allowed to run loose. Was lifted on to it, and so went on for about an hour and a half on the beast's sharp back. My Sergeant-Major makes himself of indescribable service to me.'

That night they took up their quarters in a village. 'The room was full to overflowing with people. Finally they burned the house down, and after I had lain for a short while under a rafter, I had to bivouack outside without sleep. What a sum of misery! Shall I get to Vilna?'

To know just how horrible that night must have been, we must turn once more to the account of the Russian major:

'About 40,000 men with a still significant amount of artillery had managed to cross the Beresina, but how tragic was the situation of these troops! A new and violent frost finished the business completely. Now almost everyone threw away his weapons, most of them had neither shoes nor boots, but blankets, knapsacks or old hats bound around their feet. Each had hung whatever he could find around his head and shoulders in order to have at least some protection against the cold; old sacks, tattered straw matting, newly flayed hides. Happy the men who had managed to find a shred of fur somewhere! With arms hanging and heads bowed low, officers and men plodded on side by side in sullen stupefaction; the Guard was no longer distinguishable from the rest, all were ragged, starving and disarmed. All resistance was at an end; the mere cry of "Cossacks!" brought the whole column to a shambling trot. The route which the army had taken was littered with corpses, every bivouac looked like a battlefield the next morning. No sooner had a man collapsed from exhaustion than the next fell upon him and stripped him naked before he was dead. Every house and barn was burned, and among the ashes lay a heap of dead men, who had gathered round to warm themselves and had been too weak to flee from the fire. All the country roads were swarming with prisoners, of whom no one took the least notice, and here one saw scenes of horror beyond all experience. Black with smoke and filth, they flitted like ghosts among their dead comrades in the burning houses until they too fell in and died. On bare feet covered with burns some went limping onwards down the road, no longer conscious, others had lost the power of speech and many had fallen into a kind of frenzy from cold and hunger, in which they roasted corpses and gorged upon them, or gnawed their own arms and hands. Some were too weak to drag wood to the fires; they merely sat on their dead companions huddled round some small fire which they had chanced to find, and died there as these had already done. Some in a state of frenzy would of their own free will stagger into the fires and burn themselves in the illusion that they were getting warm, and others following them would meet with the same death.'

Now the Captain speaks, and his quiet voice is very terrible:

'My Sergeant, Jost, went blind tonight. I had to leave him in the most wretched circumstances. The poor soldiers meet with horrible misfortunes: blinded by smoke, fire and lack of sleep, dazed, crazed ... My own life was twice endangered by falling with the pony among the wagons.'

On December 1st they reached their division and he bivouacked once more with Dr Amman. The bivouac was horrible, but at least they were able to roast a chicken. In the midst of all this he still remembered Mina. 'It was a year ago today that they told me she had to die. The memory has cost me many tears.' He told Amman about it; he had to tell someone.

The next day there was a small amelioration of their sufferings, for the sun shone. Also they had a little to eat, for 'Vogel and I pilfered a loaf of bread yesterday evening and this morning a copper saucepan. Overmastering need! We had to do as all the rest did!'

Somehow they had managed to concoct themselves a pea soup, and he tells how he wrapped some slices of fat pork in paper and took them with him, but the sudden surfeit of food upset his starving stomach. However, 'I made a good seven or eight hours and reached a village to the left of the road without knowing that the Division was in it. Only Vogel was with me. I had to cling for support to several Poles, who were lodged with me in the barn, of which, however, the French broke up the greater part for firewood. They made such a tumult that I was heartily glad when we could go upon our way again at four o'clock in the morning after another sleepless night.'

Just before the town of Moldzieczno, where the division was bivouacked, although they did not know it, they took shelter in a small copse by the roadside, where a party of 'uncouth Württembergers' lost no time in stealing his horse and two saddle-bags. 'Now I shall have to get to Vilna poor and like a beggar, with my sack of bread on my back. We wandered on, for we could find no place to sleep.'

They entered Moldzieczno, and once again deliverance came when all seemed lost: 'Plodding on with Vogel at one o'clock in the morning, behold! In one of the streets we came upon Dietrich! What joy that the honest fellow is still alive! He had my second writing case with him too, and had managed to get my valise over the long bridge!'

Cheered by this meeting, they quartered themselves once more 'at an inn by the roadside, and slept well enough until daylight'. Two days later they lost Dietrich once more in the crowd. Only Vogel never left the Captain's side. 'He always kept an eye on me; it was for him that I shouted through the crowds. How often I fell upon the icy roads; how often I could not walk at all without clinging to him, for my legs were weak and stiff and my shoes studded with nails after the fashion of soldiers. This man has endured all things for me.'

But even Vogel could not find food where no food was to be found, nor could he cook without a pot. On December 8th the Captain wrote: 'We were unable to prepare any food and walked on until the afternoon, when we came upon some barrels of biscuits, which were being rifled by those who passed by. Naturally we helped ourselves and took a supply for eight days. I ate without reflection and another biscuit pottage was made that evening. It was too nourishing for me and resulted in terrible diarrhoea, which made it very hard for me to go on.'

But go on he must, for 'the Russian advance guard is forever at the heels of our insignificant rearguard, and our stragglers fall into their hands. The Cossacks, however, have now taken to plundering them completely and letting them go.' There was no need to kill a destitute man in that cold. Eventually the Cossacks caught up with the Captain himself and once more he had a miraculous escape. They came upon him in a somewhat undignified situation for, owing to his indisposition, he had retired to the bushes by the roadside, when the troop of horsemen rode up. A Guards officer was something of a prize, and they lost no time in stripping him of his fur coat. Then, to his boundless astonishment, the Cossack stopped short, staring at one of the decorations on his chest. He summoned the others, who gathered round looking closely at the ribbon. 'They treated me with moderation,' he wrote afterwards, and this was true, for after they had relieved him of a little money and some pages of his diary which they found in his pockets, they mounted their horses and rode off. Somewhat dazed, the Captain stepped forth from the bushes to shout for the trembling Vogel:

'Lord love us, sir,' said that worthy, emerging from his hiding place, 'I thought they had you that time, sir! Why did they let you go?'

The Knight of the Hessian Order of Merit swayed unsteadily against the shoulder of his Sergeant-Major:

'They thought it was the Order of Vladimir,' he said, 'It has the same ribbon. They thought I had been decorated by their own Czar! Now, Vogel, now I really begin to believe that it must be God's will that we should get to Vilna!'

That night he records a curiously trivial incident in a scene of horror. 'The cords and rosettes were stolen from my hat, when a sudden cry of "Fire!" flung into activity all the men who were packed like herrings into a single room. Our cooking pot was stolen at the same time. By turning aside from the main street of a little village I had been so fortunate as to get shelter in a room, but the usual story was repeated. The house was set alight, either because a fire had been made on the deal floor of the outhouse, or those who had not been able to get into the room had bivouacked outside and lit their fire too close. The room, in which the bake oven had been heated up, instantly became so full of smoke that anyone remaining there for a moment would have been suffocated. There was nothing left for night quarters but to fling oneself on the ground as soon as one came upon a vacant space. At least one had earth to sleep on and air to breathe.'

And there in that merciless carnage he knew what he had never known before, that somewhere there was God, and God was merciful.

VICTOR HUGO
(1802–1885)
Russia 1812

The snow fell, and its power was multiplied.
For the first time the Eagle bowed its head –
dark days! Slowly the Emperor returned –
behind him Moscow! Its onion domes still burned.
The snow rained down in blizzards – rained and froze.
Past each white waste a further white waste rose.

None recognized the captains or the flags.
Yesterday the Grand Army, today its dregs!
No one could tell the vanguard from the flanks.
The snow! The hurt men struggled from the ranks,
hid in the bellies of dead horse, in stacks
of shattered caissons. By the bivouacs,
one saw the picket dying at his post,
still standing in his saddle, white with frost,
the stone lips frozen to the bugle's mouth!
Bullets and grapeshot mingled with the snow,
that hailed . . . The Guard, surprised at shivering, march
in a dream now; ice rimes the grey moustache.
The snow falls, always snow! The driving mire
submerges; men, trapped in that white empire,
have no more bread and march on barefoot – gaps!

They were no longer living men and troops,
but a dream drifting in a fog, a mystery,
mourners parading under the black sky.
The solitude, vast, terrible to the eye,
was like a mute avenger everywhere,
as snowfall, floating through the quiet air,
buried the huge army in a huge shroud.
Could anyone leave this kingdom? A crowd –
each man, obsessed with dying, was alone.
Men slept – and died! The beaten mob sludged on,
ditching the guns to burn their carriages.
Two foes. The North, the Czar. The North was worse.
In hollows where the snow was piling up,
one saw whole regiments fallen asleep.
Attila's dawn, Cannaes of Hannibal!
The army marching to its funeral!
Litters, wounded, the dead, deserters – swarm,
crushing the bridges down to cross a stream.

They went to sleep ten thousand, woke up four.
Ney, bringing up the former army's rear,

hacked his horse loose from three disputing Cossacks . . .
All night, the *qui vive?* The alert! Attacks;
retreats! White ghosts would wrench away our guns,
or we would see dim, terrible squadrons,
circles of steel, whirlpools of savages,
rush sabring through the camp like dervishes.
And in this way, whole armies died at night.

The Emperor was there, standing – he saw.
This oak already trembling from the axe,
watched his glories drop from him branch by branch:
chiefs, soldiers. Each one had his turn and chance –
they died! Some lived. These still believed his star,
and kept their watch. They loved the man of war,
this small man with his hands behind his back,
whose shadow, moving to and fro, was black
behind the lighted tent. Still believing, they
accused their destiny of *lèse-majesté*.
His misfortune had mounted on their back.
The man of glory shook. Cold stupefied
him, then suddenly he felt terrified.
Being without belief, he turned to God:
'God of armies, is this the end?' he cried.
And then at last the expiation came,
as he heard someone call him by his name,
someone half-lost in shadow, who said, 'No,
Napoleon.' Napoleon understood,
restless, bareheaded, leaden, as he stood
before his butchered legions in the snow.

Translated from the French by Robert Lowell.

PRIVATE WHEELER
The Letters of Private Wheeler

Private Wheeler, like Sergeant Lawrence, was a humble but literate man who decided to go for a soldier in the unusual circumstances of the Napoleonic Wars. By 1815 he was a veteran of the Peninsular War and his regiment, the 51st (later the 1st Battalion, King's Own Yorkshire Light Infantry), had been posted to Wellington's army stationed in Belgium in the aftermath of Napoleon's abdication as Emperor to the French in 1814, and his exile to Elba. When Napoleon escaped from Elba and collected a new French army to challenge the powers that had defeated him – Britain, Austria, Prussia and Russia – Wheeler's regiment was among those marshalled by Wellington to oppose Napoleon's advance out of France into the Low Countries. At the Battle of Waterloo on 18 June 1815, the 51st held the extreme right of the line.

*   *   *

The morning of the 18th June broke upon us and found us drenched with rain, benumbed and shaking with the cold. We stood to our arms and moved to a fresh spot to get out of the mud. You [his parents] often blamed me for smoking when I was at home last year but I must tell you if I had not had a good stock of tobacco this night I must have given up the Ghost. Near the place we moved to were some houses, these we soon gutted and what by the help of doors, windows, shutters and furniture, we soon made some good fires. About 8 o'clock our brigade went into position on the right of the line, on high ground that commanded the farm of Hougumont. The Regiment was commanded by Lt. Col. Rice, Colonel Mitchel having the command of the brigade. Major Keyt commanded the light troops in advance, consisting of Capt. Phelps' Company 51st, The Light Companys of the 23rd and 14th Regt.

About 9 o'clock three field pieces were discharged from our position and Capt. McRoss' company was ordered down to reinforce the advance, who were warmly engaged. A quarter of an hour had not elapsed, before four more of our companys were ordered to the front, the company I belong to was one. We soon saw what was up. Our advance was nearly surrounded by a large body of the enemy's Lancers. Fortunately the 15th Huzzars was at hand and rendered assistance. Our appearance altered the state of affairs and ere we could make them a present of three rounds each, the Lancers were glad to get off. We were now exposed to a heavy fire of grape, and was obliged to push across a large space of fallow ground to cover ourselves from their fire. Here we found a deep cross road that ran across our front, on the opposite side of this road the rye was as high as our heads. We remained here some time, then retired back to the ground we had advanced from, the 15th Huzzars were in column on our left.

I shall here endeavour to describe to you how matters stood where we were. On the hill behind us were posted some 20 or 30 guns blazing away over our heads at the enemy. The enemy on their side with a battery of much the same force were returning the compliment, grape and shells were dropping about like hail, this was develish annoying. As we could not see the enemy, altho they were giving us a pretty good sprinkling of musketry, our buglers sounded to lie down. At this moment a man near me was struck and as I was rising to render assistance I was struck by a spent ball on the inside of my right knee, exactly on the place I was hit at Lezaca. Like that it was a glance and did no harm, only for the moment cause a smart pain. A shell now fell into the column of the 15th Huzzars and bursted. I saw a sword and scabbard fly out from the column. It was now time to shift our ground to a place of shelter, the Huzzars moved to the left and we advanced again to the cross road under a sharp shower of shells. One of the shells pitched on the breast of a man some little distance on my right, he was knocked to atoms.

We gained the cross road and was then under good shelter, this was my position the remainder of the day. This road was opposite the Observatory where it is said the Emperor with his staff were posted. On our left a main road ran direct into the enemys lines, on this road was an arch that crossed the deep road we were in. I was ordered to go to this place with a message to Lt. Colonel Keyt. I now found our left communicated with about 300 of the Brunswick Lt. Infantry, and saw that the bridge was blocked up with trees. A little to the front and to the left stood the farm house of Hougomont, on which the enemy was pouring a destructive fire of shot, shell and musketry. The house was soon on fire and the Battle increased with double fury. Never was a place more fiercely assaulted, nor better defended, it will be a lasting honor and glory to the troops who defended it. So fierce was the combat that a spectator would imagine a mouse could not live near the spot, but the Guards, who had the honor to be posted there not only kept possession but repulsed the enemy in every attack. The slaughter was dreadful, but I must speak of this when I come to the close of the action.

I was ordered with two men to post ourselves behind a rock or large stone, well studed with brambles. This was somewhat to our right and in advance. About an hour after we were posted we saw an officer of [French] Huzzars sneaking down to get a peep at our position. One of my men was what we term a dead shot, when he was within point blank distance. I asked him if he could make sure of him. His reply was 'To be sure I can, but let him come nearer if he will, at all events his death warrant is signed and in my hands, if he should turn back.' By this time he had without perceiving us come up near to us. When Chipping fired, down he fell and in a minute we had his body with the horse in our possession behind the rock.

p.s. I omitted to say that Captain John Ross' company had a very narrow escape of being made prisoners at the commencement.

No. 75. Camp Cato plains, 23rd June 1815.

I have finished one letter this morning. I shall get on with this in continuation of the last. We had a rich booty, forty double Napoleons and had just time to strip the lace of the clothing of the dead Huzzar when we were called in to join the skirmishers. The battle was now raging with double fury. We could see most of the charges made by the cavalry of both armies. I never before witnessed such large masses of cavalry opposed together, such a length of time.

I am at a loss which to admire most, the cool intrepid courage of our squares, exposed as they often were to a destructive fire from the French Artillery and at the same time or in less than a minute surrounded on all sides by the enemy's Heavy Cavalry, who would ride up to the very muzzles of our men's firelocks and cut at them in the squares. But this was of no use, not a single square could they brake, but was always put to the rout, by the steady fire of our troopes. In one of those charges made by the enemy a great many over charged themselves and could not get back without exposing themselves to the deadly fire of the infantry. Not choosing to return by the way they came they took a circuitous rout and came down the road on our left. There were nearly one hundred of them, all Cuirassiers. Down they rode full gallop, the trees thrown across the bridge on our left stopped them. We saw them coming and was prepared, we opened our fire, the work was done in an instant. By the time we had loaded and the smoke had cleared away, one and only one, solitary individual was seen running over the brow in our front. One other was saved by Capt. Jno. Ross from being put to death by some of the Brunswickers.

I went to see what effect our fire had, and never before beheld such a sight in as short a space, as about an hundred men and horses could be huddled together, there they lay. Those who were shot dead were fortunate for the wounded horses in their struggles by plunging and kicking soon finished what we had began. In examining the men we could not find one that would be like to recover, and as we had other business to attend to we were obliged to leave them to their fate.

Either the noise of our fire or the man who escaped informed the enemy of our lurking place, for we were soon informed by a Fedet [vedette – mounted sentry or scout] that the enemy were marching down on us with Cavalry, Artillery and Infantry. Hougomont had been in flames some time and the tremendous fire of guns and Howitzers on the place seemed to increase. The news brought by the Fedet caused us to move and form square. In a short time we were obliged to shift more to our left to get out of the range of some cannon the enemy opened on us.

Lord Hill now paid us a visit and asked for water, he was very much fatigued. While his Lordship was drinking out of one of our men's wooden canteens an eight pounder picked out four of our men. We were then ordered to shift our ground a little further. The enemy did not make their appearance.

We remained here until dusk when we discovered a large column of cavalry coming down on us from our rear. Their commander saw we were ready to receive them, rode down to us. When we found they were Prussians, they passed us to the front and we followed. At this time the enemy were in full retreat, we marched into an orchard belonging to the farm where we halted for the night. This place was full of dead and wounded Frenchmen. I went to the farm house, what a sight. Inside the yard the Guards lay in heaps, many who had been wounded inside or near the building were roasted, some who had endeavoured to crawl out from the fire lay dead with their legs burnt to a cinder. It was now certain the enemy was off in good earnest.

I managed to make up a supper, wrapped myself in my blanket and slept very comfortably until daylight, then marched to Nivelle[s]. Our loss is but trifling considering the heavy fire we were under, but we have to thank the deep road and the field of Rye for it. Killed 1 bugler and 8 rank and file, wounded Captain Beardsley, Lieutenant Tyndale, 1 Serjeant and 34 rank and file.

DUKE OF WELLINGTON
Wellington's Waterloo Despatch

The official despatch by the commander of the British Army
describing the Battle of Waterloo contrasts interestingly with
Private Wheeler's view of the fighting from the ranks. Despite
its dispassionate tone, however, the despatch is the work of a
close eyewitness. Wellington was in the front line throughout
the battle, riding his horse Copenhagen from one threatened
spot to another throughout the day and sometimes having to
take refuge inside a square of infantry when the French cavalry
charged. He saw the beginning of the action at Hougoumont,
the large walled farmhouse on the British right, which the
French attacked before noon. It was held by the 2nd (now
Coldstream) and 3rd (now Scots) Guards in bitter, eventually
hand-to-hand combat, even though the French succeeded at
one stage in breaking into the courtyard and despite the fact
that some of the buildings were set ablaze while the British
were trying to defend them. Wellington was also close to La
Haye Sainte, the other walled farmhouse, near the centre of
the line, when the Light Battalion of the King's German Legion,
from George III's possessions in Hanover, was forced to
evacuate the strongpoint after running out of ammunition.
Finally, he was on the right of the line when the Imperial Guard
attacked in the early evening and it was he who gave the crucial
orders to the 52nd Light Infantry, and to the 1st (now Grenadier)
Guards, to open the fire that drove the French back. After
their retreat, he knew that the day was won and, as dusk began
to fall, he rode forward across the battlefield – on which lay
the bodies of 40,000 dead and dying soldiers and 10,000 stricken
horses – to meet his Prussian ally, Marshal Blücher, whose
arrival with his army in the nick of time had clinched the
victory. They met at the inn called La Belle Alliance, close to

the place from which Napoleon had commanded his side of the battle. After the two Allied commanders had congratulated each other – speaking in French, their only common language – Wellington rode back to the village of Waterloo, where he gave up his bed to one of his staff officers, who was dying from his wounds. Next morning he was up early to begin writing the despatch that follows.

*   *   *

Wellington's official despatch
The *London Gazette* Extraordinary.
DOWNING-STREET, JUNE 22, 1815.
Major the Hon. H. Percy arrived late last night with a despatch from field-marshal the Duke of Wellington KG, for the Earl Bathurst his Majesty's principal secretary of state for the war department, of which the following is a copy:–

MY LORD,

WATERLOO, JUNE 19, 1815,

Bonaparte having collected the 1st, 2d, 3d, 4th and 6th corps of the French army and the imperial guards, and nearly all the cavalry on the Sambre, and between that river and the Meuse, between the 10th and 14th of the month, advanced on the 15th, and attacked the Prussian posts at Thuin and Lobbes, on the Sambre, at daylight in the morning.

I did not hear of these events till the evening of the 15th, and I immediately ordered the troops to prepare to march; and afterwards to march to their left, as soon as I had intelligence from other quarters, to prove that the enemy's movement upon Charleroi was the real attack.

The enemy drove the Prussian posts from the Sambre on that day; and general Ziethen, who commanded the corps which had been at Charleroi, retired upon Fleurus; and Marshal Prince Blucher concentrated the Prussian army upon Sombreffe, holding the villages in front of his position of St. Amand and Ligny.

The enemy continued his march along the road from Charleroi towards Brussels, and on the same evening, the 15th attacked a brigade of the army of the Netherlands under the prince of Weimar, posted at Frasnes, and forced it back to the farmhouse on the same road, called Les Quatre Bras.

The prince of Orange immediately reinforced this brigade with another of the same division, under general Perponcher and in the morning early regained part of the ground which had been lost, so as to have the command of the communication leading from Nivelles to Brussels, with marshal Blucher's position.

It the mean time I had directed the whole army to march up on Les Quatre Bras, and the 5th division, under lieutenant general Sir Thomas Picton, arrived at about half-past two in the day followed by the corps of troops under the duke of Brunswick and afterwards by the contingent of Nassau.

At this time the enemy commenced an attack upon prince Blucher with his whole force excepting the 1st and 2d corps, and a corps of cavalry under general Kellerman with which he attacked our post at Les Quatre Bras.

The Prussian army maintained their position with their usual gallantry and perseverance, against a great disparity of numbers, as the 4th corps of their army, under general Bulow, had not joined, and I was not able to assist them as I wished, as I was attacked myself, and the troops, the cavalry in particular, which had a long distance to march, had not arrived.

We maintained our position also, and completely defeated and repulsed all the enemy's attempts to get possession of it. The enemy repeatedly attacked us with a large body of infantry and cavalry, supported by a numerous and powerful artillery; he made several charges with the cavalry upon our infantry but all were repulsed in the steadiest manner. In this affair his royal highness the prince of Orange, the duke of Brunswick, and lieutenant general sir Thomas Picton, and major-general sir James Kempt and sir Denis Pack, who were engaged from the commencement of the enemy's attack, highly distinguished themselves as well as lieutenant-general Charles baron Alten, major-general sir C. Halket, lieutenant-general Cooke, and major-generals Maitland and Byng, as they successively arrived. The troops of the 5th division, and those of the Brunswick corps, were long and severely engaged and conducted themselves with the utmost gallantry. I must particularly mention the 18th, 42d, and 92d regiments and the battalion of Hanoverians.

Our loss was great, as your lordship will perceive by the inclosed return; and I have particularly to regret his Serene Highness the duke of Brunswick, who fell, fighting gallantly at the head of his troops.

Although marshal Blucher had maintained his position at Sombreffe, he still found himself much weakened by the severity of the contest in which he had been engaged and as the 4th corps had not arrived, he determined to fall back, and concentrate his army upon Wavre; and he marched in the night after the action was over.

This movement of the marshal's rendered necessary a corresponding one on my part; and I retired from the farm of Les Quatre Bras upon Genappe and thence upon Waterloo the next morning the 17th at ten o'clock.

The enemy made no effort to pursue marshal Blucher. On the contrary, a patrol which I sent to Sombreffe in the morning found all quiet, and the enemy's vedettes fell back as the patrol advanced. Neither did he attempt to molest our march to the rear, although made in the middle of the day, except by following with a large body of cavalry, brought from his right the cavalry under the Earl of Uxbridge.

This gave Lord Uxbridge an opportunity of charging them with the 1st life-guards upon their debouché from the village of Genappe upon which occasion his lordship has declared himself to be well satisfied with that regiment.

The position which I took up in front of Waterloo crossed the high roads from Charleroi and Nivelles, and had its right thrown back to a ravine near Merbe Braine, which was occupied and its left extended to a height above the hamlet Tier la Haie which was likewise occupied. In front of the right centre, and near the Nivelles road, we occupied the house and garden of Hougoumont which covered the return of that flank; and in front of the left centre, we occupied the farm of La Haye Sainte. By our left we communicated with marshal prince Blucher, at Wavre, through Ohain; and the marshal had promised me that in case we should be attacked, he would support me with one or more corps, as might be necessary.

The enemy collected his army, with the exception of the 3d and 4th corps, which had been sent to observe marshal Blucher, on a range of heights in our front, in the course of the night of the 17th, and yesterday morning: and at about ten o'clock he commenced a furious attack upon our post at Hougoumont. I had occupied that post with a detachment from general Byng's brigade of guards which was in position in its rear; and it was for some time under the command of lieutenant-colonel Macdonnel and afterwards of colonel Home; and I am happy to add, that it was maintained throughout the day with the utmost gallantry by these brave troops, notwithstanding the repeated efforts of large bodies of the enemy to obtain possession of it.

This attack upon the right of our centre was accompanied by a very heavy cannonade upon our whole line, which was destined to support the repeated attacks of cavalry and infantry occasionally mixed, but sometimes separate, which were made upon it. In one of these the enemy carried the farm-house of La Haye Sainte, as the detachment of the light battalion of the (King's German) Legion which occupied it, had expended all its ammunition, and the enemy occupied the only communication there was with them.

The enemy repeatedly charged our infantry with his cavalry, but these attacks were uniformly unsuccessful, and they afforded opportunities to our cavalry to charge, in one of which lord E. Somerset's brigade, consisting of the Life-guards, Royal Horseguards, and 1st Dragoon Guards, highly distinguished themselves, as did that of major-general sir W. Ponsonby, having taken many prisoners and an eagle [French regimental standard].

These attacks were repeated till about seven in the evening, when the enemy made a desperate effort with the cavalry and infantry, supported by the fire of artillery, to force our left centre near the farm of La Haye Sainte, which, after a severe contest, was defeated. And having observed that the troops retired from the attack in great confusion, and that the march of general Bulow's corps by Frischemont upon Plancenois and La Belle Alliance had begun to take effect, and as I could perceive the fire of his cannon, and as marshal prince Blucher had joined in person, with a corps of his army to the left of our line by Ohain, I determined to attack the enemy, and immediately advanced the whole line of infantry, supported by the cavalry and artillery. The attack succeeded in every point; the enemy was forced from his position on the heights, and fled in the utmost confusion, leaving behind him, as far as I could judge, one hundred and fifty pieces of cannon, with their ammunition which fell into our hands. I continued the pursuit till long after dark, and then discontinued it only on account of the fatigue of our troops, who had been engaged during twelve hours, and because I found myself on the same road with marshal Blucher, who assured me of his intention to follow the enemy, throughout the night; he has sent me word this morning that he had taken sixty pieces of cannon belonging to the imperial guard, and several carriages, baggage, etc., belonging to Bonaparte, in Genappe.

I propose to move, this morning, upon Nivelles and not to discontinue my operations. Your lordship will observe, that such a desperate action could not be fought, and such advantages could not be gained, without great loss; and I am sorry to add that ours has been immense. In lieutenant-general sir Thomas Picton his Majesty has sustained the loss of an officer who had frequently distinguished himself in his service, and he fell gloriously leading his division to a charge with bayonets, by which one of the most serious attacks made by the enemy on our position was defeated. The earl of Uxbridge, after having successfully got through this arduous day, received a wound by almost the last shot fired, which will, I am afraid, deprive his Majesty for some time of his services.

His royal highness the prince of Orange distinguished himself by his gallantry and conduct, till he received a wound from a musket-ball through the shoulder which obliged him to quit the field.

It gives me the greatest satisfaction to assure your lordship, that the army never, upon any occasion, conducted itself better. The division of guards under lieutenant-general Cooke, who is severely wounded, major-general Maitland, and major-general Byng, set an example which was followed by all; and there is no officer, nor description of troops, that did not behave well.

I must, however, particularly mention, for his royal highness's approbation, lieutenant-general sir H. Clinton, major-general Adam, lieutenant-general Charles Baron Alten, severely wounded; major-general sir Colin Halket, severely wounded; colonel Ompteda, colonel Mitchell, commanding a brigade of the 14th division, major-general sir James Kempt and sir Denis Pack, major-general Lambert, major-general lord E. Somerset, major-general sir W. Ponsonby, major-general sir C. Grant, major-general sir H. Vivian, major-general sir O. Vandeleur, and major-general count Dornberg. I am also particularly indebted to general lord Hill, for his assistance and conduct upon this, as upon all former occasions.

The artillery and engineer departments were conducted much to my satisfaction by colonel sir G. Wood and colonel Smyth; and I had every reason to be satisfied with the conduct of the adjutant-general, major-general Barnes, who was wounded, and of the quarter-master-general, colonel Delancy, who was killed by a cannonshot in the middle of the action. This officer is a serious loss to his Majesty's service, and to me at this moment. I was likewise much indebted to the assistance of lieutenant-colonel lord Fitzroy Somerset, who was severely wounded, and of the officers composing my personal staff who have suffered severely in this action. Lieutenant-colonel the Hon. sir Alexander Gordon, who has died of his wounds, was a most promising officer, and is a serious loss to his Majesty's service.

General Kruse, of the Nassau service, likewise conducted himself much to my satisfaction; as did general Trip, commanding the heavy brigade of cavalry, and general Vanhope, commanding a brigade of infantry of the King of the Netherlands.

General Pozzo di Borgo, general baron Vincent, general Muffling and general Alava were in the field during the action, and rendered me every assistance in their power. Baron Vincent is wounded, but I hope not severely; and general Pozzo di Borgo received a contusion.

I should not do justice to my feelings, or to marshal Blucher and the Prussian army, if I did not attribute the successful result of this arduous day to the cordial and timely assistance I received from them.

The operation of general Bulow, upon the enemy's flank, was a most decisive one; and even if I had not found myself in a situation to make the attack, which produced the final result, it would have forced the enemy to retire, if his attacks should have failed, and would have prevented him from taking advantage of them, if they should unfortunately have succeeded.

I send, with this despatch, two eagles, taken by the troops in this action, which major Percy will have the honour of laying at the feet of his royal highness.

I beg leave to recommend him to your lordship's protection.

I have the honour, etc. (Signed)
WELLINGTON.

JAMES BODELL
A Soldier's View of Empire

James Bodell had enlisted in the 59th Regiment at the age of sixteen, giving his age as seventeen. In 1849, when the regiment (later the 2nd Battalion, East Lancashire Regiment) was stationed in Ireland, it was ordered for foreign service. Such an order was dreaded by a regiment on a home station, for the army refused to allow wives who were not 'on the strength' to accompany their husbands, even though the men might be away for many years and despite the absence of any official provision for the families left behind. Only a proportion of non-commissioned officers and a very few soldiers were allowed wives 'on the strength'. The remainder, when a regiment left for imperial duty, in this case in Hong Kong, had to fend for themselves.

Parting was therefore heartbreaking. It was, however, the common lot of ordinary soldiers and their womenfolk in all long-service armies of the period. The soldier, once enlisted, surrendered almost all his civil rights and became little better than a chattel of the state. Bodell, a man of unusual spirit as well as intelligence, deeply resented the deprivations and, within the limits allowed by military law, was adamant in sticking up for his rights. His manner impressed his superior officers, who came to respect him enough not to trifle with him. After a successful military career, Bodell settled in New Zealand, where he became mayor of the town in which he took up residence. His is a rare voice from the ranks of an army which, ruled by the lash, conquered and garrisoned an enormous empire for Victorian Britain.

*   *   *

At last the Six Companies for foreign Service got the route for Cork, and off we went, the Band playing us out of Fermoy. Going through Watergrasshill a house on the Road Side wanted to charge us half Penny each for a drink of Water but the men took the bucket and got several buckets of water, and paid nothing. It certainly looked one time as if some damage would be done. This trying to extort vexed the men. Only for the officers I believe they would have wrecked the house. All the Women and Children rode on the Baggage Waggons. I was surprised at seeing so many Women on the Carts belonging to the Regt. On our arrival in Cork as usual like many others, I was off to look for lodgings. Winny [his wife] had procured these as she learnt by Fermoy the last in the field have to put up with the worst rooms. Here we stopped about three Weeks, when it was read out in Regimental Orders we should embark on Steamer, by Companies, lying at the quay in Cork to be conveyed on board HM Troopship *Apollo* at anchor in the Cove of Cork. Many were the inquiries about our destination, but no one knew. Some said the West Indies etc. The last night in Cork and the next day, I shall never forget. One good thing all Women left behind had their Passage paid home, whether in Ireland, England or Scotland, and our Colour Sergt. was good enough to put Winny['s] name down. I was up all night, Winny would not leave me, and on parading next morning in column of Companies, and the Speech of our Colonel, about going on Foreign Service and if we met an enemy of our Country, that we should show ourselves as British Soldiers. Our Colonel was a fine specimen of a Soldier, six feet two inches in his Vamps, and stout broad shouldered in proportion weighing I should say 15 to 16 stone. I thought he looked grand that morning with the Waterloo Medal on his Breast. (He proved otherwise.) At last off we went the Band playing The Girl I Left Behind Me.

When we arrived at the quay, and as we were told off in Companies to go on board of the Steamer, I shall never forget that day. I could not keep Winny off my neck. I was so bothered I forgot to unfix my bayonet. The Colonel was standing one Side of the gangway, and through Winny clinging to me, my bayonet went straight to the Colonel's Chest. He let [out] a roar (and he could roar) which brought me to my senses, and Winny was gently requested to return to the Shore. This was the last time I touched Winny. After getting on Board, we stopped at the quay about 20 minutes, and the sight was something awful. Women and children screaming, young Girls fainting, others half mad in a frenzied Condition. I assure you Women with 3, 4 & 6 children each was left behind. Poor things it was a sad Sight to see them. Some of them expected shortly to bring another into the World. This was the 8th day of June 1849. At last off we went, and on getting on Board we were shown our living or I should say sleeping Place. 500 of us was told off for the lower deck and no Ventilation but the Hatches and round Scuttle holes, and one hammock for two men. All our Meals we had to take on upper deck with the Sky or clouds for a ceiling. We remained at anchor till June the 12th, even now we did not know our destination. The four days we lay at anchor was even worse than leaving the quay in Cork. All day long boats crowded with Women and Children, Winny amongst them, pulling around the Ship. Those four days were really miserable ones. At last we commenced our Voyage. I forgot to mention several men deserted and one man shot himself in Cork Barracks before they would leave their Wives and families.

HENRY CLIFFORD
Clifford in the Crimea

Turkey, by the middle of the nineteenth century, was no longer the conquering power that had struck terror into Byzantium and the Christian lands of Southern Europe. The turn of the tide had come at the Treaty of Carlowitz in 1699, when for the first time the Turkish Sultan had conceded defeat to a Christian power, the Habsburg Emperor of Austria. During the eighteenth century the champions of the effort to regain Christian land from the 'Great Turk' had been the Russian emperors. By the 1850s they had succeeded in advancing their power to the northern shore of the Black Sea, to the Caucasus mountains and almost to the Danube in Balkan Europe. Russian assistance also helped the Serbs and the Greeks to win their independence from Turkish rule.

Russia's success at Turkey's expense then had the paradoxical effect of bringing Britain and France to the Sultan's side. Neither power, with interests of their own in the Balkans and the Eastern Mediterranean, wished to see Russia replace Turkey as the dominant state in those regions. When Russia opened a campaign against Turkey in the Balkans in 1853, they took alarm. In 1854, when Russia invaded Bulgaria, then one of Turkey's possessions, they declared war, organized an expeditionary force and despatched it, under the protection of a strong naval force, into the Black Sea to threaten Russia's main southern base at Sebastopol in the Crimea.

Thus opened the Crimean War (1854–6), familiar to the English-speaking world for the siege of Sebastopol, and the battles around that port at the Alma, Balaclava and Inkerman, that followed the initial landing on 13 September 1854.

Henry Clifford, an officer in the Rifle Brigade, was a member of one of England's old Catholic families that had declined to accept the Reformation in the sixteenth century. Like many of the 'recusants', his family maintained a strong military tradition and he proved a brave soldier, winning the newly instituted Victoria Cross (now Britain's highest decoration for valour) in the course of the war. This letter home describes two of the most famous episodes of the Crimean War, the valiant conduct of the 93rd Highlanders, the 'thin red line', at Balaclava, and the Charge of the Light Brigade of cavalry against the Russian guns in the 'Valley of Death' that followed. The epic of the Light Brigade inspired Tennyson to write one of the most famous poems in the English language:

Half a league, half a league,
Half a league onward,
All in the Valley of Death
Rode the six hundred.
'Forward the Light Brigade!
Charge for the guns!' he said:
Into the Valley of Death
Rode the six hundred.

'Forward, the Light Brigade!'
Was there a man dismayed?
Not though the soldier knew
Some one had blundered:
Their's not to make reply,
Their's not to reason why,
Their's but to do and die:
Into the Valley of Death
Rode the six hundred.

Cannon to right of them,
Cannon to left of them,
Cannon in front of them
Volleyed and thundered.
Stormed at with shot and shell,
Boldly they rode and well,
Into the jaws of Death,

Into the mouth of Hell,
Rode the six hundred.

When can their glory fade?
O the wild charge they made!
All the world wondered.
Honour the charge they made!
Honour the Light Brigade,
Noble six hundred!

The Light Brigade, which consisted of the 3rd Hussars, the 4th Hussars, the 8th Hussars, the 11th Hussars, the 13th Hussars and the 17th Lancers, numbered 673 officers and men at the beginning of the charge. During it, 247 men and 497 horses were lost. The charge of the Heavy Brigade, under General Scarlett, which preceded it, mentioned by Clifford, succeeded in its object with very little cost. Bosquet, a French general who witnessed the Light Brigade's charge, remarked, 'It is magnificent but it is not war' ('C'est magnifique, mais ce n'est pas la guerre'). His words have become as celebrated as Tennyson's.

*   *   *

27th October [1854]

With the very best of intentions it is not in my power to write every day. Three days have passed during which so much has taken place, that constant duty has not left me a moment to handle my pen. Before I attempt to give you some idea of the most important events, it may be well to mention that the siege of Sebastopol continues much as usual. Firing from our siege guns, Lancasters, 32 pounders, and 24 pound Rockets and shells, begins at daybreak every morning from our Batteries, and continues till dark, without intermission, pouring into the town a most destructive fire. All but the troops essential for its defence have left it, and they are to be seen at daylight, before the firing begins, lying under the walls close to the sea coast, as far from our guns as possible.

From the nature of the ground in front of our batteries, we have not been able to make close approaches, but the French, who have better ground to work on, are closing on the Town, their foremost works being only 600 yards from it. The Russian batteries keep up a good fire, but the deserters and prisoners say they suffer much. In five days from this it is hoped we shall be able to make an assault; a combined one.

Our losses in the trenches and batteries have been few. Poor young Maule, the Adjutant of the 88th Regt, lost his left arm by a round shot in the trenches yesterday – it has been amputated and he is doing well. So far for our front.

The events I am about to acquaint you with, and which are full of the greatest interest, have taken place on our rear in front of Balaclava, and yesterday on our right. As I am obliged to make use of I so often you must understand that it is only in my power to tell you what I see and hear, but as this is only written for you, you will not think it brought in too often – as it is with no wish to bring myself before you more than possible – but only to try and give you a better idea of events, difficult to make clear, even with its frequent use.

On the morning of the 24th the siege went on as usual, but in the evening of that day a report came from the French Division in rear, that a large Russian force had made its appearance, on the road leading into Balaclava and had taken up its position on the high ground to the right of that road and the entrance of the first plain in front of Balaclava. This first plain is separated from the second by a ridge of low hills, on which at intervals of two or three hundred yards, redoubts or earthworks had been thrown up, and six or nine guns belonging to the English placed in the three works to the right of Balaclava. (I call the right of Balaclava the side to the right hand as you come from the sea). These works, six in number, were made by the Turkish troops, under the direction of our Engineer Officers and given over to them to defend. A few, four I think, privates of the Artillery, were left to show them any difference there might be between our guns and theirs. After the defence of Silistria etc., such confidence was placed in their troops that it was not thought necessary to place any but Turks in these redoubts.

The constant appearance of a large Russian force in front of Balaclava, prevented any very serious notice being taken of the report. We had our troops in Balaclava on their posts, the Turks on their earthworks were supplied with ammunition, and on the morning of the 25th of October, the 4th Division under Sir G. Cathcart, the other two Regiments of Highlanders under Sir Colin Campbell (Commandant of Balaclava), a Battn of Guards and the First Battalion Rifle Brigade, with all the Cavalry under Lord Lucan, and the Light Brigade under Lord Cardigan, were behind the range of earthworks, about 3,000 yards in rear and to the left of Balaclava.

I went at daybreak to the heights in rear of our camp, by order of Sir G. Brown, to see what was going on. It was a dull cloudy morning and mist lay in the plain below, but it was evident from what little we could see of the enemy, he was preparing for an attack.

The heights on which I stood are held by the French troops, and I had the advantage of standing by the side of General Brite, Brigadier commanding under General Bosquet, and we watched the movements of the Russian Army most closely. At about 8 o'clock the first cannon shot was fired by the Turks at the Russians, from the right of our earthworks, and shortly after we saw a large force of Cavalry and Infantry, about 20,000 men, moving towards us. When they came within range of our (the French) guns in position in front of us, on the heights on which we stood, we opened fire on them, and they retired in good order to the front of the three guns, in the earthworks, worked by the Turks. Why or wherefor I don't know, but though not a shot had been fired at them, and no attack was being made on them, the Turks on the other three earthworks deserted them, and fell back upon our troops, close upon Balaclava. The whole Russian army then made an attack upon the three earthworks with our guns in them on the right. These works were so strong that had they been occupied by our troops, I am confident we should have held them. After a short resistance during which our Artillery men say the conduct of the Turkish Officers was most disgraceful, the Turkish troops fled, leaving our guns in the hands of the enemy; but fortunately our Artillery men spiked them first.

A great part of the Russian cavalry then made a charge all along the range of little hills, passing the abandoned forts, one after the other, till they came to the last which was closest to our troops. Here a small party of Turks opened fire on them, and I saw about 20 fall out of their saddles. As they rose to the crest of the hill, they saw the English troops formed up ready to receive them; they halted to a man.

One of the Highland regiments (the 93rd) opened a heavy fire of musketry on their right but the centre stood fast. But it was only for a few minutes. The Scots Greys and the Enniskillen Dragoons, advanced in a slow, steady trot towards them, the Russians looked at them as if fascinated, unable to move. The distance between the two Cavalries at last decreased to about 50 yards, and the shrill sound of the trumpet, ordering the charge, first broke the awful silence.

Like a shot from a cannon ball our brave fellows went at the astounded enemy like one man, and horses and men were seen struggling on the ground in every direction. The Russians fled in the greatest disorder, our splendid Cavalry not leaving them till they got under the protection of their artillery.

A pause then ensued, during which the Russian army reformed its Cavalry with guns protecting its front across the plain and a line of infantry lying down in front of the guns to protect them, their infantry in rear of their cavalry on the left, on the heights in the earthworks they had taken from us, and where they had placed some of their field pieces, and on the high ground to their right, where they also had some artillery, and a few cavalry.

On our side, the whole of our Cavalry advanced along the range of low hills supported by Artillery and Infantry, till they came to the earthwork next to that in possession of the Russians. The two forces were about three quarters of a mile, or a short mile, apart. Shots from the field-pieces were exchanged, but with little effect, on either side.

For the first time that morning I saw Lord Raglan and Staff ride up to the troops in position, and General Canrobert with his Staff [as Lord Fitzroy Somerset, Raglan, the British Commander-in-Chief in the Crimea, was mentioned in Wellington's despatch after Waterloo as wounded; he lost an arm]. Two Squadrons of the Chasseurs D'Afrique, and about 700 Chasseurs de Vincennes and Zouaves, came out from the same direction on our right, the Chasseurs d'Afrique taking position in the plain in rear, and to the left of our Cavalry, and the French Infantry remaining in support of ours.

It was evident a consultation was taking place, which resulted in one of the greatest disasters, and the most useless and shocking sacrifice of the lives of hundreds of brave men that was ever witnessed.

I must tell you that but little confidence has been placed in the commanding powers of Lord Lucan commanding the Cavalry, and long and loud have been the feuds on public grounds, between his Lordship and Lord Cardigan (than whom, a braver soldier never held a sword) who commands the Light Brigade; and it was thought if a verbal order was sent to Lord L. it might be misunderstood, or not carried out. A written order was, therefore, sent from Lord Raglan by Captain Nolan, General Airey's ADC (formerly my brother ADC in the Light Division) desiring his Lordship 'to charge.' 'To charge what?' said Lord Lucan very naturally. 'Here are your orders,' said poor Nolan, pointing to the paper, 'and there,' pointing to the Russian army, 'is the enemy,' and shouting 'Come on' to the Light Brigade of Cavalry, he dashed forward. He was wrong, poor fellow; in doing so, he forgot his position, and his conduct was most insulting to Lord Lucan and Lord Cardigan, who at the head of his Brigade, pale with indignation, shouted to him to stop, that he should answer for his words and actions before Lord Raglan, but he was called to a higher tribunal, a shell struck him in the chest, and in a few minutes he was a mangled corpse. Lord Lucan then ordered the Light Brigade of Cavalry between 600 and 700 to charge the Russian Army, 30,000 strong. This is the explanation I heard afterwards.

From the commanding position in which I stood by the side of General Brite we saw the Light Brigade of Cavalry moving forward at a trot, in face of the Russian Army. 'Mon Dieu!!' said the fine old French General, 'Que vont-ils faire?' They went steadily on, as

Englishmen only go under heavy fire. Artillery in front, on the right and left. When some thousand yards from the foremost of the enemy, I saw shells bursting in the midst of the Squadrons and men and horses strewed the ground behind them; yet on they went, and the smoke of the murderous fire poured on them, hid them from my sight.

The tears ran down my face, and the din of musketry pouring in their murderous fire on the brave gallant fellows rang in my ears. 'Pauvre garçon,' said the old French General, patting me on the shoulder. 'Je suis vieux, j'ai vu des batailles, mais ceci est trop.' Then the smoke cleared away and I saw hundreds of our poor fellows lying on the ground, the Cossacks and Russian Cavalry running them through as they lay, with their swords and lances.

Some time passed, I can't say how much, but it was very long, waiting to see if any would return. Horses without riders, galloped back in numbers, and men wounded on foot and men not hurt, but their horses killed, returned on foot, and then we saw a horse or a man fall, who wounded, had come as far home as he could and then fell and died.

At length about 30 horsemen dashed through a line of Cossacks, who had reformed to interrupt their retreat, and then another larger body came in sight from the middle of the smoke and dust. Two hundred men! They were all that returned of 600 odd that charged. I don't know the names of the Officers who fell or were taken prisoner, but very few returned, and some are since dead of their wounds; one of the Officers of the 17th Lancers (his Regiment suffered most severely, I believe) told me they charged through a line of infantry, drove the gunners from their guns, but of course could not bring them away. Then through the line of Cavalry till they came to the Infantry when the handful that remained, turned about and recharged the same forces again. The Chasseurs d'Afrique also made a small charge, but when they came face to face with the Russian Infantry in square, with the exception of one or two Officers, they turned round and came back again.

It was thought an attack would be made on the Russians and the redoubts retaken. Sir G. Cathcart wished to take on his Division, but the evening passed away and the Russians remained in their positions and are there now. Yesterday morning I went into Balaclava. The Scots Greys and the Enniskilleners left good marks of the work they had done; Russian Hussars lying in all directions, with awful sabre-cuts.

I got my letters from the post, two very kind ones from Conny and two from Charles. I cannot thank you enough for letters. Such as these would be dear and valuable for the kind expressions of sincere anxiety, interest and affection they contain at any time, but to receive them here, under such circumstances, I cannot tell you what pleasure they give. The latest date is up to the 9th.

I have seen Lord Raglan's despatch and think it a very good and a very true one, the French a very paltry and insignificant one.

On my return from Balaclava I found the troops under arms. A large force of Russians, about 6,000, I thought, some said more, with about 15 pieces of cannon, were moving to attack our right. Our Division being in the centre and having the siege parties to protect we could not move in support of Sir De Lacy Evans' Division on the right, whose Pickets were engaged and fighting hard, but Sir G. Brown went up, and sent the nine pounder Battery of our Division, the Guards sent up a Brigade, and the 4th Division moved up, sending our 1st Battalion in front. The Russians advanced in 8 columns at a very quick pace, considering the ground they moved on, which was stony and covered with stunted oak-bushes, about three feet high. They were evidently bent upon mischief, and thought the attack of the day previous had weakened our force on the right to strengthen Balaclava.

They were mistaken: the pickets made a desperate resistance, and held the Russian skirmishers at bay. The advance of the column obliged them to retire, yet the Officer in command of the 48th Picket tried to hold his ground in spite of the numbers that were attacking him. He called out to his men to stand and fight longer, and only retired with them when he had killed two Russian skirmishers with his sword, and received I fear his death wound, a ball in his left breast.

On the Picket coming in, our guns, 24 in number, opened on the advancing columns and the whole of the 2nd Brigade, 2nd Division, went out as skirmishers, with the 1st Battalion Rifle Brigade, and poured in such a fire of Minié [rifle] balls, that the Russians turned and fled in disorder. Our soldiers pursued them for about two miles from camp, almost within range of the guns of Sebastopol and killed about 400 on the field, besides 62 wounded brought into camp, and many wounded taken away by the Russians.

The prisoners taken say that the sortie was occasioned by the General commanding the Army in front of Balaclava, sending into Sebastopol the news of a victory over the English the day before – that the General commanding in Sebastopol, called them together and told them it, and said that it was their turn to thrash the English. A ration of spirits was served round to each man and they asked to be led against the English. The Russian General took advantage of their enthusiasm and brought them up – only to be driven back with great loss. We have had very few men killed, and 60 wounded, in round numbers, for I have not seen the return yet.

It is said Balaclava is going to be given up as our position is too large to protect, and we are going to have the same place of embarkation and disembarkation as the French. Till Sebastopol falls we have not force enough to attack the Russian army in our rear; to take their position, we should have to lose many men, and we cannot spare one we are so weakened by illness and casualties in fight. If we take Sebastopol I have no doubt we shall pay off old scores.

I believe the Russians treated our wounded shamefully in spite of our kindness to theirs at 'Alma.' Our men will not forgive them and will pay them off in their own coin. I don't think Sebastopol will fall for five days at least, and it will be tough work when we do go at it.

Poor Mr Sheehan, our Priest, went down to Balaclava yesterday sick, so we have only one Priest in our four Divisions up here. God bless you all, continue to pray for me, and think me always your affectionate

&c. &c.
Henry.

COLIN FREDERICK CAMPBELL
Letters from Camp

The maladministration of the British army sent to the Crimea caused one of the great political scandals of the Victorian age. The arrangements for the supply, accommodation and welfare, particularly the medical welfare, of the expeditionary force broke down quickly and almost completely. While the French army, commanded by generals with recent experience of military reality gained in the conquest of Algeria, was excellently provisioned and maintained, the British froze and nearly starved in its entrenchments outside Sebastopol, while its wounded died in thousands of infection and lack of the simplest medical care. The wounded evacuated to the barracks at Scutari, the Turkish military base opposite Constantinople, were eventually taken into the care of Florence Nightingale, the pioneer of modern nursing technique. Her example, and her relentless harrying of the government at home, by private letter and public exposure of inefficiency, saved thousands of lives. Meanwhile the army left in the Crimea, entrenched around Sebastopol, continued to undergo a dreadful ordeal, ill supplied with necessities and exposed to the hardships of the Russian winter.

*   *   *

Jan. 17. [1855]

Still waiting for cattle at Koolalee, the snow having been so deep on the hills that very few have been able to get down. I quite dread to hear what will be the condition of the army when we get back to Balaclava. When I left the men were suffering terribly, and many had been frostbitten. There is a very prevalent impression in the army (which I share) that Lord Raglan does not hear the truth, and that those about him are in the habit of making things pleasant to him; but as Commander-in-Chief he should see into such matters himself, and know the state the army is in. The want of transport has destroyed more lives and caused more misery than all other mistakes put together. I have seen our men after having come back from the trenches, and having barely time to eat some biscuit and coffee, sent off to Balaclava to bring up rations, warm clothing, blankets, etc. They would return at night after their fourteen-mile tramp through the mud, and throw themselves down on the floor of their tents as if they were dead, so exhausted, that even if their dinners had been got ready for them, many of them could not have eaten a morsel. Next morning probably one third of them would be in hospital, and the remainder for the trenches the following evening.

The day after the battle of Inkerman, and even before it, every man with one grain of sense could foresee that Sebastopol would not fall for two or three months and that we must spend part of the winter, at least, there. Notwithstanding this! as far as I can see, not one single preparation was made for it. If each regiment had been furnished with two or three hundred short poles and a few entrenching tools, they could have hutted themselves in a week. The proper hut for this country is merely a long narrow hole about eight feet deep and eight feet broad, and merely wants wall enough above the surface to be able to form eaves to the roof to let the water run off.

I do not think a single mule was bought, although even in the fine weather we were very insufficiently supplied, and lived as it were from hand to mouth, never being able to bring up more than one day's rations. Yet with the whole coast of Asia Minor teeming with ponies and barley, within forty-eight hours' sail of us, and such vessels as the *Jason* and *Simla* (which could bring over 300 at each trip) lying in the harbour of Balaclava, it is scarcely credible that not one single animal was brought. Somewhere about the beginning of December 250 mules arrived, being a totally insufficient number. Our cavalry, with their dying horses, were then set to work to carry biscuit, an occupation which killed them (the horses) at the rate of about twenty a day. When one sees our commissariat mules one sees a set of half-starved dying animals savagely thrashed along by Poles, Bulgarians, Tartars, and every sort of blackguard. What a contrast in the French animals! They pass our camp in long lines of hundreds daily, they walk in a row, every mule as fat and sleek as if he were a pet, and stepping along cheerfully with a quick and rapid step. To every three mules there is a workmanlike, well-appointed Frenchman, a soldier who, when he has nothing else to do, chats to his mules as if they were his friends. This is only one of the points in which they beat us; it is the same in everything.

You will perhaps be rather surprised when I give you my opinion that the English army is virtually destroyed. They can stand behind trenches and pull a trigger, or they could again muster up endurance to fight a second battle of Inkerman; but for anything like a campaign they are utterly useless. We have not a battery of artillery in the army that could bring its guns into action. I am certain that I am not exaggerating when I say that 400 of the London mounted police would utterly overthrow our whole cavalry force. Our only stand-by is the French; they are still an army, and in first-rate order; their cavalry officers' horses have, of course, suffered greatly, as all civilized horses must do in this country; but their 2,000 Chasseurs d'Afrique are in as fine order as if they had never left Algeria. I wonder what Lord Raglan thinks when he contrasts the two armies.

If I were to set to work and try to write about all the mistakes and blunders made in our different departments here, I should fill a tolerable volume. They are endless. Those in the medical department, though not worse than others, are more dreadful in their consequences. Doctors will tell you how they have been suddenly ordered on board a ship to take charge of 300 men across the Black Sea; how the men would lie on the hard boards in every form of cholera, dysentery, and fever, with not one atom of medicine to give them, and two or three drunken pensioners to attend on them. In the morning the doctors and the pensioners would go round picking out the dead from the living, and throwing them overboard. It is not one man will tell you this story, but twenty.

If you fancy horrors, I will give you a little story about Russian prisoners, which may give you some idea of the horrors of war. I will give it to you as I heard it from Dr Franklyn as near as possible in his own words.

'I was sitting in my room one afternoon, having had rather a hard morning's work, and thinking I had now time to write a letter, when an orderly came in and said:

'"Sir, there's 168 wounded Rooshians down below, and I was to tell you you is to take care of 'em." I said, "Oh, very well;" and the orderly went away. I scratched my head a little, and at last I thought it was no use doing that, so I walked down to see the Russians. I found 168 men in a sort of team, lying like sandwiches. There were men with their legs lying across their chests, and their hands forced into their bodies and shot through the lungs, and, in fact, every mutilation of the human form you can conceive. Well, I scratched my head again, and then I went up to the invalid battalion and got a fatigue party, and we carried about half of them up to the Greek Church and laid them out there, and we got the others into some sort of a row. There were eight young surgeons landed that day from the *Prince* and put under my orders, so I sent for them. Well, they did well, those youngsters. They worked away and their knives got blunt, and they worked away all the harder, and they performed hipjoint operations, and cut out bullets a foot deep, and there weren't as many dead as you'd have thought. Well, after this there was nothing for these wretches to lie on and nothing to give them to eat, and so they died very fast; and the gangrene broke out among those in the barn, and they all died but some five-and-forty in the Greek Church pulled through, and we sent them away yesterday, and on the whole I think we did pretty well.'

I have shortened this story as much as possible, in order to get it within the bounds of a letter; but do you not think that there is material in it for the Invalide Russe to write a pretty strong article touching English barbarians, etc.?

W. H. FITCHETT
The Relief of Lucknow

Immediately after the termination of the war in the Crimea, the British army was involved in a new crisis in India. In May 1857, a regiment of the East India Company's army in Bengal mutinied and killed several of its officers. The mutineers then fled to Delhi, capital of the former Mogul Empire, and proclaimed the last emperor's descendant, Bahadur Shah, as ruler. The example of the mutineers was quickly followed by most of the other regiments in Bengal. Many Europeans were killed and the survivors took refuge under the protection of the East India Company's few European regiments and those of the British army stationed in India.

A main centre of resistance to what became known as the Indian Mutiny, or Great Mutiny, was at Lucknow, capital of the province of Oudh, from which many of the mutineers originated. In the Residency, seat of the British administration, Brigadier-General Sir Henry Lawrence organized a defence with a scratch force of Europeans and loyal Indian troops. It was greatly outnumbered by the besieging mutineers, who subjected the improvised fortifications to constant bombardment; Lawrence was killed by one of the mutineers' shells. The survival of the Residency's defenders turned on the success of a relief operation. A small relief column, commanded by Major-General Sir Henry Havelock, managed to break through on 25 September but it was not strong enough to drive the besiegers away. In November Major-General Sir James Outram, who had taken command, learnt that a second relief column, under Lieutenant-General Sir Colin Campbell, was approaching. Outram decided it was essential to send someone out of the city to meet Campbell and guide his soldiers to a spot where they might successfully penetrate the mutineers' lines. The man chosen was a clerk, Kavanagh, who was decorated with the Victoria Cross for his bravery, one of the few awards of this highest of all honours to a civilian. Colin Campbell, to whom he delivered the vital intelligence, was one of the most famous fighting soldiers in the British army. He had commanded the 93rd Highlanders at the Battle of Balaclava in the Crimea in 1854, where its steadfastness earned it the title of the 'thin red line'; one of the rare senior Victorian soldiers to rise from the ranks, he died Field Marshal Lord Clyde.

Campbell's column, led by pipers playing 'The Campbells Are Coming', managed to break through to the Residency on 16 November and to evacuate its occupants to safety on the 22nd, although Havelock died from dysentery that morning. With strong reinforcements, Campbell recaptured the city in March 1858, by which time the Mutiny was in its last throes. By June of that year it was effectively ended. The British government then dissolved the East India Company and imposed direct rule on the sub-continent, which was exercised through a Viceroy until the grant of independence in 1947. The Mutiny had, however, shaken British confidence, and henceforth the Viceregal government was always supported by a European garrison large enough to hold the Indian Army in check.

*   *   *

On November 12 Campbell had reached the Alumbagh [one of Lucknow's gates] and, halting there, decided on the line of his advance to the Residency. Instead of advancing direct on the city, and fighting his way through loopholed and narrow lanes, each one a mere valley of death, he proposed to swing round to the right, march in a wide curve through the open ground, and seize what was known as the Dilkusha Park, a great enclosed garden, surrounded by a wall 20 feet high, a little over two miles to the east of the Residency. Using this as his base, he would next move round to the north of the city, forcing his way through a series of strongposts, the most formidable of which were the Secundrabagh and the Shah Nujeef, and so reach the Residency. And the story of the fighting at those two points makes up the tragedy and glory of the Relief of Lucknow.

Outram, of course, was not the man to lie inertly within his defences while Campbell was moving to his relief. He had already sent plans of the city and its approaches, with suggestions as to the best route, to Campbell by means of a spy, and he was prepared to break out on the line by which the relieving force was to advance. But if Campbell could be supplied with a guide, who knew the city as he knew the palm of his own hand, this would be an enormous advantage; and exactly such a guide at this moment presented himself. A civilian named Kavanagh offered to undertake this desperate mission.

Kavanagh was an Irishman, a clerk in one of the civil offices, and apparently possessed a hundred disqualifications for the business of making his way, disguised as a native, through the dark-faced hordes that kept sleepless watch round the Residency, and through the busy streets of Lucknow beyond. He was a big-limbed, fair man, with aggressively red hair, and uncompromisingly blue eyes! By what histrionic art could he be 'translated', in Shakespeare's sense, into a spindle-shanked, narrow-shouldered, dusky-skinned Oude peasant? But Kavanagh was a man of quenchless courage, with a more than Irish delight in deeds of daring, and he had a perfect knowledge of native dialect and character. He has left a narrative of his adventure.

A spy had come in from Campbell, and was to return that night, and Kavanagh conceived the idea of going out with him, and acting as guide to the relieving force. Outram hesitated to permit the attempt to be made, declaring it to be too dangerous; but Kavanagh's eagerness for the adventure prevailed. He hid the whole scheme from his wife, and, at half-past seven o'clock that evening, when he entered Outram's headquarters, he was so perfectly disguised that nobody recognized him. He had blackened his face, neck, and arms with lampblack, mixed with a little oil. His red hair, which even lamp-black and oil could hardly subdue to a colder tint, was concealed beneath a huge turban. His dress was that of a budmash, or irregular native soldier, with sword and shield, tight trousers, a yellow-coloured chintz sheet thrown over the shoulders, and a white cummerbund.

A little after eight o'clock Kavanagh, with his native guide, crept to the bank of the Goomtee, which ran to the north of the Residency entrenchment. The river was a hundred yards wide, and between four feet and five feet deep. Both men stripped, crept down the bank, and slipped, as silently as otters, into the stream. Here for a moment, as Kavanagh in his narrative confesses, his courage failed him. The shadowy bank beyond the black river was held by some 60,000 merciless enemies. He had to pass through their camps and guards, and through miles of city streets beyond. If detected, he would certainly perish by torture. 'If my guide had been within my reach,' he says, 'I should perhaps have pulled him back and abandoned the enterprise.' But the guide was already vanishing, a sort of crouching shadow, into the blackness of the further bank, and, hardening his heart, Kavanagh stole on through the sliding gloom of the river.

Both men crept up a ditch that pierced the riverbank to a cluster of trees, and there dressed; and then, with his tulwar [sword] on his shoulder and the swagger of a budmash, Kavanagh went boldly forward with his guide. A matchlock man first met the adventurous pair and peered suspiciously at them from under his turban. Kavanagh in a loud voice volunteered the remark that 'the night was cold,' and passed on. They had to cross the iron bridge which spanned the Goomtee, and the officer on guard challenged them lazily from the balcony of a two-storeyed house. Kavanagh himself hung back in the shade, while his guide went forward and told the story of how they belonged to a village some miles distant, and were going to the city from their homes.

They were allowed to pass, ran the gauntlet of many troops of Sepoys, re-crossed the Goomtee by what was called the stone bridge, and passed unsuspected along the principal street of Lucknow, jostling their way through the crowds, and so reached the open fields beyond the city. 'I had not been in green fields,' writes Kavanagh, 'for five months. Everything around us smelt sweet, and a carrot I took from the roadside was the most delicious thing I had ever tasted!' But it was difficult to find their way in the night. They wandered into the Dilkusha Park, and stumbled upon a battery of guns, which Kavanagh, to the terror of his guide, insisted upon inspecting.

They next blundered into the canal, but still wandered on, till they fell into the hands of a guard of twenty-five Sepoys, and Kavanagh's guide, in his terror, dropped in the dust of the road the letter he was carrying from Outram to Campbell. Kavanagh, however, kept his coolness, and after some parleying he and his guide were allowed to pass on. The much-enduring pair next found themselves entangled in a swamp and, waist-deep in its slime and weeds, they struggled on for two hours, when they reached solid ground again. Kavanagh insisted on lying down to rest for a time. Next they crept between some Sepoy pickets which, with true native carelessness, had thrown out no sentries, and finally, just as the eastern sky was growing white with the coming day, the two adventurers heard the challenge, 'Who comes there?' from under the shadow of a great tree!

It was a British cavalry picket, and Kavanagh had soon the happiness of pouring into Sir Colin Campbell's ears the messages and information he brought, while a flag, hoisted at twelve o'clock on the summit of the Alumbagh, told Outram that his messenger had succeeded, and that both the garrison and the relieving force had now a common plan. It is difficult to imagine a higher example of human courage than that supplied by 'Lucknow' Kavanagh, as he was afterwards called, and never was the Victoria Cross better won.

LIEUTENANT-COLONEL FREMANTLE
The Fremantle Diary

By 1863 the American Civil War, which had broken out two years earlier, had attracted a gallery of official observers from foreign armies, whose mission was to accompany the warring armies and report on their operations to headquarters at home.

The role of the foreign observer was widely accepted. General George McClellan, of the United States Army, had been present as an observer with the Anglo-French expeditionary force at the siege of Sebastopol during the Crimean War.

Lieutenant-Colonel James Fremantle was an officer of the (British) Coldstream Guards who, in mid-1863, was attached to the Confederate Army of Northern Virginia and accompanied it on its march to Gettysburg in July. In these extracts from his personal diaries, he describes the layout of the battle line on the second day of the fighting there, 2 July. The Northern troops of the Army of the Potomac were defending a ridge some three miles long, running from Cemetery Hill to the Little Round Top. The Confederates lined the ridge opposite. Fremantle's description of the topography, which remains today much as it did at the time of the battle, is strikingly accurate.

He was on excellent terms with the Confederate generals – Robert E. Lee, in supreme command, and Longstreet, A. P. Hill and Ewell, commanding I, II and III Corps respectively. His report reveals him to be an acute observer of the Confederate general staff at work, Longstreet wanting to outflank the Union position, Lee to make a frontal attack. Lee had his way but Longstreet was probably in the right. The Union position was indeed weak on the left flank and, but for a hasty reinforcement of Little Round Top, might well have been turned.

Lee, far from home in central Pennsylvania, appears to have thought that his best hope was to press forward with all his strength and break the Union centre. On the third day, 3 July, he ordered just such an effort, the attack known as Pickett's Charge. It was defeated with heavy loss and, on 4 July Lee was forced to lead his army back in defeat to Virginia.

*   *   *

2d July (Thursday) – We all got up at 3:30 A.M., and breakfasted a little before daylight. Lawley insisted on riding, notwithstanding his illness. Captain and I were in a dilemma for horses; but I was accommodated by Major Clark (of this staff), whilst the stout Austrian was mounted by Major Walton. The Austrian, in spite of the early hour, had shaved his cheeks and ciréd [waxed] his mustaches as beautifully as if he was on parade at Vienna.

Colonel Sorrell, the Austrian, and I arrived at 5 A.M. at the same commanding position we were on yesterday, and I climbed up a tree in company with Captain Schreibert of the Prussian Army. Just below us were seated Generals Lee, Hill, Longstreet, and Hood, in consultation – the two latter assisting their deliberations by the truly American custom of whittling sticks. General Heth was also present; he was wounded in the head yesterday, and although not allowed to command his brigade, he insists upon coming to the field.

At 7 A.M. I rode over part of the ground with General Longstreet, and saw him disposing of M'Laws's division for today's fight. The enemy occupied a series of high ridges, the tops of which were covered with trees, but the intervening valleys between their ridges and ours were mostly open, and partly under cultivation. The cemetery was on their right, and their left appeared to rest upon a high rocky hill. The enemy's forces, which were now supposed to comprise nearly the whole Potomac army, were concentrated into a space apparently not more than a couple of miles in length.

The Confederates inclosed them in a sort of semicircle, and the extreme extent of our position must have been from five to six miles at least. Ewell was on our left, his headquarters in a church (with a high cupola) at Gettysburg; Hill in the center; and Longstreet on the right. Our ridges were also covered with pine woods at the tops, and generally on the rear slopes.

The artillery of both sides confronted each other at the edges of these belts of trees, the troops being completely hidden. The enemy was evidently intrenched, but the Southerners had not broken ground at all. A dead silence reigned till 4:45 P.M., and no one would have imagined that such masses of men and such a powerful artillery were about to commence the work of destruction at that hour.

Only two divisions of Longstreet were present today – M'Laws's and Hood's – Pickett being still in the rear. As the whole morning was evidently to be occupied in disposing the troops for the attack, I rode to the extreme right with Colonel Manning and Major Walton, where we ate quantities of cherries and got a feed of corn for our horses. We also bathed in a small stream, but not without some trepidation on my part, for we were almost beyond the lines, and were exposed to the enemy's cavalry.

At 1 P.M. I met a quantity of Yankee prisoners who had been picked up straggling. They told me they belonged to Sickles's corps (3d, I think), and had arrived from Emmetsburg during the night. About this time skirmishing began along part of the line, but not heavily.

At 2 P.M. General Longstreet advised me, if I wished to have a good view of the battle, to return to my tree of yesterday. I did so, and remained there with Lawley and Captain Schreibert during the rest of the afternoon. But until 4:45 P.M. all was profoundly still, and we began to doubt whether a fight was coming off today at all.

At that time, however, Longstreet suddenly commenced a heavy cannonade on the right. Ewell immediately took it up on the left. The enemy replied with at least equal fury, and in a few moments the firing along the whole line was as heavy as it is possible to conceive. A dense smoke arose for six miles. There was little wind to drive it away, and the air seemed full of shells – each of which appeared to have a different style of going, and to make a different noise from the others. The ordnance on both sides is of a very varied description.

Every now and then a caisson [ammunition wagon] would blow up – if a Federal one, a Confederate yell would immediately follow. The Southern troops, when charging, or to express their delight, always yell in a manner peculiar to themselves. The Yankee cheer is much more like ours; but the Confederate officers declare that the Rebel yell has a particular merit, and always produces a salutary and useful effect upon their adversaries. A corps is sometimes spoken of as a 'good yelling regiment'.

As soon as the firing began, General Lee joined Hill just below our tree, and he remained there nearly all the time, looking through his fieldglass – sometimes talking to Hill and sometimes to Colonel Long of his staff. But generally he sat quite alone on the stump of a tree. What I remarked especially was, that during the whole time the firing continued, he only sent one message, and only received one report. It is evidently his system to arrange the plan thoroughly with the three corps commanders, and then leave to them the duty of modifying and carrying it out to the best of their abilities.

When the cannonade was at its height, a Confederate band of music, between the cemetery and ourselves, began to play polkas and waltzes, which sounded very curious, accompanied by the hissing and bursting of the shells.

At 5:45 all became comparatively quiet on our left and in the cemetery; but volleys of musketry on the right told us that Longstreet's infantry were advancing, and the onward progress of the smoke showed that he was progressing favorably. About 6:30 there seemed to be a check, and even a slight retrograde movement. Soon after 7, General Lee got a report by signal from Longstreet to say 'We are doing well.'

A little before dark the firing dropped off in every direction, and soon ceased altogether. We then received intelligence that Longstreet had carried everything before him for some time, capturing several batteries, and driving the enemy from his positions; but when Hill's Florida brigade and some other troops gave way, he was forced to abandon a small portion of the ground he had won, together with all the captured guns, except three. His troops, however, bivouacked during the night on ground occupied by the enemy this morning.

Everyone deplores that Longstreet will expose himself in such a reckless manner. Today he led a Georgian regiment in a charge against a battery, hat in hand, and in front of everybody. General Barksdale was killed and Semmes mortally wounded; but the most serious loss was that of General Hood, who was badly wounded in the arm early in the day. I heard that his Texans are in despair. Lawley and I rode back to the General's camp, which had been moved to within a mile of the scene of action. Longstreet, however, with most of his staff, bivouacked on the field.

Major Fairfax arrived at about 10 P.M. in a very bad humor. He had under his charge about 1,000 to 1,500 Yankee prisoners who had been taken today; among them a general, whom I heard one of his men accusing of having been 'so G—d d—d drunk that he had turned his guns upon his own men'. But, on the other hand, the accuser was such a thundering blackguard, and proposed taking such a variety of oaths in order to escape from the US Army, that he is not worthy of much credit. A large train of horses and mules, &c., arrived today, sent in by General Stuart [J. E. B. Stuart, noted Confederate cavalry commander], and captured, it is understood, by his cavalry, which had penetrated to within 6 miles of Washington.

STEPHEN CRANE
(1871–1900)
War is Kind

Do not weep, maiden, for war is kind.
Because your lover threw wild hands toward the sky
And the affrighted steed ran on alone,
Do not weep.
War is kind.

Hoarse, booming drums of the regiment,
Little souls who thirst for fight,
These men were born to drill and die.
The unexplained glory flies above them,
Great is the Battle-God, great, and his Kingdom –
A field where a thousand corpses lie.

Do not weep, babe, for war is kind.
Because your father tumbled in the yellow trenches,
Raged at his breast, gulped and died,
Do not weep.
War is kind.

Swift blazing flag of the regiment,
Eagle with crest of red and gold,
These men were born to drill and die.

Point for them the virtue of slaughter,
Make plain to them the excellence of killing
And a field where a thousand corpses lie.

Mother whose heart hung humble as a button
On the bright splendid shroud of your son,
Do not weep.
War is kind.

ELIZABETH B. CUSTER
General Custer

By the last quarter of the nineteenth century relations between
the native inhabitants of North America and the European
invaders of their continent were approaching a final crisis.
The original settlers had, in many cases, tried to reach an
accommodation with the Indians by treaty and land purchase.
Such arrangements usually broke down over misunderstandings
or allegations of bad faith, justified in one direction or the
other. After the British victory over the French in 1763, the
government in London reserved the territory to the west of
the Appalachian Chain to the Indian inhabitants but, following
the winning of the United States' independence, and under
pressure from the waves of settlement crossing the Appalachians to the richer lands beyond, the idea of a general
territorial reserve collapsed. During the early nineteenth century
the United States government began resettling Indians from
the forest lands east of the Mississippi into the western plains,
in the belief that whites would not choose to enter a region
believed in Washington to be inhospitable to farmers. That
belief proved futile. After the Civil War the 'move west' became
an unstoppable flood and the small United States Army found
itself committed to protecting the passage of the wagon trains

towards California and Oregon, to defending those settlers who chose to stake claims on the Great Plains themselves, and to persuading the nomadic Indian tribes of the region to accept confinement to new reservations defined for them by the Federal government.

Many of the tribes agreed to reside peacefully on the reserved land. Some intransigents would not. The least co-operative proved to be the Sioux, of what is now Montana, and some of their Cheyenne allies. In 1876 Colonel George Custer, commanding the 7th Cavalry and permitted the courtesy of being addressed by his Civil War rank of general, was one of the officers given the task of rounding up the recalcitrant Sioux who, under the influence of such warrior chiefs and medicine men as Sitting Bull and Crazy Horse, were threatening war to the death against the white men and refusing to leave the open range to take up life in the reservation.

Custer was a Civil War hero who had never reconciled himself to the tamer life of the peacetime army. Dramatic in appearance – flowing locks, cavalier boots – and married to an adoring and beautiful wife, he approached the mission of pacification on the Great Plains in the same spirit that had inspired him as a cavalry leader against the Confederates in the Civil War. In June 1876, he led the 7th Cavalry in a sweep on the Powder, Tongue, Rosebud and Big Horn Rivers that was intended to curtail the movements of the Sioux and Cheyenne and shepherd them back to the assigned reservation. The response of the Indians was hostile – Sitting Bull, though he did not take part in the actions to follow, had preached war – and on 25 June 1876, when Custer found a large Indian encampment on the Little Big Horn, he determined to encircle the braves and inflict a punishing defeat.

He had, intemperately, underestimated both their numbers and their mood. Galloping ahead of his supports, who escorted the vital ammunition reserves, he attempted to encircle the Indians, only to find that they were encircling him. In the few hours of desperate combat that followed, his command was surrounded and overwhelmed. He and all his 212 officers and men were killed.

*   *   *

Official report of the engagement with Indians on the 4th and 11th ultimo

At early dawn the next day (the 11th instant), the Indians appeared in strong force on the river bank opposite us, and opened a brisk fire upon us from their rifles. No attention was paid to them until encouraged by this they had collected at several points in full view, and within range of our rifles, when about thirty of our best marksmen, having posted themselves along the bank, opened a well-directed fire upon the Indians and drove them back to cover.

In the mean time strong parties of Indians were reported by our pickets to be crossing the river below and above us, their ponies and themselves being so accustomed to the river as to render this operation quite practicable for them. Captain French, commanding the right wing, was directed to watch the parties crossing below, while Colonel Hart, commanding the left wing, posted a force to discharge this duty with regard to parties crossing above. It would have been possible, perhaps, for us to have prevented the Indians from effecting a crossing, at least when they did, but I was not only willing but anxious that as many of them should come over as were so disposed. They were soon reported as moving to the bluffs immediately in rear of us from the river. Lieutenant Brush was directed to employ his [Indian] scouts in watching and reporting their movements – a duty which they discharged in a thorough manner.

While this was transpiring I had mounted my command and formed it in line close under the bluffs facing from the river, where we quietly waited the attack of the Indians in our front. The sharpshooting across the river still continued, the Indians having collected some of their best shots – apparently armed with long-range rifles – and were attempting to drive our men back from the water's edge. It was at this time that my standing orderly, Private Tuttle, of 'E' troop, 7th Cavalry, one of the best marksmen in my command, took a sporting Springfield rifle and posted himself, with two other men, behind cover on the river bank, and began picking off the Indians as they exposed themselves on the opposite bank. He had obtained the range of the enemy's position early in the morning, and was able to place his shots wherever desired. It was while so engaged that he observed an Indian in full view near the river. Calling the attention of his comrade to the fact, he asked him 'to watch him drop that Indian', a feat which he succeeded in performing. Several other Indians rushed to the assistance of their fallen comrade, when Private Tuttle, by a skilful and rapid use of his breech-loading Springfield, succeeded in killing two other warriors. The Indians, enraged no doubt at this rough handling, directed their aim at Private Tuttle, who fell pierced through the head by a rifle-bullet. He was one of the most useful and daring soldiers who ever served under my command.

About this time Captain French, who was engaged with the Indians who were attacking us from below, succeeded in shooting a warrior from his saddle, while several ponies were known to be wounded or disabled. The Indians now began to display a strong force in our front on the bluffs. Colonel Hart was ordered to push a line of dismounted men to the crest, and prevent the further advance of the enemy towards the river. This duty was handsomely performed by a portion of Captain Yates's squadron. Colonel Hart had posted Lieutenant Charles Braden and twenty men on a small knoll which commanded our left. Against this party the Indians made their first onslaught. A mounted party of warriors, numbering nearly two hundred, rode boldly to within thirty yards of Lieutenant Braden's position, when the latter and his command delivered such a well-directed fire that the Indians were driven rapidly from that part of the field, after having evidently suffered considerable loss.

Unfortunately Lieutenant Braden received a rifle-ball through the upper part of the thigh, passing directly through the bone, but he maintained his position with great gallantry and coolness until he had repulsed the enemy. Hundreds of Indians were now to be seen galloping up and down along our front, each moment becoming bolder, owing to the smallness of our force which was then visible.

Believing the proper time had arrived to assume the offensive, orders to this effect were accordingly sent to Colonel Hart and Captain French, the two wing commanders. Lieutenant Weston was directed to move his troop 'L' up a deep ravine on our left, which would convey him to the enemy's position, and as soon as an opportunity occurred he was to charge them, and pursue the Indians with all the vigor practicable. Immediately after, Captain Owen Hale was directed to move his squadron, consisting of 'E' and 'K' troops, in conjunction with 'L' troop, and the three to charge simultaneously. Similar dispositions were ordered in the centre and right. Lieutenant Custer [Tom, the general's brother], commanding 'B' troop, was ordered to advance and charge the Indians in front of our centre, while Captains Yates and Moylan moved rapidly forward in the same direction. Before this movement began, it became necessary to dislodge a large party of Indians posted in a ravine and behind rocks in our front, who were engaged in keeping up a heavy fire upon our troops while the latter were forming. It was at this point that the horse of Lieutenant Hiram H. Ketchum, Acting assistant adjutant-general of the expedition, was shot under him. My own horse was also shot under me within a few paces of the latter.

The duty of driving the Indians engaged in sharp shooting was intrusted to Lieutenant Charles A. Varnum, 7th Cavalry, with a detachment of 'A' troop, 7th Cavalry, who soon forced the Indians back from their cover.

Everything being in readiness for a general advance, the charge was ordered, and the squadrons took the gallop to the tune of 'Garryowen', the band being posted immediately in rear of the skirmish line. The Indians had evidently come out prepared to do their best, and with no misgivings as to their success, as the mounds and high bluffs beyond the river were covered with groups of old men, squaws, and children, who had collected there to witness our destruction. In this instance the proverbial power of music to soothe the savage breast utterly failed, for no sooner did the band strike up the cheery notes of 'Garryowen', and the squadrons advance to the charge, than the Indians exhibited unmistakable signs of commotion, and their resistance became more feeble, until finally satisfied of the earnestness of our attack they turned their ponies' heads and began a disorderly flight. The cavalry put spurs to their horses and dashed forward in pursuit, the various troop and squadron commanders vying with one another as to who should head the advance. The appearance of the main command in sight, down the valley at this moment, enabled me to relieve Captain French's command below us, and he was ordered to join in the pursuit. Lieutenant McIntosh commanding 'G' troop, moved his command up the valley at a gallop, and prevented many of the Indians from crossing. The chase was continued with the utmost vigor until the Indians were completely dispersed, and driven a distance of nine miles from where the engagement took place, and they were here forced back across the Yellowstone [river], the last pony killed in the fight being shot fully eight miles from the point of attack.

The number of Indians opposed to us has been estimated by the various officers engaged as from eight hundred to a thousand. My command numbered four hundred and fifty, including officers and men. The Indians were made up of different bands of Sioux, principally Uncpapas, the whole under command of 'Sitting Bull', who participated in the second day's fight, and who for once has been taught a lesson he will not soon forget.

A large number of Indians who fought us were fresh from their reservations on the Missouri River. Many of the warriors engaged in the fight on both days were dressed in complete suits of the clothes issued at the agencies to Indians. The arms with which they fought us (several of which were captured in the fight) were of the latest improved patterns of breech-loading repeating rifles, and their supply of metallic rifle-cartridges seemed unlimited, as they were anything but sparing in their use. So amply have they been supplied with breech-loading rifles and ammunition that neither bows nor arrows were employed against us. As an evidence that these Indians, at least many of them, were recently from the Missouri River agencies, we found provisions, such as coffee, in their abandoned camps, and cooking and other domestic utensils, such as only reservation Indians are supplied with. Besides, our scouts conversed with them across the river for nearly an hour before the fight became general, and satisfied themselves as to the identity of their foes. I only regret that it was impossible for my command to effect a crossing of the river before our presence was discovered, and while the hostile village was located near at hand, as I am confident that we could have largely reduced the necessity for appropriation for Indian supplies the coming winter . . .

The losses of the Indians and ponies were particularly heavy, while we know their losses in killed and wounded were beyond all proportion to that which they were enabled to inflict upon us, our losses being one officer badly wounded, four men killed, and three wounded; four horses killed and four wounded.

Careful investigation justifies the statement that including both days' battles, the Indians' losses will number forty warriors, while their wounded on the opposite bank of the river may increase this number.

Respectfully submitted.

(Signed) G. A. CUSTER, Lieutenant-colonel 7th Cavalry, Brevet-major-general, U.S.A. [Army] commanding

Letters from the second expedition to the Yellowstone, 1876, by General Custer to his wife

POWDER RIVER, ABOUT TWENTY MILES ABOVE ITS MOUTH, JUNE 9, 1876.

. . . We are now in a country heretofore unvisited by white men. Reynolds, who had been guiding the command, lost his way the other day, and General Terry did not know what to do about finding a road from O'Fallon's Creek across to Powder River. I told him I thought I could guide the column. He assented; so Tom [Custer], 'Bos', and I started ahead, with company D and the scouts as escort, and brought the command to this point, over what seems to be the only practicable route for miles on either side, through the worst kind of Bad Lands. The general did not believe it possible to find a road through. When, after a hard day's work, we arrived at this river by a good, easy road, making thirty-two miles in one day, he was delighted and came to congratulate me.

Yesterday I finished a Galaxy [magazine] article, which will go in the next mail; so, you see, I am not entirely idle. Day before yesterday I rode nearly fifty miles, arose yesterday morning, and went to work at my article, determined to finish it before night, which I did, amidst constant interruptions. It is now nearly midnight, and I must go to my bed, for reveille comes at three.

As a slight evidence that I am not very conceited regarding my personal appearance, I have not looked in a mirror or seen the reflection of my beautiful (?) countenance, including the fine growth of auburn whiskers, since I looked in the glass at Lincoln.

ON YELLOWSTONE, AT MOUTH OF POWDER RIVER,
JUNE 11TH — 10.30 P.M.

. . . This morning we left our camp on Powder River, I acting again as guide. The expedition started to make its way through unknown Bad Lands to the mouth of the river. General Terry felt great anxiety in regard to the trip, as he feared that we could not get through with the wagons. He had been down the river to its mouth with cavalry, and he and those with him said that wagons could not make the march in a month, and the Bad Lands looked still more impracticable. He came to my tent before daylight, and asked me if I would try to find the road. He seems to think I have a gift in that way, and he hoped that we might get within ten miles of the river's mouth today. What rendered our condition more embarrassing was that the men had only rations for one day left.

I started with one company and the scouts, and in we 'plunged boldly'. One company had been sent out the previous day to look for a road, and their failure to return the same day increased the anxiety. I thought it likely they had lost their way and had slept in the Bad Lands. Sure enough we found them about 10 A.M.

After passing through some perfectly terrible country I finally struck a beautiful road along a high plateau, and instead of guiding the command within ten miles of here we have all arrived and the wagon-train beside.

If you will look on the map near my desk you will find the mouth of the Powder River and our present location on the Yellowstone, almost due west from Lincoln. Follow up the Yellowstone a short distance, and the first stream you come to is the Tongue River, to which point we will move after resting three or four days. We will there be joined by the six companies of the regiment now absent on a scout, and I shall then select the nine companies to go with me . . .

The steamer Far West leaves for Fort Buford tomorrow . . . As I was up at three this morning, and have had a hard day's march, and as it is now going on to twelve, I must hie to bed to get a little rest and slumber . . .

MONDAY, JUNE 12TH — BEFORE BREAKFAST.

. . . I rose early this morning, without waiting to be called to breakfast, in order that I might write my letter. The Yellowstone is very high; steamers loaded to their utmost capacity can go up some distance above the mouth of the Big Horn. I wanted to send you a letter that I wished you to read and afterwards re-mail, had I not thought you might have found an opportunity to come up the river in the Josephine. The new supplies for our mess – of onions, potatoes, and dried apples – have just come from the boat.

'Tuck' [the general's favourite dog] regularly comes when I am writing, and lays her head on the desk, rooting up my hand with her long nose until I consent to stop and notice her. She and Swift, Lady and Kaiser sleep in my tent.

You need not be anxious about my leaving the column with small escorts; I scarcely hunt any more . . .

MOUTH OF TONGUE RIVER, JUNE 17TH.

. . . I fear that my last letter, written from the mouth of Powder River, was not received in very good condition by you. The mail was sent in a row-boat from the stockade to Buford, under charge of a sergeant and three or four men of the 6th Infantry. Just as they were pushing off from the Far West the boat capsized, and mail and soldiers were thrown into the rapid current; the sergeant sank and was never seen again. The mail was recovered, after being submerged for five or ten minutes. Captain Marsh and several others sat up all night and dried it by the stove. I was told that my letter to you went off all right, also my Galaxy article. The latter was recognized by a young newspaper reporter and telegraph operator who came up on the train with us from St Paul, and he took special pains in drying it.

With six companies of the 7th, the Gatling battery, the scouts, and the pack-mules, I left the mouth of Powder River Thursday morning, leaving all our wagons behind, and directing our march for this point, less than forty miles distant. General Terry and staff followed by steamer. We marched here in about one and a quarter days. The boat arrived yesterday evening ... The officers were ordered to leave their tents behind. They are now lying under tentflies or in shelter-tents. When we leave here I shall only take a tentfly. We are living delightfully. This morning we had a splendid dish of fried fish, which Tom, 'Bos', and I caught a few steps from my tent last evening.

The other day, on our march from Powder River, I shot an antelope. That night, while sitting round the camp fire, and while Hughes was making our coffee, I roasted some of the ribs Indian fashion, and I must say they were delicious. We all slept in the open air around the fire, Tom and I under a fly, 'Bos' and Autie Reed on the opposite side. Tom pelted 'Bos' with sticks and clods of earth after we had retired. I don't know what we would do without 'Bos' to tease ...

Yesterday Tom and I saw a wild-goose flying overhead quite high in the air. We were in the bushes and could not see each other. Neither knew that the other intended to fire. Both fired simultaneously, and down came the goose, killed. Don't you think that pretty good shooting for rifles?

On our march here we passed through some very extensive Indian villages – rather the remains of villages occupied by them last winter.

I was at the head of the column as we rode through one, and suddenly came upon a human skull lying under the remains of an extinct fire. I halted to examine it, and lying near by I found the uniform of a soldier. Evidently it was a cavalry uniform, as the buttons on the overcoat had 'C' on them, and the dress-coat had the yellow cord of the cavalry uniform running through it. The skull was weatherbeaten, and had evidently been there several months. All the circumstances went to show that the skull was that of some poor mortal who had been a prisoner in the hands of the savages, and who doubtless had been tortured to death, probably burned.

We are expecting the Josephine to arrive in a day or two. I hope that it will bring me a good long letter from you, otherwise I do not feel particularly interested in her arrival – unless, by good luck, you should be on board; you might just as well be here as not . . . I hope to begin another Galaxy article, if the spirit is favorable . . .

MOUTH OF ROSEBUD, JUNE 21, 1876.

. . . Look on my map and you will find our present location on the Yellowstone, about midway between Tongue River and the Big Horn.

The scouting-party has returned. They saw the trail and deserted camp of a village of three hundred and eighty (380) lodges. The trail was about one week old. The scouts reported that they could have overtaken the village in one day and a half. I am now going to take up the trail where the scouting-party turned back. I fear their failure to follow up the Indians has imperilled our plans by giving the village an intimation of our presence. Think of the valuable time lost! But I feel hopeful of accomplishing great results. I will move directly up the valley of the Rosebud. General Gibbon's command and General Terry, with steamer, will proceed up the Big Horn as far as the boat can go . . . I like campaigning with pack-mules much better than with wagons, leaving out the question of luxuries. We take no tents, and desire none.

I now have some Crow scouts with me, as they are familiar with the country. They are magnificent-looking men, so much handsomer and more Indian-like than any we have ever seen, and so jolly and sportive; nothing of the gloomy, silent red-man about them. They have formally given themselves to me, after the usual talk. In their speech they said they had heard that I never abandoned a trail; that when my food gave out I ate mule. That was the kind of a man they wanted to fight under; they were willing to eat mule too.

I am going to send six Ree scouts to Powder River with the mail; from there it will go with other scouts to Fort Buford . . .

JUNE 22D — 11 A.M.

. . . I have but a few moments to write, as we move at twelve and I have my hands full of preparations for the scout . . . Do not be anxious about me. You would be surprised to know how closely I obey your instructions about keeping with the column. I hope to have a good report to send you by the next mail . . . A success will start us all towards Lincoln . . .

I send you an extract from General Terry's official order, knowing how keenly you appreciate words of commendation and confidence, such as the following: 'It is of course impossible to give you any definite instructions in regard to this movement; and were it not impossible to do so, the Department Commander places too much confidence in your zeal, energy, and ability to wish to impose upon you precise orders, which might hamper your action when nearly in contact with the enemy.'

Garrison life [Fort Lincoln]

There were about forty in our garrison circle, and as we were very harmonious we spent nearly every evening together. I think it is the general belief that the peace of an army post depends very much upon the example set by the commanding officer. My husband, in the six years previous, had made it very clear, in a quiet way, that he would much prefer that there should be no conversation detrimental to others in his quarters. It required no effort for him to refrain from talking about his neighbors, but it was a great deprivation to me occasionally. Once in a while, when some one had brought down wrath upon his or her head by doing something deserving of censure, the whole garrison was voluble in its denunciation; and if I plunged into the subject also and gave my opinion, I soon noticed my husband grow silent and finally slip away. I was not long in finding an excuse to follow him and ask what I had done. Of course I knew him too well not to divine that I had hurt him in some manner. Then he would make a renewed appeal to me beginning by an unanswerable plea, 'if you wish to please me,' and imploring me not to join in discussions concerning any one. He used to assure me that in his heart he believed me superior to such things. In vain I disclaimed being of that exalted order of females, and declared that it required great self-denial not to join in a gossip. The discussion ended by his desiring me to use him as a safety-valve if I must criticize others. From motives of policy alone, if actuated by no higher incentive, it seemed wise to suppress one's ebullitions of anger. In the States it is possible to seek new friends if the old ones become tiresome and exasperating, but once in a post like ours, so far removed, there is no one else to whom one can turn. We never went away on leave of absence, and heard ladies in civil life say emphatically that they did not like some person they knew, and 'never would', without a start of terror. I forgot that their lives were not confined to the small precincts of a territorial post, where such avowed enmity is disastrous.

I had very little opportunity to know much of official matters; they were not talked about at home. Instinct guided me always in detecting the general's enemies, and when I found them out, a struggle began between us as to my manner of treating them. My husband urged that it would embarrass him if others found out that I had surmised anything regarding official affairs. He wished social relations to be kept distinct, and he could not endure to see me show dislike to any one who did not like him. I argued in reply that I felt myself dishonest if I even spoke to one whom I hated. The contest ended by his appealing to my good sense, arguing that as the wife of the commanding officer I belonged to every one, and in our house I should be hospitable upon principle. As every one visited us, there was no escape for me, but I do not like to think now of having welcomed any one from whom I inwardly recoiled.

I was not let off on such occasions with any formal shake of the hand. My husband watched me, and if I was not sufficiently cordial he gave me, afterwards, in our bedroom, a burlesque imitation of my manner. I could not help laughing, even when annoyed to see him caricature me by advancing coldly, extending the tips of his fingers, and bowing loftily to some imaginary guest. His raillery, added to my wish to please him, had the effect of making me shake hands so vigorously that I came near erring the other way and being too demonstrative, and thus giving the impression that I was the best friend of some one I really dreaded.

As I was in the tent during so many summers, and almost constantly in my husband's library in our winter quarters, I naturally learned something of what was transpiring. I soon found, however, that it would do no good if I asked questions in the hope of gaining further information. As to curiosity ever being one of my conspicuous faults, I do not remember, but I do recollect most distinctly how completely I was taken aback by an occurrence which took place a short time after we were married. I had asked some idle question about official matters, and was promptly informed in a grave manner, though with a mischievous twinkle of the eye, that whatever information I wanted could be had by application to the adjutant-general. This was the stereotyped form of endorsement on papers sent up to the regimental adjutant asking for information. One incident of many comes to me now, proving how little I knew of anything but what pertained to our own home circle. The wife of an officer once treated me with marked coldness. I was unaware of having hurt her in any way, and at once took my grievance to that source where I found sympathy for the smallest woe. My husband pondered a moment, and then remembered that the husband of my friend and he had had some slight official difficulty, and the lady thinking I knew of it was taking her revenge on me.

When I first entered army life I used to wonder what it meant when I heard officers say, in a perfectly serious voice, 'Mrs — commands her husband's company.' It was my good fortune not to encounter any such female grenadiers. A circumstance occurred which made me retire early from any attempt to assume the slightest authority. One of the inexhaustible jokes that the officers never permitted me to forget was an occurrence that happened soon after the general took command of the 7th Cavalry. A soldier had deserted, and had stolen a large sum of money from one of the lieutenants. My sympathy was so aroused for the officer that I urged him to lose no time in pursuing the man to the nearest town, whither he was known to have gone. In my interest and zeal I assured the officer that I knew the general would be willing, and he need not wait to apply for leave through the adjutant's office. I even hurried him away. When the general came in I ran to him with my story, expecting his sympathy and that he would endorse all that I had done. On the contrary, he quietly assured me that he commanded the regiment, and that he would like me to make it known to the lieutenant that he must apply through the proper channels for leave of absence. Thereupon I ate a large piece of humble pie, but was relieved to find that the officer had shown more sense than I, and had not accepted my proffered leave, but had prudently waited to write out his application. Years afterwards, when my husband told me what a source of pride it was to him that others had realized how little I knew about official affairs, and assured me that my curiosity was less than that of any woman he had ever known, I took little credit to myself. It would have been strange, after the drilling of military life, if I had not attained some progress.

The general planned every military action with so much secrecy that we were left to divine as best we could what certain preliminary movements meant. One morning when it was too cold for anything but important duty, without any explanations he started off with a company of cavalry and several wagons. As they crossed the river on the ice, we surmised that he was going to Bismarck. It seemed that the general had been suspicious that the granaries were being robbed, and finally a citizen was caught driving off a loaded wagon of oats from the reservation in broad daylight. This was about as high-handed an instance of thieving as the general had encountered, and he quietly set to work to find out the accomplices. In a little while it was ascertained that the robbers had concealed their plunder in a vacant store in the principal street of Bismarck.

The general determined to go himself directly to the town, thinking that he could do quickly and without Opposition what another might find difficult. The better class of citizens honored him too highly to oppose his plan of action, even though it was unprecedented for the military to enter a town on such an errand. The general knew the exact place at which to halt, and drew the company up in line in front of the door. He demanded the key, and directed the men to transfer the grain to the wagons outside. Without a protest, or an exchange of words even, the troops marched out of the town as quietly as they had entered. This ended the grain thefts.

It was a surprise to me that after the life of excitement my husband had led, he should grow more and more domestic in his tastes. His daily life was very simple. He rarely left home except to hunt, and was scarcely once a year in the sutler's store [the sutler was a civilian supply contractor], where the officers congregated to play billiards and cards. If the days were too stormy or too cold for hunting, as they often were for a week or more at a time, he wrote and studied for hours every day. We had the good fortune to have a billiard-table loaned us by the sutler, and in the upper room where it was placed, my husband and I had many a game when he was weary with writing.

RUDYARD KIPLING
(1865–1936)
Tommy

I went into a public-'ouse to get a pint o' beer,
The publican 'e up an' sez, 'We serve no red-coats here.'
The girls be'ind the bar they laughed an' giggled fit to die,
I outs into the street again an' to myself sez I:
O it's Tommy this, an' Tommy that, an' 'Tommy, go away';

But it's 'Thank you, Mister Atkins,' when the band begins to
     play –
The band begins to play, my boys, the band begins to play,
O it's 'Thank you, Mister Atkins,' when the band begins to
     play.

I went into a theatre as sober as could be,
They gave a drunk civilian room, but 'adn't none for me;
They sent me to the gallery or round the music-'alls;
But when it comes to fightin', Lord! they'll shove me in the
     stalls!
For it's Tommy this, an' Tommy that, an' 'Tommy wait
     outside';
But it's 'Special train for Atkins' when the trooper's on the
     tide –
The troopship's on the tide, my boys, the troopship's on the
     tide,
O it's 'Special train for Atkins' when the trooper's on the tide.

Yes, makin' mock o' uniforms that guard you while you sleep
Is cheaper than them uniforms, an' they're starvation cheap;
An' hustlin' drunken soldiers when they're goin' large a bit
Is five times better business than paradin' in full kit.
Then it's Tommy this, an' Tommy that, an' 'Tommy, 'ow's yer
     soul?'
But it's 'Thin red line of 'eroes' when the drums begin to roll –
The drums begin to roll, my boys, the drums begin to roll,
O it's 'Thin red line of 'eroes' when the drums begin to roll.

We aren't no thin red 'eroes, nor we aren't no blackguards too,
But single men in barricks, most remarkable like you;
An' if sometimes our conduck isn't all your fancy paints,
Why, single men in barricks don't grow into plaster saints;
While it's Tommy this, an' Tommy that, an' 'Tommy, fall
     be'ind,'
But it's 'Please to walk in front, sir,' when there's trouble in the
     wind –

There's trouble in the wind, my boys, there's trouble in the
    wind,
O it's 'Please to walk in front, sir,' when there's trouble in the
    wind.

You talk o' better food for us, an' schools, an' fires, an' all:
We'll wait for extry rations if you treat us rational.
Don't mess about the cook-room slops, but prove it to our
    face
The Widow's Uniform is not the soldier-man's disgrace.
For it's Tommy this, an' Tommy that, an' 'Chuck him out, the
    brute!'
But it's 'Saviour of 'is country' when the guns begin to shoot;
An' it's Tommy this, an' Tommy that, an' anything you please;
An' Tommy ain't a bloomin' fool – you bet that Tommy sees!

ISANDHLWANA AND RORKE'S DRIFT, 1879
Four eyewitness accounts

Africa south of the Sahara had, until the beginning of the
nineteenth century, scarcely been penetrated by foreigners. A
few coastal enclaves were held by slave-traders, European on
the west coast, Arab on the east, while a Dutch colony, set up
to service ships sailing to the East Indies, had been established
at the Cape of Good Hope in the seventeenth century. The
rest of black Africa remained in the possession of its native
peoples.

Following the end of the Napoleonic Wars, however, Europeans arrived to settle the richer and more accessible land in
and around Cape Colony, while the Dutch of the Cape, irked
by the imposition of British rule, instituted a Great Trek into
the interior, to set up what would become the settlements of

the Orange Free State and the Transvaal. The Africans of the south were not untouched by these disturbances. One small tribe, the Zulus, began to develop an aggressive military system of their own and, under the leadership of Shaka (reigned 1819–28), created a local empire in south-eastern Africa which imposed its power over its African neighbours and successfully confronted the Dutch trekkers for control of the region.

The Boers, and later the British, eventually succeeded in confining the Zulus to what would become known as Zululand, part of the province of Natal, but within it the Zulus ruled supreme. Border disputes continued. In 1879 the British in South Africa demanded that Cetewayo, the Zulu king, should grant them a protectorate over his kingdom, which would effectively have reduced it to the status of a British possession. He refused and war followed.

The Zulu military system was formidable. It had no equivalent among any of the so-called 'primitive' peoples fought by European armies in any of the imperial campaigns of the nineteenth century. Shaka's system, sustained and improved by Cetewayo, required all young Zulu males to join an 'age' regiment of contemporaries. Once enlisted, they were forbidden to marry – marriage was a reward for proven warriors that came later in their careers – and were hardened to combat by privation and tests of endurance. Shaka's regiments had been armed with the assegai, a short stabbing spear, alone. By the time of Cetewayo, his *impis* (armies) had also acquired firearms. The strength of the Zulu army, however, derived from its tactics. Trained to run for long distances, the impis formed crescent-shaped formations on encountering the enemy, the centre advancing to attack, the wings enveloping the flanks. The Zulus pressed forward to hand-to-hand combat, accepting losses however heavy, and overwhelmed the foe by terror and weight of numbers.

In the 1879 war, the Zulus had an early success at Isandhlwana on 22 January, when they discovered part of the invading British army encamped. Attacking off the line of march, 10,000 Zulus swamped the British force 1,800 strong – consisting partly of one battalion of the 24th Regiment (later the South Wales Borderers) and partly of 'native' troops under European officers, with a smattering of other units – whose ammunition resupply broke down in the heat of action. Almost all the force were killed. One of the survivors from the encampment was Lieutenant Horace Smith-Dorrien, whose account of his escape to Rorke's Drift follows. He survived the Zulu War to command II Corps of the British Expeditionary Force in France, in 1914–15.

Private Henry Hook of the 24th actually came from Gloucestershire, an English county bordering Wales. He was twenty-eight in 1879, and serving as a cook in the small field hospital set up by his regiment at Rorke's Drift to care for its sick and wounded, and those of other corps in the Zululand field force; Rorke's Drift was also a supply post for the British column. For his bravery in the defence of the post, which, on the day and night following Isandhlwana, was successfully defended by eighty-five able-bodied soldiers against a Zulu force estimated to number five thousand, he was awarded the Victoria Cross. Eleven Victoria Crosses were awarded for the defence altogether, the largest number ever given for a single action. The other recipients were Lieutenant John Chard, Royal Engineers, commanding, Commissary James Dalton, Corporal Friedrich Schiess, a Swiss citizen serving in the Natal Native Contingent, Surgeon-Major James Reynolds, Lieutenant Gonville Bromhead of the 24th, and, besides Hook, five other soldiers of the regiment. Seventeen soldiers of the garrison had been killed in the defence and about four hundred Zulus. The epic is today commemorated both by the lineal descendants of the South Wales Borderers, the Royal Regiment of Wales, and the Zulu nation, who meet to honour each other on the battleground.

*   *   *

Account of Isandhlwana by a Zulu warrior

On this day [according to the Zulu], the army which had been marching hitherto in single column divided into two, marching parallel to and in sight of each other, that on the left consisting of the uNokenke, umCijo, and uNodwengu Regiments, under the command of Tshingwayo, the other commanded by Mavumeng-wana. There were a few mounted men belonging to the chief Sihayo, who were made use of as scouts. On the 20th [January 1879] we moved across the open country and slept by the Isipezi Hill. We saw a body of mounted white men on this day, to our left. On the 21st, keeping away to the eastward, we occupied a valley running north–south under the spurs of the Nqutu Hill, which concealed the Isandhlwana Hill, distant from us about four miles, and nearly due west of our encampment. We had been well fed during our whole march, our scouts driving in cattle and goats, and on that evening we lit up our camp fires as usual. Our scouts also reported to us that they had seen the videttes of the English force at sunset on some hills west-south-west of us. Our order of encampment on the 21st January was as follows: on the extreme right was the uNodwengu, uNokenke, and umCijo; the centre was formed by the inGobamakh-osi and uMbonambi; and the left of the Undi corps and the uDloko.

On the morning of the 22nd January there was no intention whatever of making any attack on account of a superstition regarding the state of the moon, and we were sitting resting when firing was heard on our right, which we at first imagined was the inGobamakhosi engaged, and we armed and ran forward in the direction of the sound. We were soon told, however, it was the white troops fighting with Matyana's people, some ten miles away to our left front, and returned to our original position. Just after we sat down again, a small herd of cattle came past our line from our right, being driven down by some of our scouts, and just when they were opposite to the umCijo regiment, a body of mounted men on the hill to the west, galloping evidently trying to cut them off. When several hundred yards off they perceived the umCijo and, dismounting, fired one volley at them and retired. The umCijo at once jumped up and

charged, an example which was taken up by the uNokenke and
uNodwengu on their right, and by the inGobamakhosi and uMbonambi on their left; while the Undi corps and uDloko formed a
circle and remained where they were. With the latter were the two
commanding officers, Mavumengwana and Tshingwayo, and several
of the King's brothers, who with these two corps bore away to the
northwest, after a short pause, and keeping on the northern side of
Isandhlwana performed a turning movement on the right without
any opposition from the whites, who from the nature of the ground
could not see them.

Thus the original Zulu left became their extreme right, while their
right became their centre, and the centre the left. The two regiments
which formed the latter, the inGobamakhosi and uMbonambi, made
a turning along the front of the camp towards the English right, but
became engaged long before they could accomplish it; and the uVe
regiment, a battalion of the inGobamakhosi, was repulsed and had
to retire until reinforced by the other battalion, while the uMbonambi
suffered very severely from the artillery fire. Meanwhile, the centre,
consisting of the umCijo on the left centre, and the uNokenke and
uNodwengu higher up on the right under the hill, were making an
attack on the left of the camp. The umCijo suffered very severely,
both from artillery and musketry fire, the uNokenke from musketry
fire alone; while the uNodwengu lost least. When we at last carried
the camp our regiments became mixed up; a portion pursued the
fugitives down to the Buffalo River, and the remainder plundered
the camp; while the Undi and uDloko Regiments made the best of
their way to Rorke's Drift to plunder the post there, in which they
failed and lost very heavily, after fighting all the afternoon and night.

Another warrior describes the battle

We were lying in the hills up there, when one of our scouting parties
came back followed by a number of mounted men; they were most
of them natives, but some were whites. They fired upon us. Then
the whole impi became very excited and sprang up. When the
horsemen saw how numerous we were they began to retreat. We

formed up in rank and marched towards the camp. At the top of the last hill we were met by more horsemen, but we were too many for them and they retreated. Here, where we are standing . . . there were some parties of soldiers in red coats who kept up a heavy fire upon us as we came over. My regiment was here and lost a lot of men; they kept tumbling over one upon another. [The narrator became quite excited . . . and indulged in much gesticulation, illustrating the volleys by cracking his fingers.]

Then the inGobamakhosi regiment, which formed the left horn of the impi, extended and swept round on the south of the conical kopje so as to outflank the soldiers, who, seeing this, fell back and took cover in the donga [dry watercourse], and fired upon us from there. By that time the inGobamakhosi had got among the rockets [besides two field guns, the British force also had a rocket battery] and killed the horses, and were circling round so as to shut in the camp on the side of the river, but we could not advance because the fire from the donga was too heavy. The great indunas [chiefs] were on the hill to the north of the camp, and just below them a number of soldiers were engaging the umCijo regiment, which was being driven back, but one of the chiefs of the umCijo ran down from the hill and rallied them, calling out that they would get the whole impi beaten and must come on. Then they all shouted 'Usutu!' and waving their shields charged the soldiers with great fury. The chief was shot through the forehead and dropped down dead, but the umCijo rushed over his body and fell upon the soldiers, stabbing them with their assegais and driving them right in among the tents.

My regiment and the uDloko formed the centre of the impi. When the soldiers in the donga saw that the umCijo were getting behind them, they retreated upon the camp, firing at us all the time. As they retreated we followed them. I saw several white men on horseback galloping towards the neck, which was the only point open; then the uNokenke and uNodwengu regiments, which had formed the right horn of the impi, joined with the inGobamakhosi on the neck. After that there was so much smoke that I could not see whether the white men had got through or not. The tumult and the firing was wonderful, every warrior shouted 'Usutu!' as he killed anyone, and the sun got very dark like night. The English fought long and hard, there were so many of our people in front of me that I did not get into the thick of the fight until the end. The warriors called out that all the white men had been killed, and then we began to plunder the camp. We found tywala [drink] in the camp, and some of our men got very drunk. We were so hot and thirsty that we drank everything liquid we found, without waiting to see what it was. Some of them found some black stuff in bottles [ink], it did not look good, so they did not drink it; but one or two who drank some paraffin, thinking it was tywala, were poisoned. We took as much plunder as we could carry, and went away home to our kraals [villages]. We did not reassemble and march back to Ulundi [the King's kraal]. At first we had not intended attacking the camp that day, as the moon was wrong, but as the whites had discovered our presence the indunas said we had better go on.

Account of Isandhlwana by Lieutenant Horace Smith-Dorrien

Since I wrote the first part of my letter a dreadful disaster has happened to us. It seems to me a pure miracle that I am alive to tell you about it. On the 21st January an order came to me, then stationed at Rorke's Drift, to go out to advanced camp [at Isandhlwana] to escort a convoy of twenty-five waggons from there to Rorke's Drift and bring them back loaded with supplies. Accordingly I slept in camp. At about three a.m. on the morning of the 22nd the General sent for me and told me not to take the waggons, but to convey a dispatch to Colonel Durnford, who was at Rorke's Drift, with about 500 mounted black fellows, as a battle was expected. He (Colonel Durnford) accordingly started off with his men to join the camp. I did not return with him, but came out an hour afterwards by myself.

When I arrived in camp, I found the greater part of the column gone out with the General [Lord Chelmsford] to meet the Zulu force, so that there was really only a caretaking force left in the camp – viz., five companies of the 1st Battalion of the 24th, two guns, about 600 Native Contingent, and a few servants looking after the tents; the Army Hospital Corps (thirteen men), and the sick in the hospital tents. The first Zulu force appeared about six o'clock in the morning. Two companies of the 24th were sent out after them. The Zulus seemed to retire, and there was firing kept up at long ranges. At about ten-thirty the Zulus were seen coming over the hills in thousands. They were in most perfect order, and seemed to be in about twenty rows of skirmishers one behind the other. They were in a semi-circle round our two flanks and in front of us and must have covered several miles of ground. Nobody knows how many there were of them, but the general idea is at least 20,000.

Well, to cut the account short, in half an hour they were right up to the camp. I was out with the front companies of the 24th handing them spare ammunition. Bullets were flying all over the place, but I never seemed to notice them. The Zulus nearly all had firearms of some kind and lots of ammunition. Before we knew where we were they came right into the camp, assegaiing everybody right and left. Everybody then who had a horse turned to fly. The enemy were going at a kind of very fast half-walk and half-run. On looking round we saw that we were completely surrounded and the road to Rorke's Drift was cut off. The place where they seemed thinnest was where we all made for. Everybody went pell-mell over ground covered with huge boulders and rocks until we got to a deep spruit or gulley. How the horses got over I have no idea. I was riding a broken-kneed old crock which did not belong to me, and which I expected to go on its head every minute. We had to go bang through them at the spruit. Lots of our men were killed there. I had lots of marvellous escapes, and was firing away at them with my revolver as I galloped along. The ground there down to the river was so broken that the Zulus went as fast as the horses, and kept killing all the way. There were very few white men; they were nearly all mounted blacks of ours flying. This lasted till we came to a kind of precipice down to the River Buffalo.

I jumped off and led my horse down. There was a poor fellow of the mounted infantry (a private) struck through the arm, who said as I passed that if I could bind up his arm and stop the bleeding he would be all right. I accordingly took out my handkerchief and tied up his arm. Just as I had done it, Major Smith of the Artillery came down by me wounded, saying, 'For God's sake get on, man, the Zulus are on the top of us.' I had done all I could for the wounded man and so turned to jump on my horse. Just as I was doing so the horse went with a bound to the bottom of the precipice, being struck with an assegai. I gave up all hope, as the Zulus were all round me, finishing off the wounded, the man I had helped and Major Smith among the number. However, with the strong hope that everybody clings to that some accident would turn up, I rushed off on foot and plunged into the river, which was little better than a roaring torrent.

I was being carried down the stream at a tremendous pace, when a loose horse came by me and I got hold of his tail and he landed me safely on the other bank; but I was too tired to stick to him and get on his back. I got up again and rushed on and was several times knocked over by our mounted blacks, who would not get out of my way, then up a tremendous hill with my wet clothes and boots full of water. About twenty Zulus got over the water and followed us up the hill, but I am thankful to say they had not their firearms. Crossing the river, however, the Zulus on the opposite side kept firing at us as we went up the hill and killed several of the blacks round me. I was the only white man to be seen until I came to one who had been kicked by his horse and could not mount. I put him on his horse and lent him my knife. He said he would catch me a horse. Directly he was up he went clean away. A few Zulus followed us for about three miles across the river, but they had no guns and I had a revolver, which I kept letting them know. Also the mounted blacks stopped a little and kept firing at them. They did not come in close, and finally stopped altogether.

Well, to cut it short, I struggled into Helpmakaar, about twenty miles off, at nightfall, to find a few men who had escaped, about ten or twenty, with others who had been entrenched in a waggon laager [literally, camp, but meaning a defensive circle of wagons]. We sat up all night, momentarily expecting attack. The next day there was a dense fog all day, nearly as bad as night, and we could not make out what had happened to everybody. I was dead beat of course, but on the 24th I struggled down to Rorke's Drift, my former headquarters, which had been so gallantly defended for a whole night against the Zulus by a single company, to find that the General and remainder of the column had arrived all right. I am there now in a laager. We keep a tremendous look-out, and sit up all night expecting attack. It has been raining for the last three hours, and did so all last night. We have not a single thing left. The men have no coats or anything, all being taken by the Zulus. We shall have another dreadful night of it tonight, I expect, lying on the wet ground. I have just had to drop this for a minute for one of our numerous alarms. I have no time for more now. What we are to do for transport I have not the faintest idea, the Zulus having captured 107 waggons and about 2,000 oxen, mules, horses, etc. However, we must begin to work again to get fresh transport together. I thank God I am alive and well, having a few bruises. God bless you.

P.S. We are expecting pestilence to break out here, to add to our enemies, what with the rain and the air tainted with dead bodies, as there were about 350 Zulus killed here and some are buried in the ruins.

Account of Rorke's Drift by Private Henry Hook, VC

Everything was perfectly quiet at Rorke's Drift after the column [Durnford's force] had left, and every officer and man was going about his business as usual. Not a soul suspected that only a dozen miles away the very men that we had said 'Goodbye', and 'Good luck' to were either dead or standing back-to-back in a last fierce fight with the Zulus. Our garrison consisted of B company of the 2/24th under Lieutenant Bromhead, and details which brought the total number of us up to 139. Besides these, we had about 300 men of the Natal Native Contingent; but they didn't count, as they bolted in a body when the fight began. We were all knocking about, and I was making tea for the sick, as I was hospital cook at the time.

Suddenly there was a commotion in the camp, and we saw two men galloping towards us from the other side of the river, which was Zululand. Lieutenant Chard of the Engineers was protecting the pontoons over the river and, as senior officer, was in command at the drift [ford]. The pontoons were very simple affairs, one of them being supported on big barrels, and the other on boats. Lieutenant Bromhead was in the camp itself. The horsemen shouted and were brought across the river, and then we knew what had happened to our comrades. They had been butchered to a man. That was awful enough news, but worse was to follow, for we were told that the Zulus were coming straight on from Isandhlwana to attack us. At the same time a note was received by Lieutenant Bromhead from the Column to say that the enemy was coming on, and that the post was to be held at all costs.

For some little time we were all stunned, then everything changed from perfect quietness to intense excitement and energy. There was a general feeling that the only safe thing was to retire and try and join the troops at Helpmakaar. The horsemen had said that the Zulus would be up in two or three minutes; but luckily for us they did not show themselves for more than an hour. Lieutenant Chard rushed up from the river, about a quarter of a mile away, and saw Lieutenant Bromhead. Orders were given to strike the camp and make ready to go, and we actually loaded up two waggons. Then Mr Dalton, of the Commissariat Department, came up and said that if we left the drift every man was certain to be killed. He had formerly been a sergeant-major in a line [infantry] regiment and was one of the bravest men that ever lived. Lieutenants Chard and Bromhead held a consultation, short and earnest, and orders were given that we were to get the hospital and storehouse ready for defence, and that we were never to say die or surrender.

Not a minute was lost. Lieutenant Bromhead superintended the loop-holing and barricading of the hospital and storehouse, and the making of a connection of the defences between the two buildings with walls of mealie-bags and waggons. The mealie-bags were good big heavy things, weighing about 200 pounds each, and during the fight many of them were burst open by assegais and bullets, and the mealies (Indian corn) were thickly spread about the ground.

The biscuit boxes contained ordinary biscuit. They were big square wooden boxes, weighing about a hundredweight each. The meat boxes, too, were very heavy, as they contained tinned meat. They were smaller than the biscuit boxes. While these preparations were being made, Lieutenant Chard went down to the river and brought in the pontoon guard of a sergeant and half-a-dozen men, with the waggons and gear. The two officers saw that every soldier was at his post, then we were ready for the Zulus when they cared to come.

They were not long. Just before half past four we heard firing behind the conical hill at the back of the drift, called Oskarsberg Hill, and suddenly about five or six hundred Zulus swept round, coming for us at a run. Instantly the natives – Kaffirs who had been very useful in making the barricade of waggons, mealie-bags and biscuit boxes around the camp – bolted towards Helpmakaar, and what was worse their officer and a European sergeant went with them. To see them deserting like that was too much for some of us, and we fired after them. The sergeant was struck and killed. Half-a-dozen of us were stationed in the hospital, with orders to hold it and guard the sick. The ends of the building were of stone, the side walls of ordinary bricks, and the inside walls or partitions of sun-dried bricks of mud. These shoddy inside bricks proved our salvation, as you will see. It was a queer little one-storeyed building, which it is almost impossible to describe; but we were pinned like rats in a hole, because all the doorways except one had been barricaded with mealie-bags, and we had done the same with the windows. The interior was divided by means of partition walls into which were fitted some very slight doors. The patients' beds were simple rough affairs of boards, raised only about half a foot above the floor. To talk of hospital and beds gives the idea of a big building, but as a matter of fact this hospital was a mere little shed or bungalow, divided up into rooms so small that you could hardly swing a bayonet in them. There were about nine men who could not move, but altogether there were about thirty. Most of these, however, could help to defend themselves.

As soon as our Kaffirs bolted, it was seen that the fort as we had first made it was too big to be held, so Lieutenant Chard instantly reduced the space by having a row of biscuit-boxes drawn across the middle, above four feet high. This was our inner entrenchment, and proved very valuable. The Zulus came on at a wild rush, and although many of them were shot down they got to within about fifty yards of our south wall of mealie-bags and biscuit boxes and waggons. They were caught between two fires, that from the hospital and that from the storehouse, and were checked; but they gained the shelter of the cookhouse and ovens, and gave us many heavy volleys. During the fight they took advantage of every bit of cover there was, anthills, a tract of bush that we had not had time to clear away, a garden or sort of orchard which was near us, and a ledge of rock and some caves (on the Oscarsberg) which were only about a hundred yards away. They neglected nothing, and while they went on firing large bodies kept hurling themselves against our slender breastworks. But it was the hospital they assaulted most fiercely. I had charge with a man that we called Old King Cole of a small room with only one patient in it. Cole kept with me for some time after the fight began, then he said he was not going to stay. He went outside and was instantly killed by the Zulus, so that I was left alone with the patient, a native whose leg was broken and who kept crying out, 'Take my bandage off, so that I can come.' But it was impossible to do anything except fight, and I blazed away as hard as I could. By this time I was the only defender of my room. Poor Old King Cole was lying dead outside and the helpless patient was crying and groaning near me. The Zulus were swarming around us, and there was an extraordinary rattle as the bullets struck the biscuit boxes, and queer thuds as they plumped into the bags of mealies. Then there were the whizz and rip of the assegais, of which I had experience during the Kaffir Campaign of 1877–8. We had plenty of ammunition, but we were told to save it and so we took careful aim at every shot, and hardly a cartridge was wasted. One of my comrades, Private Dunbar, shot no fewer than nine Zulus, one of them being a chief.

From the very first the enemy tried to rush the hospital, and at last they managed to set fire to the thick grass which formed the roof. This put us in a terrible plight, because it meant that we were either to be massacred or burned alive, or get out of the building. To get out seemed impossible; for if we left the hospital by the only door which had been left open, we should instantly fall into the midst of the Zulus. Besides, there were the helpless sick and wounded, and we could not leave them. My own little room communicated with another by means of a frail door like a bedroom door. Fire and dense choking smoke forced me to get out and go into the other room. It was impossible to take the native patient with me, and I had to leave him to an awful fate. But his death was, at any rate, a merciful one. I heard the Zulus asking him questions, and he tried to tear off his bandages and escape.

In the room where I now was there were nine sick men, and I was alone to look after them for some time, still firing away, with the hospital burning. Suddenly in the thick smoke I saw John Williams, and above the din of battle and the cries of the wounded I heard him shout, 'The Zulus are swarming all over the place. They've dragged Joseph Williams out and killed him.' John Williams had held the other room with Private William Horrigan for more than an hour, until they had not a cartridge left. The Zulus then burst in and dragged out Joseph Williams and two of the patients, and assegaied them. It was only because they were so busy with this slaughtering that John Williams and two of the patients were able to knock a hole in the partition and get into the room where I was posted. Horrigan was killed. What were we to do? We were pinned like rats in a hole. Already the Zulus were fiercely trying to burst in through the doorway. The only way of escape was the wall itself, by making a hole big enough for a man to crawl through into an adjoining room, and so on until we got to our inmost entrenchment outside. Williams worked desperately at the wall with the navvy's pick, which I had been using to make some of the loop-holes with.

All this time the Zulus were trying to get into the room. Their assegais kept whizzing towards us, and one struck me in front of the helmet. We were wearing the white tropical helmets then. But the helmet tilted back under the blow and made the spear lose its power, so that I escaped with a scalp wound which did not trouble me much then, although it has often caused me illness since. Only one man at a time could get in at the door. A big Zulu sprang forward and seized my rifle, but I tore it free and, slipping a cartridge in, I shot him point-blank. Time after time the Zulus gripped the muzzle and tried to tear the rifle from my grasp, and time after time

I wrenched it back, because I had a better grip than they had. All this time Williams was getting the sick through the hole into the next room, all except one, a soldier of the 24th named Conley, who could not move because of a broken leg. Watching for my chance I dashed from the doorway, and grabbing Conley I pulled him after me through the hole. His leg got broken again, but there was no help for it. As soon as we left the room the Zulus burst in with furious cries of disappointment and rage.

Now there was a repetition of the work of holding the doorway, except that I had to stand by a hole instead of a door, while Williams picked away at the far wall to make an opening for escape into the next room. There was more desperate and almost hopeless fighting, as it seemed, but most of the poor fellows were got through the hole. Again I had to drag Conley through, a terrific task because he was a very heavy man. We were now all in a little room that gave upon the inner line of defence which had been made. We (Williams and Robert Jones and William Jones and myself) were the last men to leave the hospital, after most of the sick and wounded had been carried through the small window and away from the burning building; but it was impossible to save a few of them, and they were butchered. Privates William Jones and Robert Jones during all this time were doing magnificent work in another ward which faced the hill. They kept at it with bullet and bayonet until six of the seven patients had been removed. They would have got the seventh, Sergeant Maxfield, out safely, but he was delirious with fever and, although they managed to dress him, he refused to move. Robert Jones made a last rush to try and get him away like the rest, but when he got back into the room he saw that Maxfield was being stabbed by the Zulus as he lay on his bed. Corporal Allen and Private Hitch helped greatly in keeping up communication with the hospital. They were both badly wounded, but when they could not fight any longer they served out ammunition to their comrades throughout the night. As we got the sick and wounded out they were taken to a verandah in front of the storehouse, and Dr Reynolds under a heavy fire and clouds of assegais, did everything he could for them. All this time, of course, the storehouse was being valiantly defended by the rest of the garrison. When we got into the inner fort, I took my post at a place where two men had been shot. While I was there another man was shot in the neck, I think by a bullet which came through the space between two biscuit boxes that were not quite close together. This was at about six o'clock in the evening, nearly two hours after the opening shot of the battle had been fired. Every now and then the Zulus would make a rush for it and get in. We had to charge them out. By this time it was dark, and the hospital was all in flames, but this gave us a splendid light to fight by. I believe it was this light that saved us. We could see them coming, and they could not rush us and take us by surprise from any point. They could not get at us, and so they went away and had ten or fifteen minutes of a war-dance. This roused them up again, and their excitement was so intense that the ground fairly seemed to shake. Then, when they were goaded to the highest pitch, they would hurl themselves at us again.

SIR HENRY NEWBOLT
(1862–1938)
Vitaï Lampada

There's a breathless hush in the Close tonight –
Ten to make and the match to win –
A bumping pitch and a blinding light,
An hour to play and the last man in.
And it's not for the sake of a ribboned coat,
Or the selfish hope of a season's fame,
But his Captain's hand on his shoulder smote –
'Play up! play up! and play the game!'

The sand of the desert is sodden red, –
Red with the wreck of a square that broke; –
The Gatling's jammed and the Colonel dead,

And the regiment blind with dust and smoke.
The river of death has brimmed his banks,
And England's far, and Honour a name,
But the voice of a schoolboy rallies the ranks:
'Play up! play up! and play the game!'

This is the word that year by year,
While in her place the School is set,
Every one of her sons must hear,
And none that hears it dare forget.
This they all with a joyful mind
Bear through life like a torch in flame,
And falling fling to the host behind –
'Play up! play up! and play the game!'

PART III

This third section is a collection of extracts about war in our own age, from the beginning to the end of the twentieth century. War in our century has been dominated by the power of technology, that of machine-guns, heavy artillery and the tank on land, and of the armoured ship and the submarine at sea; in mid-century the aircraft, of which the first practicable example was flown by the Wright brothers in December 1903, became a significant weapon of war, against both land targets and, through the development of the aircraft carrier, targets at sea as well. Many of the extracts that follow testify to the power of military technology on the battlefields, on land, sea and in the air, of the twentieth century.

Yet, despite the encroaching dominance of the technological factor, the importance of the warrior spirit remains. Weapons do not bring victory unless there is the will to use them resolutely. In the face of superior technology, moreover, there is a human instinct to shift the focus of conflict to ground where ingenuity and evasion can nullify some or all of its effects. Hence the rise to importance of the guerrilla and of his strategy of subversion and delay. It has proved effective even in a world led by the logic of modern war to develop the ultimate instrument of destruction, the nuclear bomb.

STEPHEN GRAHAM

How the News of War Came to a Village on the Chinese frontier

The most important military development in Europe in the second half of the nineteenth century was the establishment of universal conscription, military training and reserve duty for all males in the Continent's major states. The success and efficiency of the – largely conscript – Prussian army in the wars of 1866 against Austria and of 1870–1 against France, prompted those countries, but also Russia and Italy, to imitate its system. That required each fit male, at the age of about twenty, to report for two or more years of military duty, after which, on return to civilian life, he remained a member of the reserve, with the obligation to undergo refresher training in subsequent years and to return to the army on the declaration of mobilization for war.

This 'Prussian' system produced the enormous numbers of trained soldiers which filled the mobilization depots in August 1914 at the outbreak of the First World War. The French army, with a strength of 600,000 in peacetime, recalled 3 million reservists, the German army, 800,000 strong, over 4 million. Russia, with a million men under arms, had the largest peacetime army, and recalled 4 million reservists also.

Among them were its Cossack 'hosts', whose mobilization is described here by the travel writer Stephen Graham. The Russian Empire had an unusual relationship with the Cossacks, the horse peoples of Russia's frontier with the Central Asian steppe. Originally fugitives from serfdom who had made their way to the open land beyond the imperial government's rule, their life was modelled on that of the Muslim nomads – Tartars, Turks, Mongols – who were the Tsar's historic enemies. They kept large herds of horses, lived under arms and subsisted by freebooting. They remained, however, Christian and because of their skills in fighting the Tsar's nomadic Muslim enemies, were gradually drawn back into his service. By the time of Napoleon's invasion in 1812 they were supplying large contingents of light cavalry to the imperial army – Captain Roeder describes (p. 155) their attacks against the Grand Army during its retreat to the Beresina – and by 1914 Cossacks formed several of its cavalry divisions.

Unlike the ordinary subjects of the Tsar, however, the Cossacks did not serve as individuals but as members of their particular community or 'host'. It was the head (*ataman*) of a host who had the duty of mustering his Cossacks, with their horses, arms and equipment, and bringing them to the mobilization centre when called upon to do so. Stephen Graham was a witness to such a muster, in which echoes can be found of the preparation for war of Attila's Huns, Genghis Khan's Mongols and of the Turkish nomads who, between the battles of Manzikert in 1071 and the siege of Constantinople in 1453, overthrew the last outpost of the Roman Empire.

*   *   *

I was staying in an Altai Cossack village on the frontier of Mongolia when the war broke out, 1,200 versts south of the [Trans-]Siberian railway, a most verdant resting-place with majestic fir forests, snow-crowned mountains range behind range, green and purple valleys deep in larkspur and monkshood. All the young men and women of the village were out on the grassy hills with scythes; the children gathered currants in the wood each day, old folks sat at home and sewed furs together, the pitch-boilers and charcoal-burners worked at their black fires with barrels and scoops, and athwart it all came the message of war.

At 4 a.m. on 31 July 1914 the first telegram came through; an order to mobilize and be prepared for active service. I was awakened that morning by an unusual commotion, and, going into the village street, saw the soldier population collected in groups, talking excitedly. My peasant hostess cried out to me, 'Have you heard the news?' There is war.' A young man on a fine horse came galloping down the street, a great red flag hanging from his shoulders and flapping in the wind, and as he went he called out the news to each and every one, 'War! War!'

Horses out, uniforms, swords! The village *feldscher* [headman] took his stand outside our one government building, the *volostnoe pravlenie*, and began to examine horses. The Tsar called on the Cossacks; they gave up their work without a regret and burned to fight the enemy.

Who was the enemy? Nobody knew. The telegram contained no indications. All the village population knew was that the same telegram had come as came ten years ago, when they were called to fight the Japanese [the Russo-Japanese War of 1904–5]. Rumours abounded. All the morning it was persisted that the yellow peril had matured, and that the war was with China. Russia had pushed too far into Mongolia, and China had declared war.

The village priest, who spoke Esperanto and claimed that he had never met anyone else in the world who spoke the language, came to me and said:

'What think you of Kaiser William's picture?'

'What do you mean?' I asked.

'Why, the yellow peril!'

Then a rumour went round, 'It is with England, with England.' So far away these people lived they did not know that our old hostility had vanished. Only after four days did something like the truth come to us, and then nobody believed it.

'An immense war,' said a peasant to me. 'Thirteen powers engaged – England, France, Russia, Belgium, Bulgaria, Serbia, Montenegro, Albania, against Germany, Austria, Italy, Romania, Turkey.'

Two days after the first telegram a second came, and this one called up every man between the ages of eighteen and forty-three. Astonishing that Russia should at the very outset begin to mobilize its reservists 5,000 versts from the scene of hostilities!

Flying messengers arrived on horses, breathless and steaming, and delivered packets into the hands of the *Ataman*, the head-man of the Cossacks – the secret instructions. Fresh horses were at once given them, and they were off again within five minutes of their arrival in the village. The great red flag was mounted on an immense pine-pole at the end of our one street, and at night it was taken down and a large red lantern was hung in its place. At the entrance of every village such a flag flew by day, such a lantern glowed by night.

The preparations for departure went on each day, and I spent much time watching the village vet certifying or rejecting mounts. A horse that could not go fifty miles a day was not passed. Each Cossack brought his horse up, plucked its lips apart to show the teeth, explained marks on the horse's body, mounted it bare-back and showed its paces. The examination was strict; the Cossacks had a thousand miles to go to get to the railway at Omsk. It was necessary to have strong horses.

On the Saturday night there was a melancholy service in the wooden village church. The priest, in a long sermon, looked back over the history of Holy Russia, dwelling chiefly on the occasion when Napoleon defiled the churches of 'Old Mother Moscow', and was punished by God. 'God is with us,' said the priest. 'Victory will be ours.'

Sunday was a holiday, and no preparations were made that day. On Monday the examination of horses went on. The Cossacks brought also their uniforms, swords, hats, half-*shubas* [groundsheets], overcoats, shirts, boots, belts – all that they were supposed to provide in the way of kit, and the *Ataman* checked and certified each soldier's portion.

On Thursday, the day of setting out, there came a third telegram from St Petersburg. The vodka shop, which had been locked and sealed during the great temperance struggle which had been in progress in Russia, might be opened for one day only – the day of mobilization. After that day, however, it was to be closed again and remain closed until further orders.

What scenes there were that day!

All the men of the village had become soldiers and pranced on their horses. At eight o'clock in the morning the holy-water basin was taken from the church and placed with triple candles on the open, sun-blazed mountain side. The Cossacks met there as at a rendezvous, and all their women-folk, in multifarious bright cotton dresses and tear-stained faces, walked out to say a last religious goodbye.

The bare-headed, long-haired priest came out in vestments of violent blue, and behind him came the old men of the village carrying the icons and banners of the church; after them the village choir, singing as they marched. A strange mingling of sobbing and singing went up to heaven from the crowd outside the wooden village, this vast irregular collection of women on foot clustered about a long double line of stalwart horsemen.

The consecration service took place, and only then did we learn the almost incredible fact that the war was with Germany. It made the hour and the act and the place even more poignant. I at least understood what it meant to go to war against Germany, and the destiny that was in store.

'God is with you,' said the priest in his sermon, the tears running down his face the while. 'God is with you; not a hair of your heads will be lost. Never turn your backs on the foes. Remember that if you do, you endanger the eternal welfare of your souls. Remember, too, that a letter, a postcard – one line – will be greedily read by all of us who remain behind . . . God bless His faithful slaves!'

THOMAS HARDY
(1840–1928)
In Time of 'The Breaking of Nations'

I

Only a man harrowing clods
In a slow silent walk
With an old horse that stumbles and nods
Half asleep as they stalk.

II

Only thin smoke without flame
From the heaps of couch-grass;
Yet this will go onward the same
Though Dynasties pass.

III

Yonder a maid and her wight
Come whispering by:
War's annals will cloud into night
Ere their story die.

1915

ERWIN ROMMEL
Infantry Attacks

Far away from the Altai Mountains of Central Asia in August
1914, the young Lieutenant Erwin Rommel, future field marshal
and commander of the Afrikakorps, had mobilized with his
regiment, the 124th Infantry (King William I, 6th Württemberg)
in the old monastery at Weingarten in south Germany. Rommel
was a regular officer, the son of a schoolmaster, a subject of
the King of Württemberg but, as events would shortly prove,
one of the most effective soldiers in the army of the German
Kaiser. His Swabian soldiers would also show themselves to
be hardy, brave and resolute.

The 124th Infantry, with the 123rd Grenadiers, formed 53
Brigade of the 27th Division in the Württemberg Corps, part
of the German Fifth Army. After mobilization, and a pause

for hasty refresher training of its reservists, it was moved by rail to the border of French Lorraine where, in the third week of August 1914, it deployed to begin the invasion of France. Rommel had never before been in action and was disabled in the early stages of the operation by a severe stomach upset, which he attributed to undercooked army bread. His later medical history suggests that the symptoms were psychosomatic. Though exceptionally brave, Rommel was also highly strung and, during the Second World War, succumbed to illness in stressful situations.

Nerves or not, Rommel demonstrated outstanding qualities of leadership in his first engagement. He was a keen observer of the enemy's behaviour under fire – intimates would later ascribe his success as a commander to his possession of *fingerspitzengefühl*, feeling in the end of his fingers or sixth sense – and he showed this ability to detect enemy moral weakness in his first firefight. By close observation of the French reaction to the opening exchange of fire, he decided that they were operating by the drill book, advancing bravely but failing to take cover or to identify the source of the resistance opposite. He, with a handful of riflemen, shot men down as they rushed forward, stopped and turned the rest of the French unit back and went on to occupy the ground they defended. His little victory, the outcome of personal bravery and cool-headedness, was to set the pattern for many others that would follow, on an increasing scale, during the First and then the Second World War. In 1917 Rommel, then a company commander in the Württemberg Mountain Battalion, won the Pour le Mérite, Germany's highest decoration for bravery. In the Second World War he would become, in Western eyes at least, Germany's greatest soldier. The foundations of his glittering career were laid in this tiny action in the cornfields of Bleid.

*   *   *

To the right and above us lay Hill 325 still covered with fog. In the tall fields of grain on its southern slope, we could not recognize friend or foe. Off to the right and about half a mile ahead of us on the far side of a draw, we saw the red breeches of French infantry in company strength on the front edge of a yellow wheatfield behind fresh earthworks. (They belonged to the 7th Company of the French 101st Infantry Regiment.) In the low area to the left and below us, the fight for burning Bleid still raged. Where were our company and the 2d Battalion? Were some still in Bleid with their bulk farther to the rear? What was I to do? Since I did not wish to remain idle with my platoon, I decided to attack the enemy opposite us in the sector of the 2d Battalion. Our deployment behind the ridge, our movement into position, and the opening of fire by the platoon was carried out with the composure and precision of a peacetime maneuver. Soon the groups were in echelon, part of them in the potato field, part of them well concealed behind the bundles of oats from whence they delivered a slow and well-aimed fire as they had been taught to do in peacetime training.

As soon as the leading squads went into position, the enemy opened with heavy rifle fire. But his fire was still too high. Only a few bullets struck in front of and beside us, and we soon became accustomed to this. The only result of fifteen minutes' fire was a hole in a messkit. Half a mile to our rear we saw our own skirmish line advancing over Hill 325. This assured support for our right, and the platoon was now free to attack. We rushed forward by groups, each being mutually supported by the others, a maneuver we had practiced frequently during peacetime. We crossed a depression which was defiladed from the enemy's fire. Soon I had nearly the whole platoon together in the dead angle on the opposite slope. Thanks to poor enemy marksmanship, we had suffered no casualties up to this time. With fixed bayonets, we worked our way up the rise and to within storming distance of the hostile position. During this movement the enemy's fire did not trouble us, for it passed high over us toward those portions of the platoon that were still a considerable distance behind us. Suddenly, the enemy's fire ceased entirely. Wondering if he was preparing to rush us, we assaulted his position but, except for a few dead, found it deserted. The tracks of the enemy led off to the west through the field in which the grain was as tall as a man. Again I found myself well in advance of my own line with my platoon.

I decided to wait until our neighbors on the right came up. The platoon occupied the position they had just gained; then, together with the commander of the 1st Section, a first sergeant of the 6th Company, and Sergeant Bentele, I went off on a reconnaissance to the west to learn where the enemy had gone. The platoon maintained contact. Some four hundred yards north of Bleid we reached the road connecting Gévimont and Bleid without having encountered the enemy. The road became higher as it went to the north, passing through a cut at this point. On both sides of the road large clumps of bushes interfered with the view to the northwest and west. We used one of these clumps of bushes as an OP (observation point). Strange to say, nothing was to be seen of the retreating enemy. Suddenly, Bentele pointed with his arm to the right (north). Scarcely 150 yards away the grain was moving: and through it we saw the sun's reflection on bright cooking gear piled on top of the tall French packs. The enemy was withdrawing from the fire of our guns which were sweeping the highest portion of the ridge to the west from Hill 325. I estimated that about a hundred Frenchmen were coming straight at us in column of files. Not one of them lifted his head above the grain. (These soldiers belonged to the 6th Company of the French 101st Infantry Regiment. They had been attacked on the west slope of Hill 325 by elements of the 123d Grenadier Regiment and were now retreating toward the southwest.)

Was I to call up the remainder of the platoon? No! They could give us better support from their present position. The penetration effect of our rifle ammunition came to mind! Two or three men at this distance! I fired quickly at the head of the column from a standing position. The column dispersed into the field; then, after a few moments, it continued the march in the same direction and in the same formation. Not a single Frenchman raised his head to locate this new enemy who had appeared so suddenly and so close to him. Now the three of us fired at the same time. Again the column disappeared for a short time, then split into several parts and hastily dispersed in a westerly direction toward the Gévimont–Bleid highway.

We opened with rapid fire on the fleeing enemy. Strange to say, we had not been fired on even though we were standing upright and were plainly visible to the enemy. To the left, on the far side of the clump of bushes where we were standing, Frenchmen came running down the highway. They were easily shot down as we fired at them through a break in the bushes at a range of about ten yards. We divided our fire and dozens of Frenchmen were put out of action by the fire of our three rifles.

The 123d Grenadier Regiment was advancing up the slope to the right. I signalled my platoon to follow, and we then advanced northwards on both sides of the Gévimont–Bleid road. During our advance we encountered a number of Frenchmen in the bushes along the road. It took a lot of talking to get them out of their hiding places and make them lay down their arms. They had been taught that the Germans would behead all their prisoners. We got more than fifty men out of the bushes and grain fields, including two French officers, a captain and a lieutenant who had been slightly wounded in the arm. My men offered the prisoners cigarettes which increased their confidence.

To the right on the hill the 123d Grenadier Regiment also reached the Gévimont–Bleid road. We were being fired on from the direction of the forest-covered peak, Le Mat, which was five thousand feet high and lay northwest of Bleid. As quickly as possible I got the platoon into the cut on the right so they would be under cover, with the intention of resuming the fight with an attack on Le Mat from this point. Suddenly, however, everything went black before my eyes and I passed out. The exertions of the previous day and night; the battle for Bleid and for the hill to the north; and, last but not least, the terrible condition of my stomach had sapped the last ounce of my strength.

I must have been unconscious for some time. When I came to, Sergeant Bentele was working over me. French shell and shrapnel were striking intermittently in the vicinity. Our own infantry was retiring toward Hill 325 from the direction of the Le Mat woods. What was it, a retreat? I commandeered part of a line of riflemen, occupied the slope along the Gévimont–Bleid road, and ordered them to dig in. From the men I learned that they had sustained heavy casualties in Le Mat woods, had lost their commander, and that their withdrawal was executed on orders from a superior commander. Above all, French artillery wrought great havoc among them. A quarter of an hour later, buglers sounded 'regimental call' and 'assembly'. From all sides parts of the regiment worked their way toward the area west of Bleid. One after the other the different companies came in. There were many gaps in their ranks. In its first fight the regiment had lost twenty-five per cent of its officers and fifteen per cent of its men in dead, wounded, and missing. I was deeply grieved to learn that two of my best friends had been killed. As soon as the formations had been reordered, the battalions set off toward Gomery through the south part of Bleid.

Bleid presented a terrible sight. Among the smoking ruins lay dead soldiers, civilians, and animals. The troops were told that the opponents of the German Fifth Army had been defeated all along the line and were in retreat; yet in achieving our first victory, our success was considerably tempered by grief over the loss of our comrades. We marched south, but our progress was frequently halted, for in the distance we saw enemy columns on the march. Batteries of the 49th Artillery Regiment trotted ahead and went into position on the right of the highway. By the time we heard their first shots, the enemy columns had disappeared into the distance.

Night fell. Nearly dead from fatigue, we finally reached the village of Ruette, which was already more than filled with our own troops. We bivouacked in the open. No straw could be found, and our men were much too tired to search for it. The damp, cold ground kept us from getting a refreshing sleep. Toward morning it grew chilly – all of us were pitifully cold. During the early morning hours, my complaining stomach made me restless. Finally day dawned. Again thick fog lay over the fields.

COMPTON MACKENZIE
Gallipoli Memories

The opening campaign of the First World War in the west, in which Erwin Rommel had played a tiny but significant part, concluded in the failure of the German army to win a great victory in the open field. The outcome was the construction of a trench line, nearly five hundred miles long, running from the North Sea in Belgium to the border of Switzerland.

The frustration of the efforts by the French army, assisted by the British Expeditionary Force, to break the trench line in the winter and spring of 1914–15 persuaded the Allied high command to look for a way round. The spot on which they fixed for the outflanking manoeuvre was Gallipoli in Turkey, on the waterway separating Europe from Asia and leading to the Black Sea. The object of the original mission was to force a way up the waterway – the Bosphorus – as far as Constantinople, capital of Turkey, which had joined Germany and Austria as an ally in October 1914. The British and French governments believed that the arrival of their battle fleet at Constantinople would frighten the Turkish (Ottoman) government into making peace, so allowing supplies to be sent from the Mediterranean to their Russian allies via the ports of the Black Sea.

The attempt to force the battleships past Gallipoli failed, when several were sunk by mines and shore batteries. It was then decided to land troops, and on 25 April 1915, several divisions of British, Australian, New Zealand and French troops assaulted the beaches at the tip of the Gallipoli peninsula. Footholds were secured but the losses suffered were heavy and a Turkish counter-offensive, directed by Mustafa Kemal, later Kemal Atatürk, the founder and first President of modern Turkey, hemmed the Allies in. The Gallipoli front was stale-mated, just as that in France and Belgium had been. Neverthe-

less, Britain and France persisted with the effort to defeat
the Turkish defenders all through the summer and autumn.
Compton Mackenzie, already a well-known writer, was in 1915
an officer of the Royal Naval Division (whose sailors fought
as infantry), one of those involved in the initial landings. His
account of the trench fighting in June vividly conveys the
smallness of the battlefield, overlooked by the Turks from all
sides, and the atmosphere of the fighting, which brought a
constant, daily flow of casualties without any progress on the
ground. The Collingwood Battalion – the battalions of the
Royal Naval Division were named after famous British admirals
– had suffered particularly heavily in the first month of fighting.

*   *   *

On Monday, the seventh of June, there was a chance of crossing
again to Cape Helles [the southern point of the Gallipoli peninsula]
in a destroyer. I welcomed an excuse to escape from the stifling
atmosphere of the tent which had been full of blood-stained Turkish
notebooks for the last two days, and asked leave to visit my Divisional
Paymaster to discuss the problem of my pay and at the same time
try for a batman. The atmosphere at Divisional Headquarters was
gloomy in the extreme. I was not astonished when I heard details
of what the Division had been through last Friday. The casualties
had been very heavy. They thought that the French had let them
down completely on the right. Patrick Shaw-Stewart was seen running
along, waving his cane and shouting, 'Avancez! Avancez!' The Senegalese came out of their trenches, advanced seventeen yards, and
then bolted back into them like so many gigantic black rabbits, after
which nothing would persuade them to show themselves again. I
suppose this was after the Colonial troops and Senegalese had been
bombed out of the Haricot Redoubt which they had held for a time.
There was no disposition to put any blame for the failure of the
fourth of June on the General Staff. Any gibing was mostly directed
at Maxwell's *Peninsula Press* [a trench newspaper] which had come
out with a rosified account of our 'success', though of course it was
recognized that a daily sheet of unmitigated gloom would hardly be

worth printing and circulating. I was promised a batman; but the problem of my pay looked like being for ever insoluble, and I started to walk back. Small shells kept dropping all round me, and it seemed inevitable that I should be hit presently. There is no doubt that the sensation of being shelled when alone is most infernally unpleasant. After walking about three-quarters of a mile I felt inclined to sit down and cry with exasperation because those Turkish gunners would not realize that I really was not worth so much expensive ammunition. I wanted to argue with them personally about the futility of war. It seemed so maddeningly stupid that men should behave as impersonally and unreasonably as nature. Over to the right I saw a clump of trees and, feeling I simply must somehow get a sensation of cover, I hurried across toward them at a diagonal jog-trot. I could not have made a more foolish move, because apparently there was a well by them at which mules were watered, and at regular intervals the enemy used to spray the clump with shrapnel. I must have come in for one of those antiseptic douches, for the air was alive. I began to worry about the proofs of [his novel] *Guy and Pauline*, thinking to myself that the printer's reader would be sure to change 'tralucent' to 'translucent' and that Secker [his publisher] in the depression caused by the news of my death would never remember how much importance I attached to getting rid of that unnecessary sibilant. Why couldn't those blasted Turks up on Achi Baba shut up? And I would have turned a gerund into a participle here and there ... and probably there would be a vile *nominativus pendens* ... at this moment I heard a burst of laughter and, looking round angrily, for I thought this laughter must be meant for the way I was definitely running by now, I saw a couple of men digging opposite to one another like the gravediggers in *Hamlet* and roaring with laughter every time one of the small shells either exploded or as often happened hit the ground with a thud and nothing else. Then one of the pair dropped. The other looked first at his pal and then at me who was hurrying past with haversack, water-bottle, pistol, and glasses jogging up and down in a most undignified way.

'Beg pardon, sir! Beg pardon!' he called out.

'You can't do anything,' I snapped. 'You'd better get into cover yourself as quickly as you can.'

'No, sir, it's not that,' he whimpered as he cut across my path and forced me to stop while he saluted. 'But would you mind telling me if my friend's dead, sir, because I'm new at this job.'

'Of course, I'm not bloody well dead, you silly little cod,' shouted the friend, who was sitting up by now and rubbing his head. And I left them, remembering another occasion when the friend actually had been killed and when the survivor's comment was, 'Beg pardon, sir, you think it's funny at first, but it's very serious really.'

By the time I reached the beach, the big gun on the Asiatic side of the Straits had started to shell the shipping. There were three preliminary fountains, after which a shell hit a French transport loaded with hay. The crew at once jumped overboard, and the transport caught fire. Then two destroyers rushed up and bundled all the men back on to their ship in order to extinguish the fire, which they succeeded in doing without being shelled any more.

I think it must have been that evening I met the last surviving officer of the Collingwood Battalion. He was very young, hardly more than eighteen and, after the horror of that experience to which he had gone almost within forty-eight hours of landing at Helles, he was being sent to do some work at Imbros [island west of Gallipoli, used as an Allied base; now Imroz] in connection with the rest camp which was to be formed there. We did not talk about the battle, either then or at any other time. Oldfield was his name, and I hope he survived the battles in France later. I can hear now the tone of his voice as he said to me with a nervous little laugh:

'I'm the only officer left of the Collingwood.'

SIDNEY ROGERSON
Twelve Days

The routine of life in the trenches dominated the infantry soldier's experience of the First World War on the Western Front. Great battles, as at Verdun and on the Somme in 1916, were terrifying and deadly interludes, killing soldiers in tens of thousands. The greater part of an infantry regiment's time, however, was spent in trench garrison duty, usually divided into blocks of front-line service, rest behind the lines, return to the support or reserve lines immediately behind the front and then the relief of a sister regiment in the front trenches again. The cycle, in the British army, usually lasted twelve days.

Sidney Rogerson, a company commander in a battalion of the West Yorkshire Regiment, used the twelve-day cycle to describe his unit's ordeal in and behind the trenches of the Somme battlefield in the winter of 1916. The British army had attacked the German lines on the Somme on 1 July 1916. Strongly prepared during eighteen months of quiescence on that sector, the German trenches proved almost impenetrable. Some 20,000 British soldiers were killed on 1 July, more than 40,000 wounded or missing. In the four months that followed, the British succeeded, in a sequence of step-by-step attacks, in forcing the Germans back. At the onset of winter in November, however, the German line still held. The battlefield itself, by a process of attack and counter-attack, had become a wilderness of broken barbed wire, ruined dugouts, shell holes, swamp and trench dead-ends. The trench system was a maze, which the occupiers struggled to 'improve' by digging and draining, but which easily swallowed up the individual who lost his way in any attempt to find a route from one place to another. Connection between the front line and the rear lay through 'communication trenches'. So battered and shapeless did they become under shell fire, however, that it was often quicker, though inevitably more dangerous, for those leaving the front line to travel above ground, at the risk again of becoming lost in a landscape without landmarks.

Sidney Rogerson's account of leaving his trench sector at the end of his regiment's spell of front-line duty vividly conveys what he calls 'the nightmare' of a trench relief – personal disorientation, misunderstanding between brother officers, the overwhelming fatigue of the men and their leaders, the euphoric sense of escape from all-encompassing danger when the safe area behind the lines was reached. It also conveys the arbitrary and haphazard nature of any military operation – the menace of 'friendly fire', when British artillery killed British soldiers, the hazards of the battle zone itself, ever threatening to engulf and drown the unwary in swamped shell holes or waterlogged trenches. Sidney Rogerson is one of the great soldier-writers of the First World War.

*   *   *

Account of being relieved in the line, November 1916, Somme

So that last day in the line passed under its mantle of fog. The short winter daylight passed in uneasy silence. The men slept; took their turn at a meaningless spell of duty at the sentry-posts, peering stupidly into the mist; and wrote letters or 'Whizz-Bangs' [pre-printed cards] to be posted home when we got out – 'I am well,' 'I am in hospital,' etc. 'Cross out the words which do not apply.'

Robbed of their eyes in the air [poor weather kept spotter aircraft and balloons from operating effectively], even the heavy guns were still, but one of the few shells which our field guns fired, burst short in the very mouth of A Company's sap, killing two men.

The piece of news sent me off post-haste to Fall Trench to investigate. Cropper was in the sap when I arrived. The shell that had done the damage had come from directly in rear, otherwise it could not have entered the trench as it had done. It had burst plumb in the middle of the sap-way, killing the two men instantaneously but without mutilating them much. There was little room for doubt that it was one of our own shells, and that was removed when, after a few seconds poking about in the smoke-blackened soil, we found the nosecap. It was a British eighteen-pounder. It was only natural to curse the gunners, the rotten American ammunition, the worn guns, the inefficiency of the intelligence people who did not know where their own bloody infantry were, the staff, and every one else whom we could think of for blotting out two good Yorkshire soldiers. But, living as we were in scoops and burrows which not only were not shown on any map, but which we ourselves were frantically anxious should be difficult to detect by direct observation, we were as much at the mercy of our own as of the enemy gunners. And when we had blown off our indignation, we had to admit that the marvel was that such accidents did not happen more often. Anyway, there was nothing to be done. I promised Cropper I would report to Battalion Headquarters so that they could pass the information on to our divisional artillery. Cropper undertook to see that the victims were buried as soon as night fell, so that we might leave everything shipshape for the Worcesters.

From A Company's sap to that of B Company was only a few yards along Fall Trench, and my visit took the post there, consisting of a young lance-corporal and two men, by surprise. The corporal had not only allowed his men to take off their equipment, but was minus his own, while there was a general atmosphere of slackness. It was strictly against orders to remove any essential equipment while in the trenches, and the offence naturally became still more heinous in men on what was virtually outpost duty. Still, in view of the youth of the NCO and his good record, I was prepared to let him off this time with a good dressing-down, had he not shown a kind of familiar resentment that I should have taken exception to his indiscipline. This hint of familiarity touched me on a delicate spot. It had always seemed absurd to me to try and adhere rigidly to the conventional formalities of discipline in the trenches where officers lived cheek by jowl with their men, shared the same dangers, the same dug-outs, and sometimes the same mess-tins. Quite apart from the absurdity, I believed, and nothing I ever saw subsequently shook me in the belief, that the way to get the best out of the British soldier was for an officer to show that he was the friend of his men, and to treat them as friends. This naturally involved a relaxation of pre-war codes of behaviour, but it did not mean that an officer should rub shoulders with his men at every opportunity, or allow them to become familiar with him. It meant rather that he should step down from the pedestal on which his rank put him, and walk easily among his men, relying on his own personality and the respect he had earned from them to give him the superior position he must occupy if he wished to lead. He had consequently to steer a delicate course between treating those under him as equals in humanity if inferiors in status, and losing their respect by becoming too much one of them. He must deal with them sympathetically and at all times interpret the law in the spirit and not in the letter, but he had equally to be jealous of his position, and never to allow leniency to be looked upon as weakness, or friendship to degenerate into familiarity. He had, in short, to discriminate between the men who would appreciate his interest and those who would be foolish enough to try and impose upon his good nature.

This was a case in point. Nothing was left to me but to take disciplinary action, and sending for his platoon commander, Hall, I ordered him to be relieved of his post and brought up for punishment when we got out of the line. For that show of bad manners, he was to lose his lance-stripe.

It was but a few moments before our minds were turned to less serious thoughts. Hall and I had walked a little farther along to the right of the sector – we were standing talking in the front line when we noticed a scuffling of earth in the parados [rear crest] of the trench, and out fell a furry, fat little mole. It appeared as one of Nature's miracles that this blind, slow creature could have survived in ground so pounded and upturned. After holding him for a few minutes, and marvelling at the strength of his tiny limbs, we put him into his hole again to find his way back whence he had come. A few desperate clawings, and he had disappeared. How we wished we could dig ourselves in so easily!

As darkness drew on, the mist lifted, and Purkiss, as company cook, and the officers' servants were sent out of the line to make ready the camp against our arrival. They were a reluctant party that set off over the top, loath to leave the comparative safety of their forward position to cross that dreaded two miles of shell-swept mud separating us from the nearest approach to civilized comforts.

Darkness took the place of fog, but the effect was much the same. The day had been but a continuation of the night. The only difference was that night was the time of movement. Just as an English countryside comes to life after dark, when even the roadside hedge is quick with rustlings and squeakings, so the dead landscape of the Somme stirred into activity with nightfall. And as we listened to the now familiar far-off rumble of transport behind the enemy lines as behind our own, there was this evening the knowledge that soon we were to be relieved, to go back out of this outpost to some camp where we should have braziers and blankets and hot food. We had all been through it before, most of us more times than we could remember. We were familiar with every dreary detail of relief as we were of taking-over, yet there was no one of us, not even the most hardened campaigner, who would not confess to being seized on every such occasion with the fear that at the last moment something dreadful would happen, some disaster overtake him before he had passed through the danger zone which separated the line from the safety of the back areas. It might come by a chance sniper's bullet, striking him down noiselessly before he even got out of the trenches. It might come with a scream out of the darkness when the worst of the journey lay behind him.

So it was on this night of November 13. The marvel of three days without rain had enabled us more quickly than usual to get the trenches clean and tidy, to collect our kit, and put ourselves in readiness to depart as soon as the Worcesters should appear. The result was that the men were waiting for the relief almost as soon as night fell, and tried to keep their spirits up, after the manner of the British soldier, by singing all the most mournful ditties they knew from 'Don't Go Down the Mine, Daddy', to 'Oh My, I Don't Want to Die! I Want to Go Home'.

I had to make a final last trip round the sector, noting with a certain pride the improvements that the kindness of the elements had allowed us to make. Not only had we temporarily subjugated the mud, but we had added another brand-new piece to the great jig-saw puzzle of the trench system of the Western Front. There were also arrangements to be made with Mac about moving the company out. We had been ordered to salvage as much derelict equipment and stores as possible, and that meant putting further burdens in the shape of waterproof capes, sheets, or great-coats on to already tired and overladen men. But there was another point, and one which served to show me how very easy it is to write orders which were capable of more than one interpretation. I, for some incomprehensible reason, had read the command about Company Commanders reporting 'relief complete' at Battalion Headquarters, to mean that I, at the head of my straggling followers, was to flounder through the mud to that death-trap the Sunken Road, and, while they awaited almost certain destruction, inform the Adjutant that we had handed over properly and without incident. I waxed righteously indignant. Whatever was the Colonel thinking about, I asked Mac, giving orders so unlike him in their disregard of the men's safety! Anyway, I would do nothing of the sort. If I had to report personally I would report alone, and Mac should take the company out by the quickest and easiest way he could find. I would meet him at Ginchy cross-roads.

One, two, nearly three hours dragged by with the men getting more restive and anxious to be gone. The enemy left us alone, confining his 'frightfulness' to the valley behind. This was a temporary blessing only, causing each one of us to shrink the more from the thought of the journey we should have to make. We had become such troglodytes that to leave the shelter of a trench induced a feeling of nakedness. Not until nearly eleven o'clock did there come the welcome sounds of men approaching, and the Worcesters began filing in. They had had a rotten time in the passage of the valley, poor devils, much worse than we had had, and had lost several men. Still, favoured by the weather and undisturbed by 'Fritz' the ceremony of handing-over was accomplished with all possible haste and, shepherded by the lanky Mac, the company struggled out on their journey to Ginchy. Unlike the Devons, we left no rum behind us!

As soon as the last man was clear of the trenches, I started out for Battalion Headquarters in the Sunken Road, so confident in my ability to find it that I took no orderly with me. Alas for such presumption! I had gone no more than a few dozen paces when I began to have misgivings. Surely, I should have passed Dewdrop Trench by now. I paused to take what bearings I could, but the night was black as pitch. Landmarks there were none. A shell burst here and there, and I remember thinking what a wrong impression the ordinary war pictures gave. They always showed shells exploding with a vivid flash, but all that now happened was a scream, a thud, and a little shower of red sparks as from a blacksmith's anvil. There was not the faintest glimmer to light me on my way. I stumbled on. Doubts became anxiety. I was lost! No matter that I ought to know I could not be far away from some one; I was afraid. Throughout the war this was my worst nightmare – to be alone, and lost and in danger. Worse than all the anticipation of battle, all the fear of mine, raid, or capture, was this dread of being struck down somewhere where there was no one to find me, and where I should lie till I rotted back slowly into the mud. I had seen those to whom it had happened.

So now anxiety passed almost at once into panic. I went forward more quickly first at a sharper walk, then at a desperate blundering trot. Was it imagination? Or were more shells really beginning to fall, rushing down to sink into the soft earth and burst with smothered thuds? Yes! Little showers of red sparks were all around me. I struggled on, fell once, again, many times, tore my coat on barbed wire, cut my hand. When would a bullet from those chattering machine-guns strike me in the head or back? The nape of my neck ran cold at the thought. My heart thumped louder than ever, both from terror and effort. I was getting blown [exhausted]. I could go no further. Then I stumbled, pitched forward, slithered down several feet, caught my kit on some signal wires, sank up to my elbows in wet mud. I had reached the Sunken Road!

Breathless and shaken, I struggled to my feet. A head peered at me out of a sandbagged dug-out entrance. I asked the owner the whereabouts of the West Yorkshire headquarters. He thought there were some infantry battalion headquarters a little farther up the road on the opposite side, but whose he did not know. He was a signaller, he confessed, as if to excuse his lack of information. There was nothing for it but to find out for myself.

What a cess-pit that Sunken Road was! Over ankle-deep in slime, it was strewn with the bodies of horses and mules in varying stages of decay. Yet its battered banks afforded the only convenient cover for a wide area around, so into them had been driven dug-outs, British and German, into which were crammed all those whose duties kept them in the forward zone without taking them into the trenches. There were the headquarters of two or three battalions, the forward posts of batteries, both field and heavy, signallers, sappers, and odd details of all arms. I made shouted inquiries down two or three shafts before I pulled aside the tattered sandbag cover which hung before the right dug-out, and entered a tiny candle-lit burrow. In it were installed the Worcesters. My own regiment had gone!

This was the crowning blow. With apologies for my intrusion, I set out again into the darkness, feeling more wretched and hopeless than ever after the brief vision of light and warmth. But this time luck was with me, and hardly had I scrambled up the bank than some one said, 'Who's there?' and I recognized Hawley's voice. He was making for Ginchy with an orderly who knew the way, so we set off together. With company the whole atmosphere seemed to change. The danger remained the same, yet the presence of others banished at once the terror that had assailed me. In a few minutes we were passing battery positions and were dazzled by the stabbing, lemon flashes of the guns as they fired towards us. Then we struck the duck-board track which, rickety and shell-smashed though it was, led us steadily towards Ginchy. Lights again began to appear in dug-out doors and gun-positions, now far enough away from the enemy to disregard the risk of detection, and at last we were able to make out groups of men in the darkness. We had made Ginchy cross-roads, and the men were B Company in artillery formation, into which Mac had put them to lessen the danger from any chance shell. Thanks to his guidance in avoiding the bad places, they had got out without loss.

Without wasting time in marvelling at this miracle, the Company fell in and moved off. That march was a nightmare. Not till then did I realize how tired I was, nor how done the men. I had snatched less sleep than they – my total for the three days was no more than six hours – and had been more continuously 'on the go'. They, on the other hand, were overloaded with sodden kit. We had not gone far before requests were made for a halt. I turned a deaf ear. Men so weary, I argued, would only fall asleep the moment they broke rank. It would be the harder to get them on the move again. Besides, we were still in the shelled area. Requests turned to protests. Some of the younger men could hardly walk. Officers and the fresher NCOs took over rifles and packs from the most fatigued but without avail. The querulous, half-mutinous demands for rest grew more insistent. They were the cries of semi-conscious minds tortured by over-exertion and lack of sleep. Still I took no notice, vowing that not until we were in the haven of our own camp would I call a halt. 'I won't stop! I won't stop! I won't stop!' I repeated to myself with each agonizing step. My ears, deaf to all else, sung the refrain. But to no purpose. Will said, 'I won't stop'; Body argued, 'I can't go on.' And so it was. After what seemed hours of tramping I could go no farther. All my determination was thrown to the winds. 'Halt!' I could do no more than speak the command. It was enough. The word was scarce uttered before every one, myself included, had thrown himself down on to the bricks and rubbish at the roadside. There we lay, silent, exhausted. Will began to reassert itself. This was no good. The longer we stayed the more difficult it would be to go on. Somehow, how I do not know, we stirred and prodded the men into movement again. Cursing and grunting, they shambled forward with the unsteady steps of sleep-walkers. I tried to square my shoulders for the long march that I felt must lie ahead. Irony of ironies! We had not moved more than a few score paces before the gruff voice of Company Sergeant-Major Scott was heard shouting 'This way, B Company.' We had fallen out less than a hundred yards from camp!

But what miles we must have marched! How many was it? The distance from Ginchy cross-roads to La Briqueterie camp was no more than three and a half miles! The front line was under five miles away. There is no doubt the Somme taught us that distance is a relative term, not to be measured in yards and feet.

Once again we were under canvas, La Briqueterie being a camp of bell-tents where once had stood the brickyard which had figured so prominently in the despatches describing the September battles around Montauban. Unlike the pitiful 'Camp 34' at Trônes Wood, La Briqueterie would not have been entirely disgraced by comparison with the real thing as seen at Aldershot or on Salisbury Plain. Tents were pitched in regular lines, and were well and truly guyed with a brazier alight and glowing in each one, and dear old Hinchcliffe, the Quartermaster, ready to issue hot soup and rum for every man. We felt we had fallen into luxury indeed. Lord! How the keen edge of appreciation of creature comforts is blunted by a life of peace! Did not a mess-tin of stew, a tot of rum or whisky and water in a tin mug, taste more like divine nectar than the best champagne drunk out of the finest cut-glass to-day? To enjoy ease, it is surely necessary to labour. To enjoy luxury, it is necessary to live hard. Since our work in these after-days is all too often sedentary, since we all too often tend to overfeed, and since we shun living hard, if we have not lost the capacity for it, it is small wonder that we are dyspeptically out of tune with life, and have to pay to have our jaded appetites whetted by manufactured thrills on stage, screen, dirt-track, or playing field. No such necessity existed in France in the war years. The enemy provided excitement enough.

However it may be, our spirits rose and our mouths watered at the delectable prospect which had opened so suddenly out of the night. Mac and George I sent to take off their equipment, staying myself to watch Company Quartermaster-Sergeant Carlton allot tents to the company, and Company Sergeant-Major Scott start to serve out the soup. Having seen things properly started, I went off to remove my burdensome equipment before returning to supervise the rum issue. As my guide led me to the officers' lines, he pointed to a great hole which yawned where the next tent to ours had stood.

'That was your tent this afternoon, sir,' he told me, 'but about two hours ago Jerry dropped an eight-inch shell there and "napoo-ed" it.' He reassured me that this had been the only shell the enemy had put into the camp. Still, that hole would have been a disquieting feature in a rest camp had it not been that I was too tired to do more than register its presence as a fact.

I had barely started to take off my kit when an orderly came to say, 'Colonel wants you in B Company lines at once, sir.' Tearing off the bulkiest items of my 'Christmas Tree', I hurried back to find the Colonel standing with Scott.

'What does this mean?' he demanded. 'Don't you know yet that an officer's place is with his men? Haven't you learnt yet that an officer has no right to think of his own comfort until he has satisfied himself that every one of his men is comfortable? Not only that, but you go away to your tent and leave a warrant officer to issue rum, which is entirely against all orders.' I struggled to explain that my offence was not so heinous as it seemed. I had not left until the men had been settled in and soup was being issued. I had had no intention of staying away, and I had given no instructions to any one to issue rum. At this the Colonel calmed down, and I recount the incident to illustrate his determination that the welfare of his men should be the first concern of every officer. I accompanied him through the lines, and that he was satisfied was soon evident as he left me with an invitation to come round to Battalion Headquarters for a drink when I had finished.

Although this was evidence of forgiveness, I rounded on Scott, asking why in Heaven he, of all people, had presumed to issue rum. His answer that he had done so quite deliberately because he had thought I had had about enough, and it was one way in which he could help me, completely disarmed me, and the stinging rebuke which I had contemplated turned to thanks.

At long last our duties were at an end. Every man was bedded down in warmth and with a belly reasonably full of soup and rum. It was nearly two o'clock. Bidding Scott good-night I turned thankfully towards the tent allotted to the Headquarters Mess. This I found brightly lit and almost inconveniently crowded. The Colonel was there, and Maclaren, and Matheson, and young Rayner, the assistant Adjutant, who like poor Skett was a recent arrival from Sandhurst and like him too, destined not to survive the war. There were cheery greetings, and Brownlow, imperturbable as ever, brought a whisky and soda. As I drank it the Colonel congratulated me on the behaviour of A and B Companies during the tour, and apologized for his testiness in 'strafing' me. He too was overtired.

His appreciation, coupled with a second whisky and soda, gave me a pleasurably warm feeling. Life took on a different complexion. Everything was good. Every one was friendly. I stayed chatting with Matheson for a few minutes, long enough to learn that the Boche had in fact put down a barrage as we were coming out – so that shelling when I lost my way was not imagination, I registered to myself – and D Company had been caught in it, though providentially without losing a man. Since we had last met two of our friends had gone West, but except for a passing reference – 'rotten luck' – their names were not mentioned. We were glad to be out, to be alive, and be together again.

I said my good-nights and sallied forth into the night air. Could I walk to my tent? Could I even stand? No! Two whiskies and soda on an empty stomach and in my exhausted physical condition had made me so drunk that I had to crawl home on all fours. Too tired and too fuddled to undress, I struggled out of jacket and boots, rolled into my flea-bag, and almost before I lay down had joined my snores to those of Mac and George.

JOHN MCCRAE
(1872–1918)
In Flanders Fields

In Flanders fields the poppies blow
Between the crosses, row on row,
That mark our place; and in the sky
The larks, still bravely singing, fly
Scarce heard amid the guns below.

We are the Dead. Short days ago
We lived, felt dawn, saw sunset glow,
Loved and were loved, and now we lie
            In Flanders fields.

Take up our quarrel with the foe:
To you from failing hands we throw
The torch; be yours to hold it high.
If ye break faith with us who die
We shall not sleep, though poppies grow
            In Flanders fields.

JOHN GLUBB
A Soldier's Diary of the Great War

The Western Front during the First World War was a great
battlefield, but also a huge civil-engineering site. The demands
of the war required the construction of camps, roads and light
(Decauville) railways on an enormous scale, as well as the sinking
of wells, laying of pipeline, sawing of timber, management of

canals and building of every sort of military installation, from airfields to ammunition dumps.

In the British army, the work fell on the Corps of Royal Engineers, whose officers and men were both soldiers and construction workers. One of them, John Glubb, in 1914 a cadet at the Royal Military Academy, Woolwich, proved also to be a gifted writer. Sent to France in November 1915, he was posted to the 7th Field Company, Royal Engineers, and soldiered at the front throughout the war, though three times wounded. His last wound was very serious, scarring his face for life, and keeping him out of battle until late in 1918. After the war his military duties took him to the Middle East, where he remained for the rest of his service, first in Iraq, then in Transjordan (now Jordan). In Transjordan he became the trusted servant of the royal government and, as Glubb Pasha, the commander of the Arab Legion. The Anglo-French invasion of Suez in 1956 led to his dismissal, though he was by then a legendary and famous figure in both the Middle East and Britain. In retirement he settled to writing, his talent for which he had shown in his First World War journal, published in 1977 as *Into Battle*. These extracts describe his work as an engineer officer, his pleasure in a brief holiday from the fighting amid the beauty of the unspoiled countryside behind the lines and his experience of being wounded. Few First World War memoirs combine the qualities of matter-of-fact reporting and lyrical evocation of landscape as his do.

*   *   *

September 23–26 1916, working on Decauville railway

23 September: The Boche has retired a little way on our Corps front and we have temporarily lost touch. We were above him on a forward slope beyond High Wood [on the Somme front], and he doubtless did not like to remain overlooked. We cannot immediately follow him up, as the country sloping down from High Wood to Eaucourt is simply a wilderness of shell-holes almost impossible to cross and is in full view all the way. Not a fly can cross the High Wood ridge without being seen by the Boche. The whole country is rolling downs, looking in the distance like a ploughed field, but in reality a continuous series of shell-holes and mounds between them.

The area is thickly dotted with specks of black and grey, lying motionless on the ground. When you approach, the black patches rise into a thick buzzing swarm of bluebottles, revealing underneath a bundle of torn and dirty grey or khaki rags, from which protrude a naked shin bone, the skeleton of a human hand, or a human face, dark grey in colour, with black eye holes and an open mouth, showing a line of snarling white teeth, the only touch of white left. When you have passed on again a few yards, the bluebottles settle again, and quickly the bundle looks as if covered by some black fur. The shell-holes contain every débris of battle, rifles, helmets, gas-masks, shovels and picks, sticking up out of the mud at all angles.

One cannot see these ragged and putrid bundles of what once were men without thinking of what they were – their cheerfulness, their courage, their idealism, their love for their dear ones at home. Man is such a marvellous, incredible mixture of soul and nerves and intellect, of bravery, heroism and love – it *cannot* be that it all ends in a bundle of rags covered with flies. These parcels of matter seem to me proof of immortality. This cannot be the end of so much.

24 September: It seems hopeless to overtake the front line by building roads across this vast waste of mud. The latest idea is to lay down a hasty 'decauville' light tramline, from the left of High Wood towards Eaucourt l'Abbaye. The company has stopped work on the roads and is to lay these tramlines. This, however, presents me with a new headache, for we have to carry the tramlines on our wagons from Bécourt to High Wood.

The big-gauge railway has now been laid up to Bécourt, where the guns were a week ago! [A series of British assaults from mid-September had resulted in the German line being pushed back in places.] Half a mile of big-gauge railway was laid in forty-eight hours, and the trains began to run before the line was ballasted.

25 September: I rode back to our mounted-section lines at Bécourt and loaded up the trestle wagons with tramlines, and took them up to High Wood after dark. 14th Corps (Lord Cavan), on our right, took Lesboeufs, Morval and Gueudecourt today.

26 September: Took some trolleys and sleepers for the tramline up to High Wood. This decauville 60-centimetre gauge track is complete with sleepers and is supposed to be laid down all ready-made, and just bolted together. In practice, however, some of the sleepers and bolts are always missing, and a good deal of improvisation is needed.

It is curious from the ridge beside High Wood, standing on an expanse ploughed up in shell-holes and strewn with rifles, helmets and corpses, to see only two miles away a peaceful landscape of downs, leafy woods, and the church steeple of Le Barque showing above the trees. This country had been out of range of our artillery until a week ago. However it is now within range and already the copses in front are becoming smashed and stripped. Soon all the trees will have disappeared, as in High Wood, leaving only an occasional splintered trunk standing, and a debris of split, torn and rotting timber on the ground. As you stand looking at this view, you hear the continuous quiet *wheu-u-u* of our big howitzer shells, which seem to spring up from the horizon behind you – or the lightning *wheut* of the whizz-bangs flying past – followed by a sudden volcano of grey on the distant hills, or the little white puffs of shrapnel among the trees.

By what tortuous build-up of evil have men become such tragic and cynical destroyers of their fellow beings, and of the glorious beauty of nature?

The broad-gauge railway had huge working parties of Indians and also a British Labour battalion. The latter were a butt for much scorn on the part of the boys. The sight of the war was to see a very tired whizz-bang shrapnel come over and burst with a ping somewhere up in the clouds. The Labour battalion would down tools like one man, officers and NCOs shouting and gesticulating to their men to take cover, as if the whole Boche army was coming at them with fixed bayonets. I remember old Corporal Cheale, who was in charge of No. 4 Section, gazing solemnly at them and saying, 'I'd like to get 'old of some of thim Labour battalion, Sir!'

Cheale is a character. He combines to an absolute fearlessness, a morbid love for the dead. Just before I joined in November 1915 at Armentières, a great many casualties from shelling had occurred in the company, which was billeted in a large school. In the morning, the big schoolroom was full of mutilated corpses. Many were new reinforcements, whose names nobody knew, making identification difficult. Cheale volunteered to help. When the job was over, he went up to the second-in-command and said, 'Excuse me, Sir, did you notice this young feller? Don't he make a lovely corpse?'

About the same time, he got married to a French girl, though he knew not a word of French, nor she of English. But they seemed to get on very well and he always went regularly on leave, and wrote to her most affectionately to

Madame Cheale,
Sentier de l'Eglise,
Nieppe.

He greatly enjoyed the work on 15 September, clearing the roads behind the infantry advance up to High Wood. The ground was thickly strewn with corpses, from brand new to skeletons. The pockets of the dead it was his self-imposed task to search, after which he would carry them away, wrapped in an affectionate embrace. He would always allude to them as property owners (owning six feet of soil), and talked cheerfully of the day when he would become one.

He had a habit, when other men were ducking or crawling around, of standing up very straight, looking peacefully around and remarking, 'I don't reckon much of these 'ere whizz-bangs!' . . .

Riding alone

28–30 April [1917]: The weather is absolutely and incredibly perfect – our first taste of spring, coinciding with our arrival at this beautiful village. The warmth and sunshine after snow and mud makes [the Battle of] Arras seem another world [Glubb's unit had been moved north from the Somme to the Arras front]. The village lies on a single road running along a narrow valley, and consists of farms and white-painted cottages with thatched roofs. On either side rise the downs.

Behind the officers' mess, in a one-storey farmhouse, stretches a little orchard and beyond it a paddock full of coarse meadow grass. At the bottom of this burbles a little stream, then a few trees and then the steep wall of the downs, crowned by a beech wood of tenderest, lightest spring green.

Lying on the grass in the orchard, one can feel the warm sun and watch the little fleecy clouds slowly moving across the blue to where the latter joins the light green of the beeches on the hill. One can hear the cocks crowing in the village and the birds singing, and an occasional *cuc-koo, cuc-koo* far away in the hills. We only expect to be out here for a week or so, so are doing little training – just basking in the loveliness and the warmth.

Baker is in hospital in Rouen, and there is no one else I care to go out with, except perhaps the 'fat boy', but he is too lazy to come. I went for many rides by myself, the country being thickly wooded, chiefly with beech woods. Through these I rode alone, down old neglected rides, while all round my head was a dazzling bower of light emerald green. Underfoot crunched the beech nuts, while the ground was everywhere carpeted with anemones and cowslips. Pulling up and sitting quietly on my horse in the heart of the forest, it was impossible to catch a sound of the outside world, except the jingling of my own bit and the murmuring of the trees.

Hooking a haversack on my saddle, containing a sponge and towel and Tolstoy's *The Cossacks*, I twice rode off into the woods, tied up my horse, and bathed in a little stream.

Wounded and operation

A BLIGHTY ONE

I set out on the evening of 21 August to go down to the dump at Hénin, to see certain stores loaded up correctly. I was riding an old black mare, most unsuitably called Geisha, for nothing in the world could be less like a dancing girl. She would not walk, but jogged endlessly, her nose stuck out, her neck as stiff as wood, and her mouth like iron. The champion tool-cart team was out that night in GS [General Service] wagons, proudly wearing their prize-winning rosettes.

I was expecting to meet some infantry wagons in Hénin, but, as they did not turn up, I rode on through St Martin-sur-Cojeul, to see my own wagons which had gone on ahead, but had been obliged to halt there. The road beyond St Martin was in view of the enemy, and it was not yet quite dark. I dismounted and sat on a stone for a short time, and then rode back through St Martin 'village', which consisted of a sea of untidy mounds of broken bricks, covered with grass.

Some long-range shells whined over, and burst about 120 yards beyond the road. It seemed to me to be a 4-inch gun at extreme range. I began to trot at first, but finding shells bursting well over I pulled back to a walk, determined not to run away. Just as I left St Martin, the shelling ceased. Here I met Driver Gowans coming up with a GS wagon, and stopped him to tell him he would have to do two trips, as the infantry wagons had failed to come. No shells had fallen for the last five minutes, since those which had passed over my head a few hundred yards back.

As I spoke to Gowans, I think I heard for a second a distant shell whine, then felt a tremendous explosion almost on top of me. For an instant I appeared to rise slowly into the air and then slowly to fall again. I seemed to have dimly heard the rattle of wagon wheels and then for a moment I saw my horse's neck in front of my face.

I dropped off to the ground and set out at a half run toward Hénin. I must have been dazed, for I remembered nothing afterwards of the wagon or of where my poor horse had gone. Scarcely had I begun to run towards Hénin, when the floodgates in my neck seemed to burst, and the blood poured out in torrents. I could actually hear the regular swish of the artery, like a firehose, but coming and going in regular floods and pauses.

I was in a kind of dazed panic, deserted by all my bravado, and I cowered down as the shells whipped by and burst all around. Then

I got up and stumbled on as quick as I could. I had a vague idea that I might be going to die, but was not alarmed by it. At the crossroads to Hénin, no traffic man was to be seen, but beyond it some artillery wagons were waiting for the shelling to cease, before trying to pass.

I could not speak, but I paused in the middle of the road, and gave one or two sobbing groans, whereupon the traffic man appeared, from where he had been crouching in a shell-hole to avoid the shells. He called to a gunner driver to watch the post, and led me a little further down the road to a dressing station in an old cellar under a mound of bricks. I could feel something long lying loosely in my left cheek, as though I had a chicken bone in my mouth. It was in reality half my jaw, which had been broken off, teeth and all, and was floating about in my mouth.

I sat on the table in the cellar, while they dressed my wound. The RAMC [Royal Army Medical Corps] orderly put some plug into my neck which stopped the bleeding. They also put a rubber tube in my wound, sticking out of the bandage. They told me there was no ambulance in Hénin and I should have to walk to Boiry-Becquerelle. We accordingly set out, I leaning on the medical orderly's arm. I was not looking forward to the long walk at all, but luckily the orderly remembered that there was some regimental medical officer, who lived in a dugout at the south end of the village. We turned in there and I sat down on a stone at the entrance to the dugout.

This doctor said it was all rubbish not getting an ambulance, and sent the orderly back to the dressing station to telephone to Héninel for one. He took my temperature and said that I was all right for the moment, but I heard him tell the orderly that it was a good thing they dressed me at once, or I should have been done for. I felt no anxiety about whether I should live or die, but I was very cold, and the broken pieces of jawbone in my mouth were unpleasant. I felt no pain.

I gave the doctor my name and unit, and told him that our camp was on the Neuville–Vitasse road, next to the battalion in reserve. I wrote down, 'Please let them know,' for I could not speak. I heard later that the doctor sent the company a telegram, saying that I was badly wounded and was not expected to live.

At last the ambulance arrived and we set off. I was horribly cold, which I conveyed to the medical orderly by signs and he put a blanket round me. We went through Boiry-Becquerelle to the main Casualty Clearing Station (No. 20 CCS) at Ficheux. Here they helped me out and into a chair, when a doctor came up and said, 'What's the matter here, old man?' and took off and redid my bandage. I was put into the 'pending operations' ward, and slept like a log till the morning.

Early next morning I was dressed for the slaughter in long woollen stockings and laid in a line of stretchers waiting for operating. Somebody gave me an injection of morphia, and then two orderlies came up and said, 'Come on, this one will do first.' So they picked up my stretcher and bore me out, along the duckboard walks, with a steady bobbing up-and-down motion. Lying on my back, I looked up at the blue sky and the white drifting clouds.

Then into a hut all white inside, with a row of white operating tables down the centre and white-aproned doctors and nurses moving about. They held up my stretcher and I crawled over on to the table and lay down. One or two of them came to look at me, and then the anaesthetist came up, and told me to breathe deeply through my nose. At a word from the surgeon, he put the mask on my face and I smelt that suffocating sickly smell of gas. Once or twice I felt I would suffocate and longed to pull it off. Then my head began to sing, and a tap of water which was dripping seemed to grow louder and louder. I began saying to myself 'I'm still awake! Yes, of course, I must be.' The tap grew louder and louder, and beat all through my head. For a moment, the man took off the mask, and a voice said, 'How old are you?' I tried to say twenty, and he put the mask back. The singing and the tap grew louder – and then nothing.

When I came round again, Dad was beside my bed. I was almost perfectly conscious at once, and I remember writing down on a piece of paper, 'This rather spoils our leave.' We were both due for leave to England and had been trying to arrange to go together on 24 August. Colonel Rathbone had telephoned II Army that morning to say I was hit.

When Dad came into the mess for lunch, his staff officers, Colonel Stevenson and Major de Fonblanque, told him they had a message for him, which they would tell him after lunch. Not suspecting anything, he had a good lunch, and then they told him that I had been hit and that they had ordered his car to be ready, knowing that he would want to come and see me. So he drove down at once, but he was not allowed to stay long.

I remained half alive for several days, lying still all day only semi-conscious. I asked for a book to read but found I could not read it. I had apparently nearly swallowed my tongue during the operation and, to prevent this, they had pierced my tongue and threaded a wire through it with a wooden rod on the end of it. This was extremely uncomfortable. A good deal of discharge came from my mouth, and I was very miserable, with my pillow always covered with blood and slime. I was later told that I looked very bad, with my mouth dragged down, discharging and filthy, and with my head and neck all bandages.

The CCS was made entirely of marquees and tents, and was comfortable, considering the circumstances. The officers' ward consisted of three or four marquees, placed side by side, with boarded floors, and rows of beds with coloured counterpanes, and looked neat and pretty.

BRIGADIER-GENERAL E. L. SPEARS
Prelude to Victory

Edward Spears was a regular officer of the British army who, because of his family background, spoke perfect French. At the outbreak in 1914 he was a lieutenant in the 11th Hussars, attached as a liaison officer to the French army. During the campaign of the Marne, he acted as an interpreter and intermediary between General Lanrezac, commanding the French Fifth

Army, and Field Marshal Sir John French, commanding the British Expeditionary Force, a passage in his life brilliantly described in his autobiographical account of the campaign, *Liaison 1914*. By 1917 he was a senior staff officer, still largely concerned with relations between the French and British commands. In the extracts from his second account of his war service, *Prelude to Victory*, that follow, he narrates first his observation of the successful British attack at the Battle of Arras in April 1917, then the disastrous French assault ('the Nivelle Offensive') on the Chemin des Dames the following month. So disastrous was the Nivelle Offensive that in the aftermath fifty divisions of the French army, half its strength, refused to return to the attack. These 'mutinies', akin to military strikes, were to throw the weight of the military effort on the Western Front on to the British for the rest of the year.

*   *   *

Two hours before zero, in the caves of Arras and the tunnels of Vimy an occasional flicker from a distant light showed faces grey and still against the inky distorted shadows. As the light went out again, the shadows jumped back into the uneven walls from which they had emerged, the faces disappeared into the rising tide of darkness on which they had seemed to float for a moment, and the great silence continued.

What were all these men thinking about now the hour that would be the last for so many was near? Doubtless just the ordinary thoughts that skim across the strange, thousand-faceted human mind. The very young were probably strained and excited, concealing with difficulty their nervousness, while the others, the older hands, grown fatalists by experience, thought little, fatigue playing pranks with fancy, conjuring up incongruous or pleasing memories, pictures flitting against a background of unreality. Some perhaps brooded as they evoked a future in which they would have no part. There may have been others who could have cried aloud as, wedged in between elbows and weapons, they conjured up pictures of open spaces, gardens, light, the sound of water and the breeze in English trees, small arms outstretched asking to be picked up, a face looking upwards with frightened eyes at the moment of parting; but the instinctive self-discipline of the race sealed all those lips, placed a mask on all those faces. The sudden flash of a torch, the light of a passing lantern revealed nothing; expressions of weariness or boredom, that was all.

There was almost complete silence save in some privileged corners where the reserves lay hidden, notably behind the Canadian lines where here and there tightly wedged groups round a candle played poker for unusually high stakes, the pessimists betting recklessly in paper chits.

In the assault trenches outside, the men, without greatcoats, huddled together for warmth, heads drooping under heavy steel helmets. Occasionally the rattle of equipment followed by the muffled call of a sergeant for 'less noise' disturbed the silence that hung like a curtain over the front lines in the short intervals of the bombardment.

A cold moon just past the full showed at moments between huge, black, fantastically shaped clouds racing each other to obscure it. Sometimes the desolate space of no-man's-land would stand out much as a muddy lane must appear to an ant, a lane fringed by a forest whose trees were pickets and whose interlaced branches were strands of wire.

Everywhere glistening mud and the sparkling mirrors of thousands of water-filled shell-holes reflected the moon. The smoke of an explosion twirling in high convolutions over the enemy's line would show amber edges, a high cloud would be outlined for a few seconds in blinding silver, then all was pitch darkness again.

The trench-mortars in the front line kept on their exasperating endless barking. Terriers of war, they yapped ceaselessly at the flitting moon. Machine-guns chattered intermittently like a man in a fever. Only very occasionally came the crash of an answering German shell.

The weather was getting colder. An icy drizzle was falling driven by a strong north-west wind. The shelling was not as heavy as it had been during the periods of intense bombardment, but nevertheless the impression of continuous pounding was frightful. A persistent bass formed a background of rumour to the cacophony nearer at hand.

To those standing on the Arras–St Pol road it seemed as if, far away, muffled drums were being beaten continuously. The constant sinister output of noise, pitched for long periods on one monotonous note, would sink periodically to an angry staccato muttering, then swell to a great onrushing volume of sound like a hurricane in a forest. To one advancing towards the assault trenches with the heavies behind him and the field guns near at hand, each individual salvo was perceptible, however rapid the succession of explosions. The gradation of sound ranged from that of distant trip-hammers to the crash of slamming doors, then swelled to the wild stridency of furiously beaten anvils, dominated and submerged every few moments by thunderclaps from nearby batteries, until close to the front line it ceased being a sound at all and became just a succession of sudden shattering blows of indescribable brutality.

Half an hour to go. The order to fix bayonets is now passed along. Much uneasy movement, shifting of position, a continuous clicking, miles of clicking, as thousands of bayonets are pressed home over rifle-barrels.

Ten minutes, five minutes, two minutes before zero. Every officer, head bent, gazes intently at his watch. Not a thought but of his job now.

Only a few seconds to go, then suddenly a complete silence, an absolute cessation of the immense roar, a stillness punctuated and emphasized by the barking of the trench-mortars up and down the lines; every gun had stopped firing.

That sudden silence was more terrifying than the most reverberating explosion. It had the effect of making men feel they were losing their balance on the edge of an abyss. No typhoon uprooting a tropical forest, no storm with the lightning crackling in the high mountains could give so complete an impression of all-embracing power. Only an earthquake with its sudden balance-wrecking movement underfoot could cause a comparable anguish.

It did not last long. At 5.30 to the second the earth shook as the mines exploded with a muffled roar and every gun on the fifteen-mile front of attack and beyond it opened fire with a clamour such as had probably never been heard in the world since mountains were raised from its molten surface. The air screamed as it was torn by a thousand shells. Miles up the great projectiles hummed their mighty drone. Lower down through each layer of air the shells flew according to their kind, until, quite low above the lines of men closing in behind the barrage, the missiles of the light mortars and the bullets of the machine-guns hissed.

Behind the infantry waves, hundreds of flashes a minute came from the supporting guns. In front the blinding many-coloured flames of explosions made the enemy's line appear to be burning like a furnace. Out of a skeleton-like wood between our heavy batteries and the front line a great flight of rooks arose circling in wild panic. The light was dim. Just enough to see your way, not clear enough to aim a rifle or a machine-gun save at the closest range.

Within three minutes of the time it took our men to form up behind the barrage, a new kind of illumination was added to the fantastic scene. For miles upon miles, all along the German lines hundred of flares went up. Red, white, orange, the distress signals shot high, falling back in sprays of sparkling multi-coloured rain. The German infantry was begging for support. The British were upon them.

At the first lift in the barrage the advance began. Our infantry moved steadily forward. The ground was broken here and there by enormous, impassable mine-craters, everywhere pitted with shell-holes full of gluey mud and water, in places two or three feet deep. Nevertheless, wave after wave clambered out of the trenches and made their way in astonishingly good order. The Lewis gunners had a particularly hard time. Their heavy weapons on their shoulders, they fell frequently, many of their guns becoming choked with mud and useless for the time being.

In spite of difficulties and obstacles the men pressed on eagerly, squelching and splashing in the water, as fast as they could drag their mud-clogged feet. So keen were they that, it was ascertained later, many casualties were caused by the leading waves rushing into their own barrage.

Over a scene of desolation, of flame and smoke such as Doré never dreamt of, the red sun rose. The jagged silhouette of Arras appeared on the right, on the left was the vague outline of the Vimy Ridge. The drizzle changed to a heavy rainstorm mixed with snow. Wisps of icy mist trailed across the trenches on which they seemed to catch for a moment like a veil. Through this, flying low, the sudden noise of their engines so close overhead that men stopped and looked up, our planes appeared making straight for the enemy lines. In the bad light the special markings of the infantry contact planes were hardly discernible, but the long fluttering streamers could easily be seen.

Few casualties were suffered from enemy shells. The German barrage was late, from eight to fifteen minutes after ours, and what there was of it was ragged and desultory; also it came down in a zone vacated by the attackers. It was evident that the opposing artillery had been completely mastered. Except for the heavies, on many parts of the front the enemy guns presently ceased firing altogether.

We now know that our bombardment and gas had prevented ammunition coming up and that the German batteries were short of shells. One of the reasons for this was that the German Command had decided against having large dumps near the guns, for fear these might cause undue expenditure of ammunition. They paid for their mistaken parsimony in lives and in munitions too, since the helpless guns were unable to hinder our counter-battery work, which time after time blew up what stores of shells their guns had left. Moreover, in the rare cases when the hostile batteries were not overwhelmed by our fire and had some munitions available, the gunners had no idea what the situation was, owing to the destruction of [field-telephone] cables.

There were very few incidents in the attack on the Black Line and hardly any mistakes: it fell to us in the main at exactly the moment prescribed, although here and there nests of Germans held out.

The only major exception occurred on the Canadian left (4th Canadian Division) where the advance of the centre was held up. Here the first waves passed over their objectives probably without recognizing them. Supports in the centre lost direction: the Bavarians emerging from their dug-outs swarmed back into their trenches, pouring a murderous fire into the Canadians and indulging in frequent counter-attacks. Nevertheless the flanks of the division gained their objective. At some other points much the same thing occurred but with less serious results.

Whenever, owing to bad light or obliterated trenches, our moppers-up failed to locate the narrow entrances to the deep dug-outs which formed part of the German front-line system, the defenders were out in a flash firing into the backs of our advancing men. When this occurred our losses were always heavy, but in most cases our moppers-up located the entrances and dealt with the occupants who did not surrender. Very few of the Germans who showed resistance survived.

The Leinsters of the 24th Division on the extreme left met with the fiercest opposition. They had attacked earlier, under terribly difficult conditions. In sleet and snow the officers guided the men by compass over incredibly bad ground towards unrecognizable objectives. The Irish suffered some casualties from our own guns, and very severe losses from the enfilade fire coming from across the Souchez valley. A Homeric hand-to-hand struggle took place on the slippery ground of the bleak snow-covered hill. The Lewis gunners wielded their weapons, rendered useless by mud, as enormous clubs. It was a fight to the finish; few prisoners were taken, but at the end the Leinsters were masters of the German first line. A few men reached the second line which was obliterated, but the Germans were holding in strength the natural cover in rear. The position was untenable and our men in the second line were later withdrawn to the first line, which was consolidated. On the greater part of the front, the less bellicose, that is the majority, of the defenders were soon streaming back towards our lines, often unaccompanied, to be collected later in the waiting [prisoner-of-war] cages. On their way they passed the oncoming troops who, mildly curious but totally devoid of animosity, pestered the poor wretches as was their habit for 'souvenirs', though they themselves had their day's work before them.

The small resistance encountered, save at some points, is not to be wondered at. The astonishing thing is that there were any men left to defend that first shattered system. The survivors, stunned and stupefied, surrendered freely. None can blame them. Rather must one wonder at the extraordinary pluck of those who showed fight. For days now in the front line even the most urgent work had had to be abandoned. Companies reduced to seventy or eighty men, holding fronts of three to four hundred yards, had spent the last six days, in most cases without relief, huddled in the bottom of insanitary dug-outs, every now and then called upon to make desperate efforts to relieve comrades trapped in shelters whose entrances had been blown in. The trenches were half full of water, the men were stiff with cold yet many were shaking with fever.

To the defenders it seemed as if the British shelling hardly ever diminished in intensity during the preparation. There was never a moment's rest nor a second's respite. Prisoners from one battery said they had been unable to leave their dug-outs or fire their guns for four days. Food was short. In ordinary times from the sheltering eastern slope of Vimy it took ration parties only fifteen minutes to reach the front line. It had taken them six hours during the last few days, and very often they disappeared, never to be heard of again. In many cases none had come up for ninety-six hours. The roads leading to Vimy Ridge were impassable, and the defenders had been all but completely cut off for days by the iron curtain of our guns. In whole sectors the front-line garrison had to fall back on their iron rations which they ate in fear under the trembling earth. It took runners hours to reach their destinations, if they did so at all. Practically every other form of communication had broken down. The Germans speak of a symphony of hell, and so it must have appeared to them. In Douai, fifteen miles away, every pane of glass was shattered by the terrific reverberations of that distant storm. The villages behind Vimy Ridge, hitherto peaceful places, were drawn into the conflict. Great shells roared over or crashed into them without ceasing. The

Germans could see at night the flash of our guns on the Lorette Ridge continuously lighting up the ruins of Mont St Éloi, Bouvigny Wood, Berthonval, Maison Blanche and the skeleton trees on the Arras road, with hardly an answering flash from their guns.

To these men zero hour came almost as a relief. It had been long awaited. Few can have expected to survive. 'When the English attack tomorrow,' wrote Lieutenant Runge of the 79th Reserve Division to a brother officer on the evening of the 8th, 'dear Hoinicke, you will see me no more.' . . .

Almost at once, or so it seemed, the immense mass of troops within sight began to move. Long thin columns were swarming towards the Aisne. Suddenly some 75s [French 75-mm field guns] appeared from nowhere, galloping forward, horses stretched out, drivers looking as if they were riding a finish. 'The Germans are on the run, the guns are advancing,' shouted the infantry jubilantly. Then it started to rain and it became impossible to tell how the assault was progressing.

I advanced farther, trying to see. I had reached a broken-down wall when there was a deafening explosion. Stones and earth flew in every direction as I flung myself or was flung down. Getting up I peered round the wall. An officer was standing there, men were running towards him, several bodies horribly mutilated lay about. The officer began to speak. He was very white and trembling. He had been talking to those others, he named them, when suddenly they had been felled like oxen. He spoke fast in a staccato voice. Suddenly blood began to drip from his sleeve and a large red stain widened on his side. His eyes all at once showed white. Two of his men caught him. I walked on, only to find that I had lost my field-glasses (the strap had been broken probably by a flying stone) and had to come back to the horrible spot and the maculated wall to look for them.

I only knew later, in some cases much later, what was happening in front and to either side of me. I never saw the heart-breaking spectacle of the Senegalese attack. We had been taught to believe theirs would be a headlong assault, a wild savage onrush. Instead, paralysed with cold, their chocolate faces tinged with grey, they reached the assault trenches with the utmost difficulty. Most of them were too exhausted even to eat the rations they carried, and their hands were too cold to fix bayonets. They advanced when ordered to do so, carrying their rifles under their arms like umbrellas, finding what protection they could for their frozen fingers in the folds of their cloaks. They got quite a long way before the German machine-guns mowed them down.

Only slowly was it possible to get some idea of what was happening. In a dug-out in which were some preoccupied senior gunners, I unfolded my map showing the lines the attack was to reach at given hours. It was obvious that the troops were far behind their timetables. Evidently they had not been able to keep up with the insane pace of the barrage.

My map showed the ten-kilometre bound the French Armies were expected to make beyond the German defensive system. An arrow showed the line of advance of the Fifth Army towards Sissonne, another that of the Sixth Army on Laon. The Tenth Army, whose troops I had seen filling every nook and cranny of the country I had left behind me, was represented by a big blob on the map. A third arrow between the two others showed the direction in which it was to advance. Attacking through and with these, it was to carry them forward by the impetus of its immense weight, so that pressure should not be relaxed for a moment, nor the enemy given time to recover owing to a halt of the first-line troops on the conquered positions. The whole plan of the attack was staring at me from the map crudely lit by the smelly and flickering flare of a naked acetylene lamp. Brimont Fort dominating Reims was marked as falling within four hours, captured by a turning movement from the north. But the point which fixed attention unconsciously, for it had gripped the imagination of both the public and the army, was that portion of the Chemin des Dames which extended from the historic farm of Heurtebise to Craonne, along the plateau on the northern slopes of which the Forest of Vauclerc spread its thick undergrowth. It was to the I Corps, which had a magnificent fighting reputation, that the honour fell of attacking Craonne, while the II Colonial Corps, largely composed of [General] Mangin's blacks, the Senegalese, was to carry Vauclerc.

The delay reported by the gunners was disturbing as well as disappointing. All the first accounts agreed that the attack had started well, and that in all cases the French troops had raced for the enemy. No trumpets or drums cheered them on with martial blare and urgent beat as at Charleroi. Instead, the infantry dashed forward to the explosive tune of the 75s, whose lashing shells seemed like whips compelling forward waves that needed no urging. The field guns were firing in such a frenzy of speed that madness and noise appeared to have become synonymous terms, and to their insane tempo the French infantry advanced, seemingly irresistible, gaining ground at great speed, save at some points where German machine-guns mowed them down to a man. This was the case of the splendid 1st Division of the I Corps, so long our neighbour on the Somme, which was stopped dead by machine-gun fire opposite Craonne. As they left their trenches, the first waves were felled on the parapets.

When every unit displayed such magnificent qualities of courage and determination, it seems invidious to pick out any division as having specially distinguished itself; nevertheless even amongst so much gallantry it would seem that by common consent the 40th Division, which literally hurled itself at the German defences in front of Mont Sapigneul, and the 14th Division, which rushed forward in a way that caused all beholders to wonder, are worthy of special mention.

The headlong pace of the advance was nowhere long maintained. There was a perceptible slowing down, followed by a general halt of the supporting troops which had been pressing steadily forward since zero hour. German machine-guns, scattered in shell-holes, concentrated in nests, or appearing suddenly at the mouths of deep dug-outs and *creutes* [chalk caves], took fearful toll of the troops now labouring up the rugged slopes of the hills.

In a very short time after the attack was launched, the I Colonial Corps on the left, facing the woods west of Anizy-le-Château, found its waves broken, thinned and delayed, as it progressed through the forest. It was impossible to keep up with the barrage. When the attacking infantry came up to the German trenches, it found them protected by felled trees intermingled with barbed wire. The French heavies had turned these obstacles into impassable obstructions. As everywhere else, the barrage was timed to move forward at far too fast a rate, and the men were handicapped in their movements by the three-days' rations with which they were loaded to sustain them in the territory to be captured beyond the hills.

Everywhere the story was the same. The attack gained ground at most points, then slowed down, unable to follow the barrage which, progressing at the rate of a hundred yards in three minutes, was in many cases soon out of sight. As soon as the infantry and the barrage became dissociated, German machine-guns were conjured as if by magic from the most unlikely places and opened fire, in many cases from both front and flanks, and sometimes from the rear as well, filling the air with a whistling sound as if of scythes cutting hay. On the steep slopes of the Aisne, the troops, even unopposed, could only progress very slowly. The ground, churned up by the shelling, was a series of slimy slides with little or no foothold. The men, pulling themselves up by clinging to the stumps of trees, were impeded by wire obstacles of every conceivable kind.

'We were faced', ran one report, 'not by barbed-wire entanglements but by a forest of wire . . . Machine-guns appeared everywhere from the hidden mouths of concealed caves, there were traps of every description, the ground was apparently impassable.'

Meanwhile the supporting troops were accumulating in the assault trenches at the rate of a fresh battalion every quarter of an hour. As the leading waves were held up, in some cases a few hundred yards and seldom as much as half to three quarters of a mile ahead, this led to congestion and in some cases to the greatest confusion. Had the German guns been as active as their machine-guns, the massacre which was going on in the front line would have been duplicated upon the helpless men in the crowded trenches and on the tracks in rear.

General Mangin, inevitably unaware of the exact situation but informed that the attack was not progressing according to plan, was using every endeavour to press the troops on. At 8.10 a.m., he ordered the guns to move forward to make up for lost time. A little later, when the VI Corps facing the Chemin des Dames had been brought to a standstill, and the Corps Commander ordered the artillery preparation to be resumed, General Mangin issued the following order: 'The resumption of the preparation is a bad solution. It proves that the troops hesitate to advance. Our artillery preparation cannot have allowed the enemy to establish a continuous line of machine-guns. You must take advantage of the gaps and pass beyond the islands of resistance.' This the doomed troops attempted to do, the survivors courageously pressing on where they could, all unconscious that their Commander thought they were 'hesitating to advance'. The General issued another order that 'where the wire is not cut by the artillery it must be cut by the infantry. Ground must be gained,' but this order did not reach the soldiers whose bodies now hung grotesquely on the German wire.

The I Colonial Corps was ordered on in the same way by the imperious Army Commander, only in the end, after reaching Laffaux Mill, to be swept back into its assault trenches.

Only one piece of good news was received from this sector: the Corps artillery commander reported that the one local success obtained, the occupation of a hill called the Mont des Singes, lost later, had been achieved thanks to the heavy and accurate fire of British guns lent to the French. This result was all the more satisfactory as, according to this report, the British batteries had been subjected to very violent shelling. At the end of the day they had only two hundred rounds left.

W. B. YEATS
An Irish Airman Foresees His Death

I know that I shall meet my fate
Somewhere among the clouds above;
Those that I fight I do not hate,
Those that I guard I do not love;
My country is Kiltartan Cross,
My countrymen Kiltartan's poor,
No likely end could bring them loss
Or leave them happier than before.
Nor law, nor duty bade me fight,
Nor public men, nor cheering crowds,
A lonely impulse of delight
Drove to this tumult in the clouds;
I balanced all, brought all to mind,
The years to come seemed waste of breath,
A waste of breath the years behind
In balance with this life, this death.

Note: Major Robert Gregory was the son of Yeats's friend and patron, Augusta, Lady Gregory. Awarded the MC and appointed to the Légion d'Honneur, Gregory was shot down and killed on the Italian front, 23 January 1918.

ROBERT GRAVES
Goodbye to All That

Robert Graves, who survived the First World War to become one of the most famous English poets and novelists of the twentieth century, served in it as an officer of the 2nd Battalion,

Royal Welch Fusiliers, the same regiment in which his friend, the poet Siegfried Sassoon, soldiered. *Goodbye to All That*, his memoir of the trenches, has become one of the most celebrated of Great War autobiographies. It was, in fact, written at great speed in order to fulfil an obligation to a publisher. Perhaps because it is a work dashed off almost without reflection, it has a directness, vitality and often morbidity missing from more deliberate compositions.

*   *   *

The troops with the worst reputation for acts of violence against prisoners were the Canadians (and later the Australians). The Canadians' motive was said to be revenge for a Canadian found crucified with bayonets through his hands and feet in a German trench. This atrocity had never been substantiated; nor did we believe the story, freely circulated, that the Canadians crucified a German officer in revenge shortly afterwards. How far this reputation for atrocities was deserved, and how far it could be ascribed to the overseas habit of bragging and leg-pulling, we could not decide. At all events, most overseas men, and some British troops, made atrocities against prisoners a boast, not a confession.

Later in the war, I heard two first-hand accounts.

A Canadian-Scot: 'They sent me back with three bloody prisoners, you see, and one started limping and groaning, so I had to keep on kicking the sod down the trench. He was an officer. It was getting dark and I felt fed up, so I thought: "I'll have a bit of a game." I had them covered with the officer's revolver and made 'em open their pockets without turning round. Then I dropped a Mills bomb in each, with the pin out, and ducked behind a traverse. Bang, bang, bang! No more bloody prisoners. No good Fritzes but dead 'uns.'

An Australian: 'Well, the biggest lark I had was at Morlancourt, when we took it the first time. There were a lot of Jerries in a cellar, and I said to 'em: "Come out, you Camarades!" So out they came, a dozen of 'em, with their hands up. "Turn out your pockets," I told 'em. They turned 'em out. Watches and gold and stuff, all dinkum. Then I said: "Now back to your cellar, you sons of bitches!"

For I couldn't be bothered with 'em. When they were all safely down I threw half a dozen Mills bombs in after 'em. I'd got the stuff all right, and we weren't taking prisoners that day.'

An old woman at Cardonette on the Somme gave me my first-hand account of large-scale atrocities. I was billeted with her in July 1916. Close to her home, a battalion of French Turcos [North African native infantry] overtook the rearguard of a German division retreating from the Marne in September 1914. The Turcos surprised the dead-weary Germans while still marching in column. The old woman went, with gestures, through the pantomime of slaughter, and ended: 'Et enfin, ces animaux leur ont arraché les oreilles et les ont mises à la poche!' ['And finally those animals cut off their ears and put them in their pockets!']

The presence of semi-civilized coloured troops in Europe was, from the German point of view, we knew, one of the chief Allied atrocities. We sympathized. Recently, at Flixécourt, one of the instructors told us, the cook of a corps headquarters mess used to be visited at the château every morning by a Turco – the orderly to a French liaison officer. The Turco used to say: 'Tommy, give Johnny pozzy,' and got his tin of plum-and-apple jam.

One day the corps had orders to shift by the afternoon, so the cook told the Turco, giving him his farewell tin: 'Oh la, la, Johnny, napoo pozzy tomorrow!'

The Turco would not believe it. 'Yes, Tommy, mate,' he insisted, 'pozzy for Johnny tomorrow, tomorrow, tomorrow!'

To get rid of him, the cook said: 'Fetch me the head of a Fritz, Johnny, tonight. I'll ask the general to give you pozzy tomorrow, tomorrow, tomorrow.'

'Right, mate,' said the Turco, 'me get Fritz head tonight, general give me pozzy tomorrow.'

That evening the mess cook of the new corps that had taken over the château found a Turco asking for him and swinging a bloody head in a sandbag. 'Here Fritz head, mate,' said the Turco, 'general give me pozzy tomorrow, tomorrow, tomorrow.'

As Flixécourt lay more than twenty miles behind the line . . .

We discussed the continuity of regimental morale. A captain in a line battalion of a Surrey regiment said: 'Our battalion has never recovered from the first battle of Ypres. What's wrong is that we have a rotten depot. The drafts are bad, and so we get a constant re-infection.' He told me one night in our sleeping hut: 'In both the last two shows I had to shoot a man of my company to get the rest out of the trench. It was so bloody awful, I couldn't stand it. That's why I applied to be sent down here.' Thus was the truth, not the usual loose talk that one heard at the Base. I felt sorrier for him than for any other man I met in France. He deserved a better regiment.

The boast of every good battalion was that it had never lost a trench; both our line [regular] battalions made it – meaning, that they had never been forced out of a trench without recapturing it before the action ended. Capturing a German trench and being unable to hold it for lack of reinforcements did not count; nor did retirement by order from headquarters, or when the battalion next door had broken and left a flank in the air. And, towards the end of the war, trenches could be honourably abandoned as being wholly obliterated by bombardment, or because not really trenches at all, but a line of selected shell-craters.

We all agreed on the value of arms drill as a factor in morale. 'Arms drill as it should be done,' someone said, 'is beautiful, especially when the company feels itself as a single being, and each movement is not a synchronized movement of every man together, but the single movement of one large creature.' I used to get big bunches of Canadians to drill: four or five hundred at a time. Spokesmen stepped forward once and asked what sense there was in sloping and ordering arms, and fixing and unfixing bayonets. They said they had come across to fight, and not to guard Buckingham Palace. I told them that in every division of the four in which I had served – the First, Second, Seventh, and Eighth – there were three different kinds of troops. Those that had guts but were no good at drill; those that were good at drill but had no guts; and those that had guts and were good at drill. These last, for some reason or other, fought by far the best when it came to a show – I didn't know why, and I didn't care. I told them that when they were better at fighting than the Guards they could perhaps afford to neglect their arms drill.

We often theorized in the mess about drill. I held that the best drill never resulted from being bawled at by a sergeant-major: that there must be perfect respect between the man who gives the order and the men that carry it out. The test of drill came, I said, when the officer gave an incorrect word of command. If the company could, without hesitation, carry out the order intended or, if the order happened to be impossible, could stand absolutely still, or continue marching, without confusion in the ranks, that was good drill . . . Some instructors regarded the corporate spirit that resulted from drilling together as leading to loss of initiative in the men drilled.

Others argued that it acted just the other way round: 'Suppose a section of men with rifles get isolated from the rest of the company, without an NCO in charge, and meet a machine-gun. Under the stress of danger this section will have that all-one-body feeling of drill, and obey an imaginary word of command. There may be no communication between its members, but there will be a drill movement, with two men naturally opening fire on the machine-gun while the remainder work round, part on the left flank and part on the right; and the final rush will be simultaneous. Leadership is supposed to be the perfection for which drill has been instituted. That's wrong. Leadership is only the first stage. Perfection of drill is communal action. Though drill may seem to be antiquated parade-ground stuff, it's the foundation of tactics and musketry. Parade-ground musketry won all the battles in our regimental histories; this war, which is unlikely to open out, and must almost certainly end with the collapse, by "attrition", of one side or the other, will be won by parade-ground tactics – by the simple drill tactics of small units fighting in limited spaces, and in noise and confusion so great that leadership is quite impossible.' Despite variance on this point we all agreed that regimental pride remained the strongest moral force that kept a battalion going as an effective fighting unit; contrasting it particularly with patriotism and religion.

Patriotism, in the trenches, was too remote a sentiment, and at once rejected as fit only for civilians, or prisoners. A new arrival who talked patriotism would soon be told to cut it out. As 'Blighty', a geographical concept, Great Britain was a quiet, easy place for getting back to out of the present foreign misery; but as a nation it included not only the trench-soldiers themselves and those who had gone home wounded, but the staff, Army Service Corps, lines of communication troops, base units, home-service units, and all civilians down to the detested grades of journalists, profiteers, 'starred' men exempted from enlistment, conscientious objectors, and members of the Government. The trench-soldier, with this carefully graded caste-system of honour, never considered that the Germans opposite might have built up exactly the same system themselves. He thought of Germany as a nation in arms, a unified nation inspired with the sort of patriotism that he himself despised. He believed most newspaper reports on conditions and sentiments in Germany, though believing little or nothing of what he read about similar conditions and sentiments in England. Yet he never underrated the German as a soldier. Newspaper libels on Fritz's courage and efficiency were resented by all trench-soldiers of experience.

Hardly one soldier in a hundred was inspired by religious feeling of even the crudest kind. It would have been difficult to remain religious in the trenches even if one survived the irreligion of the training battalion at home. A regular sergeant at Montagne, a Second Battalion man, had recently told me that he did not hold with religion in time of war. He said that the niggers (meaning the Indians) were right in officially relaxing their religious rules while fighting. 'And all this damn nonsense, sir – excuse me, sir – that we read in the papers, sir, about how miraculous it is that the wayside crucifixes are always getting shot at, but the figure of our Lord Jesus somehow don't get hurt, it fairly makes me sick, sir.' This was his explanation why, when giving practice fire-orders from the hilltop, he had shouted, unaware that I stood behind him: 'Seven hundred, half left, bloke on cross, five rounds, concentrate, FIRE!' And why, for 'concentrate', he had humorously substituted 'consecrate'. His platoon, including the two unusual 'bible-wallahs' whose letters home always began in the same formal way: 'Dear Sister in Christ' or 'Dear Brother in Christ', blazed away.

The troops, while ready to believe in the Kaiser as a comic personal devil, knew the German soldier to be, on the whole, more devout than himself. In the instructors' mess we spoke freely of God and Gott as opposed tribal deities. For Anglican regimental chaplains we had little respect. If they had shown one-tenth the courage, endurance, and other human qualities that the regimental doctors showed, we agreed, the British Expeditionary Force might well have started a religious revival. But they had not, being under orders to avoid getting mixed up with the fighting and to stay behind with the transport. Soldiers could hardly respect a chaplain who obeyed these orders, and yet not one in fifty seemed sorry to obey them. Occasionally, on a quiet day in a quiet sector, the chaplain would make a daring afternoon visit to the support line and distribute a few cigarettes, before hurrying back. But he was always much to the fore in rest-billets. Sometimes the colonel would summon him to come up with the rations and bury the day's dead; he would arrive, speak his lines, and shoot off again. The position was complicated by the respect that most commanding officers had for the cloth – though not all. The colonel in one battalion I served with got rid of four new Anglican chaplains in four months; finally he applied for a Roman Catholic, alleging a change of faith in the men under his command. For the Roman Catholic chaplains were not only permitted to visit posts of danger, but definitely enjoined to be wherever fighting was, so that they could give extreme unction to the dying. And we had never heard of one who failed to do all that was expected of him and more. Jovial Father Gleeson of the Munsters, when all the officers were killed or wounded at the first battle of Ypres, had stripped off his black badges [of the Army Chaplains Department] and, taking command of the survivors, held the line.

Anglican chaplains were remarkably out of touch with their troops. The Second Battalion chaplain, just before the Loos fighting, had preached a violent sermon on the Battle against Sin, at which one old soldier behind me grumbled: 'Christ, as if one bloody push wasn't enough to worry about at a time!' A Roman Catholic padre, on the other hand, had given his men his blessing and told them that if they died fighting for the good cause they would go straight to Heaven or, at any rate, be excused a great many years in Purgatory. When I told this story to the mess, someone else said that on the eve of a battle in Mesopotamia the Anglican chaplain of his battalion had preached a sermon on the commutation of tithes. 'Much more sensible than that Battle against Sin. Quite up in the air, and took the men's minds off the fighting.'

I felt better after a few weeks at [the base at] Harfleur, though the knowledge that this was merely a temporary relief haunted me all the time. One day I left the mess to begin the afternoon's work on the drill-ground, and passed the place at which bombing instruction went on. A group of men stood around a table where the various types of bombs were set out for demonstration. I heard a sudden crash. A sergeant of the Royal Irish Rifles had been giving a little unofficial instruction before the proper instructor arrived. He picked up a No. 1 percussion-grenade and said: 'Now lads, you've got to be careful here! Remember that if you touch anything while you're swinging this chap, it'll go off.' To illustrate the point, he rapped the grenade against the table edge. It killed him and the man next to him and wounded twelve others more or less severely.

ERNEST HEMINGWAY
Wounded (1)

Ernest Hemingway, America's greatest novelist of the twentieth century, served as a volunteer in the medical service on the Italian Front in 1918. He used what he learned during his service with the Italian army to write one of his finest novels, *Farewell to Arms*, which begins with a dramatic account of the disaster of Caporetto, in October 1917, when the Italian army in the Julian Alps was overwhelmed by an Austro-German offensive (in which the young Rommel played a leading part).

Hemingway was not at Caporetto but succeeded nevertheless in conveying the atmosphere of a military catastrophe in an unforgettable way. In this letter home he describes how he was wounded in the defensive fighting that followed the Caporetto retreat.

*   *   *

To his family, Milan, 18 August 1918

Dear Folks:

That includes grandma and grandpa and Aunt Grace. Thanks very much for the 40 lire! It was appreciated very much. Gee, Family, but there certainly has been a lot of burbles about my getting shot up!. . . I have begun to think, Family, that maybe you didn't appreciate me when I used to reside in the bosom. It's the next best thing to getting killed and reading your own obituary.

You know they say there isn't anything funny about this war. And there isn't. I wouldn't say it was hell, because that's been a bit overworked since Gen. Sherman's time, but there have been about 8 times when I would have welcomed Hell. Just on a chance that it couldn't come up to the phase of war I was experiencing. For example. In the trenches during an attack when a shell makes a direct hit in a group where you're standing. Shells aren't bad except direct hits. You must take chances on the fragments of the bursts. But when there is a direct hit your pals get spattered all over you. Spattered is literal. During the six days I was up in the Front line trenches, only 50 yds from the Austrians, I got the rep. of having a charmed life. The rep of having one doesn't mean much but having one does! I hope I have one. That knocking sound is my knuckles striking the wooden bed tray.

It's too hard to write on two sides of the paper so I'll skip.

Well I can now hold up my hand and say I've been shelled by high explosive, shrapnel and gas. Shot at by trench mortars, snipers and machine guns, and as an added attraction an aeroplane machine gunning the lines. I've never had a hand grenade thrown at me, but a rifle grenade struck rather close. Maybe I'll get a hand grenade later. Now out of all that mess to only be struck by a trench mortar and a machine gun bullet while advancing toward the rear, as the Irish say, was fairly lucky. What, Family?

The 227 wounds I got from the trench mortar didn't hurt a bit at the time, only my feet felt like I had rubber boots full of water on. Hot water. And my knee cap was acting queer. The machine gun bullet just felt like a sharp smack on my leg with an icy snow ball. However it spilled me. But I got up again and got my wounded [man he was helping back] into the dug out. I kind of collapsed at the dug out. The Italian I had with me had bled all over my coat and my pants looked like somebody had made current jelly in them and then punched holes to let the pulp out. Well the Captain who was a great pal of mine, It was his dug out said 'Poor Hem he'll be R.I.P. soon.' Rest In Peace, that is. You see they thought I was shot through the chest on account of my bloody coat. But I made them take my coat and shirt off. I wasn't wearing any undershirt, and the old torso was intact. Then they said I'd probably live. That cheered me up any amount. I told him in Italian that I wanted to see my legs, though I was afraid to look at them. So we took off my trousers and the old limbs were still there but gee they were a mess. They couldn't figure out how I had walked 150 yards with a load with both knees shot through and my right shoe punctured two big places. Also over 200 flesh wounds. 'Oh,' says I, 'My Captain, it is of nothing. In America they all do it! It is thought well not to allow the enemy to perceive that they have captured our goats!'

The goat speech required some masterful lingual ability but I got it across and then went to sleep for a couple of minutes. After I came to they carried me on a stretcher three kilometers to a dressing station. The stretcher bearers had to go over lots because the road was having the 'entrails' shelled out of it. Whenever a big one would come, Whee – whoosh – Boom – they'd lay me down and get flat. My wounds were now hurting like 227 little devils were driving nails into the raw. The dressing station had been evacuated during the attack so I lay for two hours in a stable, with the roof shot off, waiting for an ambulance. When it came I ordered it down the road to get the soldiers that had been wounded first. It came back with a load and then they lifted me in. The shelling was still pretty thick and our batteries were going off all the time way back of us and the big 250s and 350s [howitzer shells] going over head for Austria with a noise like a railway train. Then we'd hear the bursts back of the lines. Then there would come a big Austrian shell and then the crash of the burst. But we were giving them more and bigger stuff than they sent. Then a battery of field guns would go off, just back of the shed – boom, boom, boom, boom, and the Seventy-Fives or 149s would go whipping over to the Austrian lines, and the star shells going up all the time and the machines going like rivetters, tat-a-tat, tat-atat.

After a ride of a couple of kilometers in an Italian ambulance, they unloaded me at the dressing station where I had a lot of pals among the medical officers. They gave me a shot of morphine and an anti-tetanus injection and shaved my legs and took out about Twenty X shell fragments varying from [drawing of fragment] to about [drawing of fragment] in size out of my legs. They did a fine job of bandaging and all shook hands with me and would have kissed me but I kidded them along. Then I stayed 5 days in a field hospital and was then evacuated to the base Hospital here.

I sent you that cable so you wouldn't worry. I've been in the Hospital a month and 12 days and hope to be out in another month. The Italian Surgeon did a peach of a job on my right knee joint and right foot. Took 28 stitches and assures me that I will be able to walk as well as ever. The wounds all healed up clean and there was no infection. He has my right leg in a plaster splint now so that the joint will be all right. I have some snappy souvenirs that he took out at the last operation.

I wouldn't really be comfortable now unless I had some pain. The Surgeon is going to cut the plaster off in a week now and will allow me on crutches in 10 days.

I'll have to learn to walk again.

You ask about Art Newburn. He was in our section but has been transferred to II. Brummy is in our section now. Don't weep if I tell you that back in my youth I learned to play poker. Art Newburn held some delusions that he was a poker player. I won't go into the sad details but I convinced him otherwise. Without holding anything I stood pat. Doubled his openers and bluffed him out of a 50 lire pot. He held three aces and was afraid to call. Tell that to somebody that knows the game Pop. I think Art said in a letter home to the Oak Parkers that he was going to take care of me. Now Pop as man to man was that taking care of me? Nay not so. So you see that while war isn't funny a lot of funny things happen in war. But Art won the championship of Italy pitching horse shoes.

This is the longest letter I've ever written to anybody and it says the least. Give my love to everybody that asked about me and as Ma Pettingill says, 'Leave us keep the home fires burning!'

Good night and love to all.

Ernie

P. S. I got a letter today from the Helmles addressed Private Ernest H – what I am is S. Ten. or Soto Tenente Ernest Hemingway. That is my rank in the Italian Army and it means 2nd Lieut. I hope to be a Tenente or 1st Lieut. soon.

SIEGFRIED SASSOON
(1886–1967)
The Hero

'Jack fell as he'd have wished,' the Mother said,
And folded up the letter that she'd read.
'The Colonel writes so nicely.' Something broke
In the tired voice that quavered to a choke.
She half looked up. 'We mothers are so proud
Of our dead soldiers.' Then her face was bowed.

Quietly the Brother Officer went out.
He'd told the poor old dear some gallant lies
That she would nourish all her days, no doubt.
For while he coughed and mumbled, her weak eyes
Had shone with gentle triumph, brimmed with joy,
Because he'd been so brave, her glorious boy.

He thought how 'Jack', cold-footed, useless swine,
Had panicked down the trench that night the mine
Went up at Wicked Corner; how he'd tried
To get sent home, and how, at last, he died,
Blown to small bits. And no one seemed to care
Except that lonely woman with white hair.

GERALD ULOTH
Riding to War

Not all the events of the First World War were mud-stained
or waterlogged. On the fringes of the great conflict, campaigns
were fought that better belonged in spirit to the freebooting
days of tribal raiding and the imperial punitive expedition, than
to the grim business of bombardment and trench-to-trench
offensive.

One such campaign was conducted by the British in Southern
Persia which, although nominally an independent kingdom,
had been divided by them and the Russians into 'spheres of
influence'. It was, however, also used by the Germans, Austrians
and Turks as an area of special operations, in which they tried
to raise anti-Allied resistance and through which they infiltrated
agents to foment rebellion inside India and on its borders with
Afghanistan.

In an effort to control hostile activity in the region, Britain
recruited a local regiment, the South Persian Rifles, and organ-

ized the Eastern Persian Cordon, an unofficial frontier garrisoned by regiments from the (British) Indian Army. One of these was the 28th Light Cavalry, in which Gerald Uloth was then serving. His account of duty on the cordon has a timeless quality. The characters he describes might have skirmished against Alexander the Great on the campaign of conquest two thousand years earlier; their bad habits – raiding, looting and kidnapping, particularly kidnapping their enemies' women – were even older than that. (The Brigadier-General Dyer described here was later to be responsible for the killing or wounding of some 1,500 unarmed Indians at Amritsar in the Punjab in April 1919, for which he was removed from his command.) Uloth felt at home in these lawless circumstances and describes them brilliantly. He belonged to almost the last generation of British officers who were able to enjoy the cut-throat but curiously honourable war of the Indian frontiers that had occupied warriors since the beginning of recorded history.

*　　*　　*

The convoys carrying supplies to our troops in Eastern Persia travelled along the Seistan Trade Route from Nushki. Before crossing into Persia they passed through that narrow triangle of British India bordered on the north by Afghanistan and on the south by Persian Baluchistan. The convoys entered Persia at Kuh-e-Malek Siah, the point where the frontiers of Persia, Baluchistan (in India) and Afghanistan all meet. For many years in times of peace merchant caravans had passed safely along the Seistan Trade Route unescorted. The sight of our enormous military convoys of a thousand camels and more laden with all manner of delectable goods – clothing, food and, above all, arms and ammunition – offered a great temptation to the wild Baluch, the Damanis, whose name translates so picturesquely into 'The Dwellers in the Skirts of the Hills', who inhabited the mountainous country to the south of this British triangle. Their black tents could be found in all the country surrounding the great volcano, the Kuh-e-Taftan, and in the adjacent country, all of which was known as the Sirhadd. These tribesmen had owed allegiance to no government for years; they had resisted all efforts by the Persian Government to levy taxes. From their mountain fastnesses they aided gun-running parties to reach Afghanistan and made raids into the Persian plain.

In early 1916, as the result of Turco-German propaganda, they began to show unrest. Two Germans actually visited tribes to the south of the Sirhadd in 1915. The Damanis started to raid British convoys, at first only in a small way. They were very mobile. On their small fast-trotting camels they could cover great distances at speed and subsist for two or three days on a bag of dates, a little *atta* (coarse flour) and what water they could carry in a goat skin *mashak* or find in water holes and desert wells. It was only too easy for them to find a vulnerable spot in a convoy some three miles long, with an escort of only fifteen or twenty men, who were probably concentrated round some valuable part of the convoy such as treasure or arms. Their tactics were for a small party to lie in ambush and by firing a few shots to cause alarm and disorder amongst the unarmed camel drivers and, from amongst the resultant confusion, drag off a few loaded camels. They then hurried off into the heat haze and the hills, were lost to view and made for their mountain homes. Pursuit was impossible: there were no troops with which to pursue.

Success made them bolder; they soon started more extensive raids, more shooting into convoys, the killing of drivers and the carrying off of more camels. In an attack on one convoy they killed the Force Commander's new charger which had been sent out from India and looted the saddlery and blankets. Our Force Commander, Brigadier-General Dyer, was an Indian Army General of the old type, an infantryman, tough, courageous and strict. He decided that the time had now come to teach the rascals a lesson. Perhaps the loss of his charger was the deciding factor.

He collected from his small command a force of one squadron of Indian cavalry, a machine-gun section from the Indian infantry, a company of Indian infantry and two mountain guns. He augmented this force with some friendly tribesmen from the Reki tribe, some of whom lived on both sides of the border.

The force was first concentrated at Mirjawa, a British Frontier Levy Post. The General then sent friendly tribesmen into the Sirhadd with exaggerated reports of his strength. After allowing time to elapse for these reports to sink in he advanced boldly across the border from Mirjawa. His bluff was successful. Except for one small skirmish, in which the Baluch lost a few men and our force none, he reached Khash without opposition. Khash is the very heart of the Sirhadd, and the tribal centre of the Gomshadzais, who under their Chief, Jiand Khan, were the most numerous and powerful tribe in the Sirhadd. Jiand Khan and his lieutenants came in to make submission.

Shortly after this the squadron of which I was second-in-command was ordered to Khash as reinforcements. My squadron commander was a major, some ten years senior to me – in those days quite a gulf. We marched from Mirjawa to Ladis, seventeen miles. This was a lovely camping ground, so different to the arid halting places along the Seistan Trade Route. It lay where the plain met the foothills and where the hills, which had been converging on us all day on either hand, finally closed in. The Kuh-e-Taftan rose high into the blue sky ahead; the twin domes of its craters were snow-capped. From one a wisp of smoke was rising. From out of the mountains ran a broad crystal clear stream, in which darted tiny fish. The black tents of the nomad Bekis dotted the plain near the stream, camels were being brought in from grazing and children came to gaze at us as they drove their flocks of goats back to their encampments. All was peaceful and pastoral.

The next three marches were through barren hills, in places sparsely covered with scrub and an aromatic herb, the smell of which I always associate with the hills of Baluchistan. All the marches were dominated on the right by the Kuh-e-Taftan. On the fourth day we entered the Khash valley. This is a flat camel-thorn covered plain, some twenty miles in length and three to four miles wide. The northern end is blocked by the Kuh-e-Taftan. Down both sides run ranges of rocky brown hills, which converge at the southern end. Khash itself lies a little north of the centre of the plain and consisted of a crumbling fort of sun-dried brick with a crenellated circular watch tower, about a dozen mulberry trees, and some evidence of cultivation. After life in posts on the Eastern Persian Cordon, either alone or with one other British officer, we found Khash a perfect metropolis. Including my squadron commander and myself there were ten British officers.

We gathered that the Gomshadzais had submitted to the General's terms and that all was now fun and friendship. By some superhuman effort the General had got his car up there, driven by a British corporal all the way from Nushki along the Seistan Trade Route, the first motor vehicle to come that way. In the evenings he took the Chiefs for a drive round the camp. It was a most extraordinary spectacle to see Jiand Khan, the Chief of the Robbers, festooned with well-filled bandoliers and holding his rifle between his knees, sitting in the back of this open tourer as they bumped over the uneven surface of the plain.

Many minor Chiefs had come in and hundreds of long-haired Baluch were camped in their black tents nearby. *Durbars* (formal meetings) were a daily occurrence. These wild-looking men with hair down to their shoulders in soiled loosely wound white turbans, dirty white shirts hanging outside equally dirty voluminous white pyjamas, wearing sleeveless waistcoats of black or blue cloth, trimmed in some cases with an addition of gold or silver tinsel, were all armed to their teeth. They all carried firearms of sorts, Belgian, French or British and wore one or two full bandoliers. Some also had a cartridge belt round the waist. This often held a dagger or a curved sword hung from a belt. They wandered round the camp at will.

The door of my tent faced the great volcano. Every morning I was filled with wonder and pleasure at its beauty in the clear light of the new day, as the sun started to tint the snow on its twin domes. The Khash valley lies at five thousand feet, the same altitude as Quetta. The Kuh-e-Taftan rises another eight thousand feet above it and dominates the country for a hundred miles around. The silhouette was so clear against the blue sky and so symmetrical. Through the ages the two craters had ejected their streams of lava to cascade down and form two identical domes.

However, on one morning it was not the mountain which first

caught my eye as I emerged from my tent. To my astonishment nine little female figures squatted in a row on the ground facing me. They were all dressed in shabby, dusty, black or dark blue. Their head-dresses were drawn modestly across the lower part of their faces. Perhaps two or three were old, that is to say over thirty – women age quickly in the hard life of the Oriental peasant. I was embarrassed. I was not dressed to receive ladies; in fact I was stripped to the waist, as I was about to wash in my camp basin which stood on its 'X' stand outside my tent. I hastily withdrew and donned a shirt. Although they remained for the most part with the lower part of the face covered; some of the younger ones, on the pretext of adjusting a head-dress, would reveal a shapely almond-coloured forearm, adorned with a cheap glass bangle or two, and a pair of dark eyes ringed with kohl, and a comely though obviously unwashed face.

Why they chose to visit me I never discovered. Perhaps because my tent was on the outside of the two in that part of the camp, or possibly they had been told that I spoke some Persian. One of the elder women acted as spokesman. She declared in a shrill, high-pitched voice, that they were Persians and had been carried off from their villages in a recent raid by Juma Khan. No doubt that villain would have described the younger ones as good booty. Juma Khan was Chief of the Ismaelzais, a tribe second only to the Gomshadzais in numbers. He had not long ago made a raid on a grand scale into the Seistan plain, had clashed with our troops and been forced to retreat in haste into the mountains, leaving behind him a large portion of his loot, in the shape of goats and camels. Whereas Jiand Khan was a patriarchal, dignified figure, fair in complexion, with a flowing white beard, Juma Khan was dark, squat and repulsive looking. I had seen him in the camp, swaggering around loaded with weapons.

I asked how far they had come in their escape. I do not think they knew. The woman answered, '*Besiar farsakh* (many farsakh).' The *farsakh* is roughly four miles, or the distance a laden mule can do on the flat in an hour. The Persian idea of distance is not ours. They live in a land of vast distances: we live in a tiny island. If you ask a passing traveller how far it is to the next stage or well, he will often reply, '*Nazdik* (near),' when in your opinion this proves to be far from the case. He may say this from a weakness some Orientals have of saying what they think you wish to hear, or because to him *nazdik do farsakh* (near two farsakh) really does imply near, whereas to me eight miles and possibly two more hours' travel seems far.

I went to the major's tent next door. He was in the same state as I was, but on hearing of our visitors put on a shirt and came out to meet them. They were still there looking like a row of rather shabby crows. He sent for his orderly to take them to the brigade major, who was more in a position to deal with a problem of this nature. They padded off behind the orderly in single file. No doubt he had his leg pulled in the Lines [soldiers' quarters]. I learnt later that these women were all safely returned to their homes.

SIEGFRIED SASSOON
(1886–1967)
The General

'Good-morning; good-morning!' the General said
When we met him last week on our way to the line.
Now the soldiers he smiled at are most of 'em dead,
And we're cursing his staff for incompetent swine.
'He's a cheery old card,' grunted Harry to Jack
As they slogged up to Arras with rifle and pack.

*   *   *

But he did for them both by his plan of attack.

ISAAC BABEL
Red Cavalry
Crossing into Poland

Another form of cavalry warfare was fought out on the eastern flank of the First World War as it drew to a close, a warfare quite without honour or any other form of moral restraint. When the Imperial Russian army collapsed at the end of 1917, those contending for power in the deposed, soon to be murdered Tsar's empire, and they included Red revolutionaries and White reactionaries, as well as subject nationalities – Balts, Ukrainians, Armenians, Georgians – struggling for independence, were rapidly caught up in a ferocious civil war.

One of its bitterest subsidiaries was the campaign on Russia's western frontier, where a newly re-established Poland, whose independence had been recognized by the victor nations at the Paris Peace Conference in 1919, attempted to pitch its eastern border on territory that the Russian Bolsheviks would not recognize as historically Polish. In the ensuing Russo-Polish War of 1920–1, large cavalry armies locked in combat, which swayed back and forth across the plains of Poland, White Russia and the Ukraine; at one point, the Polish army threatened Kiev, the Ukrainian capital, at another, the Red Army menaced Warsaw, Poland's capital. The soldiers engaged were desperate men, on the Russian side veterans of the Imperial army, on the other the survivors from the national contingents enlisted in the Austrian, German and Russian armies who had been conscripted by Poland's new leaders in the aftermath of those armies' collapse. Brutalized by the First World War, further brutalized by the disintegration of civil order throughout their homelands, they murdered and pillaged wherever the war took them. Isaac Babel's account of the Red cavalry's invasion of Poland, though written as a short story, captures the horror of a war fought without respect for military law, or even for the rules of common humanity. There have been many other such wars in the twentieth century.

*   *   *

The Commander of the VI Division reported: Novograd-Volynsk was taken at dawn today. The Staff had left Krapivno and our baggage train was spread out in a noisy rearguard over the highroad from Brest[–Litovsk] to Warsaw built by Nicholas I upon the bones of peasants.

Fields flowered around us, crimson with poppies; at noontide the yellowing rye; on the horizon virginal buckwheat rose like the wall of a distant monastery. The Volyn's peaceful stream moved away from us in sinuous curves and was lost in the pearly haze of the birch-groves; crawling between flowery slopes, it wound weary arms through a wilderness of hops. The orange sun rolled down the sky like a lopped-off head, and mild light glowed from the cloud-gorges. The standards of the sunset flew above our heads. Into the cool of evening dripped the smell of yesterday's blood, of slaughtered horses. The blackened Zbruch roared, twisting itself into foamy knots at the falls. The bridges were down, and we waded across the river. On the waves rested a majestic moon. The horses were in to the cruppers, and the noisy torrent gurgled among hundreds of horses' legs. Somebody sank, loudly defaming the Mother of God. The river was dotted with the square black patches of the wagons, and was full of confused sounds, of whistling and singing that rose above the gleaming hollows, the serpentine trails of the moon.

Far on in the night we reached Novograd. In the house where I was billeted I found a pregnant woman and two redhaired, scraggy-necked Jews. A third, huddled to the wall with his head covered up, was already asleep. In the room I was given I discovered turned-out wardrobes; scraps of breeze played in women's fur coats on the floor, human filth, fragments of the occult crockery the Jews use only once a year, at Eastertime.

'Clear this up,' I said to the woman. 'What a filthy way to live!' The two Jews rose from their places and, hopping on their felt soles, cleared the mess from the floor. They skipped about noiselessly, monkey-fashion, like Japs in a circus act, their necks swelling and twisting. They put down for me a feather bed that had been disembowelled, and I lay down by the wall next to the third Jew, the one who was asleep. Fainthearted poverty closed in over my couch.

Silence overcame all. Only the moon, clasping in her blue hands her round, bright, carefree face, wandered like a vagrant outside the window.

I kneaded my numbed legs and, lying on the ripped-open mattress, fell asleep. And in my sleep the Commander of the VI Division appeared to me; he was pursuing the Brigade Commander on a heavy stallion, fired at him twice between the eyes. The bullets pierced the Brigade Commander's head, and both his eyes dropped to the ground. 'Why did you turn back the brigade?' shouted Savitsky, the Divisional Commander, to the wounded man – and here I woke up, for the pregnant woman was groping over my face with her fingers.

'Good sir,' she said, 'you're calling out in your sleep and you're tossing to and fro. I'll make you a bed in another corner, for you're pushing my father about.'

She raised her thin legs and rounded belly from the floor and removed the blanket from the sleeper. Lying on his back was an old man, a dead old man. His throat had been torn out and his face cleft in two; in his beard blue blood was clotted like a lump of lead.

'Good sir,' said the Jewess, shaking up the feather bed, 'the Poles cut his throat, and he begging them: "Kill me in the yard so that my daughter shan't see me die." But they did as suited them. He passed away in this room, thinking of me. – And now I should wish to know,' cried the woman with sudden and terrible violence, 'I should wish to know where in the whole world you could find another father like my father?'

WILFRED OWEN
(1893–1918)
Anthem for Doomed Youth

What passing-bells for these who die as cattle?
– Only the monstrous anger of the guns.
Only the stuttering rifles' rapid rattle
Can patter out their hasty orisons.

No mockeries now for them; no prayers nor bells;
Nor any voice of mourning save the choirs, –
The shrill, demented choirs of wailing shells;
And bugles calling for them from sad shires.

What candles may be held to speed them all?
Not in the hands of boys, but in their eyes
Shall shine the holy glimmers of goodbyes.
The pallor of girls' brows shall be their pall;
Their flowers the tenderness of patient minds,
And each slow dusk a drawing-down of blinds.

JOHN MASTERS
Bugles and a Tiger

John Masters, who was to become one of Britain's most successful novelists of adventure after the Second World War, came
from a family with a long tradition of service to Britain's empire
in India. In 1935, as a newly commissioned officer, he himself
joined an Indian regiment, the 4th Prince of Wales's Own
Gurkha Rifles.

    To be accepted as a Gurkha officer was an honour, for the

Gurkha regiments were the elite of the Indian Army. Their soldiers were not subjects of the Empire, but were recruited under treaty from the independent Kingdom of Nepal in the Himalayas. Short in stature, with the muscular physique of mountaineers, the Gurkhas were notably hardy and famously brave. They set soldierly standards only the best British officers could match. Those accepted into a Gurkha regiment remained with it for the rest of their regimental life. The apprehension John Masters describes as he makes his way to the 'home' of the 4th Gurkhas at Bakloh in the Himalayan foothills is therefore entirely authentic. So, too, is his evocation of the sense of fraternity that bound Gurkha officers to each other and to their soldiers.

John Masters served with his regiment during the Second World War in the Middle East and South-East Asia.

*   *   *

There were only two more stops before we finally did leave the town, one to pick up a goat and another hundred cubic feet of merchandise (charcoal in sacks), and one at the octroi post. In all my time in India I never saw a bus driver or anyone else actually pay any octroi, but there was an octroi post on the outskirts of almost every town, and buses always stopped there. This time the resident tax collector, who was sitting inside with his feet on the table in the pompous trance of oriental officialdom, came out and went through a slow-motion routine of inspecting us, our baggage, and the merchandise. Then he waved a finger – and we were away.

We were away through the unspeakable squalor of the outskirts of Pathankot and then on the wide road between the mango trees. The world was green, and little convoys of overloaded donkeys and gaily shawled women walked along the grass verge under the trees. We honked at tongas – built to take three passengers but habitually loaded with eight people and a huge bale of hay – and made them swerve out of the middle of the road. The bus had an electric horn, but it was disconnected to save the battery and to enable the driver to show his virtuosity on the winding mountain road, where one hand perpetually honked at the rubber bulb of the old-fashioned horn and the other changed gear – 'Look, no hands!' To the left as the road swung, the forested hills climbed up and away, rolling higher and higher till they disappeared into the hazy surge of the Himalaya. Once a gap in the trees, the line of the road, and the drift of cloud all worked together to unveil an austere, blue-white wall of ice a hundred miles away in the main chain of the greatest mountain mass on earth. A little later we rushed heedlessly under a cliff of conglomerate and the sign guarding it: DRIVE CAREFULLY, LOOKING UPWARDS.

After three hours of terrifying effort, after the radiator had boiled twice and twice been slaked with cold water, after everyone in the back of the bus had been sick many times, we stopped on the edge of a precipice and I got out. This was Tuni Hatti, and I had arrived. There was said to be a truck coming down the three miles from Bakloh to the road junction here. It would (perhaps) arrive in an hour, or (perhaps) two.

'Salaam, sahib,' said the driver. The bus roared away down a curving slope, the engine switched off, out of gear, the top load swaying, the paper-thin tyres sliding on the loose surface.

I left my bags and boxes beside the road and started walking up the hill. It was hot in the afternoon sun, but trees covered the hillside and a footpath wound up through them to Bakloh. I knew the path, for I had hurried down it in July with clattering Gurkhas all round me, on the first stage of the road to the Shadiganj riots.

A thousand feet above the metal road and five hundred feet below the ridge crest I came to the 2nd Battalion's football ground. They had hewn it out of the hill, for there is no level place in Bakloh, and its sides were steeply built up from the valley. Below it, on the left, I saw a rifle range, and on the grassy knoll above the butts, a clump of trees. They were a rare kind of date palm and looked out of place among the pines. An exactly similar clump of the same rare date trees crowned the end of the next spur, three miles up towards Dalhousie. Thick forest covered the next lower spur towards the plains, but on the end of that too, among the trees, was another clump of palms. I sat down to draw breath, and to look at the palms.

I felt the grip of the same awed fascination that had overtaken me when I first heard the legend about them.

The local hillmen said that these palms marked the line of Alexander the Great's outposts. The clumps were descended from the dates Alexander's soldiers had brought from the banks of the Tigris. Certainly the trees rose in the places any officer would have chosen for an outpost line, and history confirms that this was indeed the farthest limit of Alexander's penetration to the east. I thought: perhaps a young Macedonian officer climbed this path to inspect his posts. They were strangers here too, and the clangor of their shields on the rock echoed over twenty-two hundred years into my ears.

I went on and came to the little 2nd Battalion bazaar. The shopkeepers on their porches stood up to greet the new officer as he passed, but I could not say anything more than 'Salaam' in return, for I did not know them. I reached the crest, which was occupied by the parade ground – another major feat of landscape engineering. Then came barracks, more pines, and the road wound down among the gardens of the officers' bungalows, past the tiny church and the German trophy guns from Flanders, and on to the mess lawn, into the mess itself.

The officers were at tea. They had been at tea when I came up to be vetted. 'Midge' looked up, said, 'Hello, Masters,' and went on with his cake, but in a friendly manner. He would have stopped eating cake had it been necessary to assure me of my welcome, or had I been a guest.

Another man, small, tough, young, with bright-blue eyes, tightly curled fair hair, and a furrowed brow, leaped to his feet and cried, 'Good God, how did you get here? Where are your things? You're early.' He looked at his watch with an unbelieving frown, shook it, placed it to his ear, and began to mutter under his breath, 'Christ – damn watch. Have some tea.'

James Sinclair Henry Fairweather, the next subaltern above me, was off again. He had been deputed to meet me and had forgotten, or had forgotten the time. He frequently did. Everyone else laughed, and after a moment James looked up with a beautiful shy smile and said, 'Oh, well, you're here, anyway.'

I wasn't going to make the mistake about the cake again [he had been too polite to cut it on his first visit], and firmly helped myself to a large slice. James bustled off to make sure the local bus picked up my kit from the road junction at Tuni Hatti where I had left it, and I looked around me.

The Victorian founders of our regiment had built the mess low, of stone, and set it on the edge of the ridge, its front turned to the Himalaya. The Edwardians had glassed in the back veranda, which faced south over the edge of the ridge, towards India. And so, near the end of the century-long dream we Georgians took our ease there, as on the promenade deck of a moored airship. Beyond the glass, forests and terraced fields dropped steeply away for two thousand five hundred feet, then climbed with the same precipitousness to a ridge two miles distant and only a few hundred feet lower than ours; and so on, and down and on in dwindling rise and fall to the flat lands of the Punjab. Beyond the last low smoky ridge the khaki plains reached round and on until, in the farthest light, only the visible curve of the earth's surface served to separate them from the sky. To the right the Ravi River burst through a cleft in the foothills and wound out across the earth in a spreading flood. The low sun sent us intermittent golden flashes from unseen water. To the left the Beas River curled out of Kulu, the Valley of the Gods, and moved off to the horizon. Far, far to the left again were haze and a loom of mountains. The light hung opalescent over all – not sharp, but warm and blue-green, washed by recent rain and split by the broad shafts of the sun. The undersides of the cumulo-nimbus cloud masses showed thick and solid and dark blue to the earth, but their anvil heads towered thirty thousand feet above the plains, and shimmered with a frozen gold in the upper light, and moved steadily on towards us.

I had come to my home.

Of the officers at tea, only 'Midge' was in uniform, though they had all been working and most of them had more to do yet. In the 4th Gurkhas, as in most regiments, an officer changed into plain clothes after lunch unless he had an outdoor parade with troops. Midge had come in from machine-gun drill and wore our ordinary working uniform. His nailed ankle-boots were of black leather. His short grey-green puttees were wound in three exactly superimposed folds over the join of boot and hosetop, and each puttee was held in place by three exactly superimposed folds of broad white tape. His pale grey woollen hosetops were turned over for four inches just below the knee, and held up invisibly under the turnover by elastic garters. To the garters were sewn oblongs of coloured felt so that a square inch showed on the outside of each leg. The colour denoted the company to which a man belonged – in Midge's case, red for D (Machine-gun) Company. His wide shorts were of heavily starched khaki drill. (The traditional way to put on these shorts was to stand them on the floor and step in.) The shorts had a deep waistband with three loops to hold the Sam Browne belt, and were tailored so that the point of the crease in front was level with the upper tip of the kneecap. His shirt was of a material called 'greyback', a pearly blue-grey woollen flannel worn by the Indian Army, with plain black buttons and plain shoulder-straps of the same material, on which were sewn black cloth silhouettes of the rank badges – for Major Madge, a crown.

Around his sunburned face was the unmistakable brand of our calling, the clear white impression of his chinstrap. His hat was probably hanging on one of the pegs on the veranda, with his Sam Browne. The Sam Browne was not brown, but black. The hat consisted of two broadbrimmed felt campaign hats, sewn together to give better protection from the sun, and dimpled fore-and-aft down the middle of the crown. Several folds of very thin khaki cotton cloth were wound in a band round the outer hat and topped with a single black fold. The number of khaki folds varied with status – nine for a King's Commissioned [British] Officer, seven for a Viceroy's Commissioned [Gurkha] Officer, five for an other rank. Finally, on the left, pinned through all the folds of cloth, was the regimental badge – a great Roman IV in black metal.

I had come home. I sat back in the wicker chair with a sigh. James returned, and the duty bearer brought another pot of tea, a paunchy white pot with the regimental crest on it. I poured out tea into a cup and poured in milk from a crested silver jug. Glancing at it, I saw from the engraving that it had been presented to 'the Officers' by someone, a long time ago, 'on the occasion of his promotion'. I took a lump of sugar from a crested silver bowl – presented by someone else 'on appointment'. Soon my name would be on one of these little bowls or jugs or ashtrays. I felt good and comfortable. The sun shone level through the glass, the talk hummed lazily, china clinked; the 4th Gurkhas' private landscape lay like a dream before me.

Midge got up. 'We machine-gunners have some work to do, even if no one else has.'

James muttered something inaudible under his breath. Midge glanced sharply at him and walked out with quick, bird-like steps. Everyone drifted off. The sun went down.

James took me to a bungalow just above the mess. Half the bachelors lived there; it was called the Rabbit Warren. It had no electric light, so James lent me a small oil lamp. He went away, and I took a look at my room. I seemed to own – or be renting – a narrow bed, a chest of drawers, a wardrobe, two hard chairs, one easy chair, a table, and a bookshelf. Attached to the room was my private *ghuslkhana* with its appropriate fittings, also rented.

A *ghuslkhana* corresponds in function to a bathroom, but it is a profound misconception to think of white porcelain, taps, or water sanitation. The *ghuslkhana* is small and square and has a hole in the outer wall to let out water and let in snakes. One corner, around that outlet, is fenced in by a low parapet the height of a single brick. In this enclosure sits an oval zinc tub. Outside the parapet is a slatted wooden board to stand on, and a wooden towel horse. Ranged along the inner wall are a deal table holding an enamel basin, a soap-dish, and a jug; a chamber pot; a packet of Bromo hanging from a nail; and a wooden thing on four legs whose proper name may possibly be 'toilet', but which was never called anything but 'the thunderbox'. The thunderbox has a hole in the middle of the seat, and a hinged top. Under the hole, fitted into the structure, is a deep enamel pot called a top hat.

ALEXANDER STAHLBERG
Bounden Duty (1)

In 1935 Adolf Hitler, the new head of the German state, abrogated the article of the Versailles Treaty of 1919 that forbade Germany to have an army larger than 100,000 men and introduced general conscription. All fit young Germans at once became liable for military training. One who volunteered was Alexander Stahlberg, son of a rich industrial family with wide connections in the German aristocracy, though the family was not technically 'noble' itself. He was posted to one of the army's last horsed cavalry regiments, Reiterregiment 6, and began to learn the duties of a trooper. His account is of the greatest interest, both as a description of an historic way of military life, shortly to disappear for ever, and of the processes by which Adolf Hitler sought to turn the new German army into a body bound to him by an oath of personal loyalty.

Stahlberg would later be commissioned as a reserve officer and serve most of the war as aide-de-camp (ADC) to Field Marshal von Manstein, the Wehrmacht's leading commander of armoured troops.

*   *   *

In the 6th (Prussian) Cavalry Regiment

It was the summer of 1935 and in the newly-built barracks area of the 6th (Prussian) Cavalry Regiment (RR6) in Schwedt, the first recruits were arriving in long lines at their squadron quarters after the re-introduction of general conscription (March 1935). Sixty to seventy young men had been allocated to each of the six squadrons, all still in civilian clothes and each with a box beside him in which to send them home.

The commander of the 3rd Squadron, Captain von Lewinski, and his three lieutenants looked over the new faces. For the modern cavalry, the military district headquarters was looking for short young men, but an outsider had obviously wandered into the 3rd Squadron by mistake: over six feet three inches tall.

The captain spoke a few words of welcome and explained that from now on we were no longer Mr So-and-So, but 'Troopers'. Then he started with the tall man on the right and asked him to give his name loudly and clearly. So Trooper Stahlberg began and the sergeant-major entered the name in his thick book, which was kept ready to hand between his second and fifth uniform buttons. No one else was allowed to leave a button undone.

'Hands up anyone who has completed their school education. Elementary school first, then secondary school, and university entrance last.' Result: one – he wore abnormally thick glasses – had not finished his elementary schooling, no secondary school pupils, one university entrant.

Then came the ages: all the recruits were under twenty, only the tall one at the end was almost twenty-three. He was asked how he had arrived in this squadron. 'I don't know, Sir!' 'Ah, are you the volunteer?' 'Yes, Sir.' The sergeant-major made a note. Then the captain asked: 'Shouldn't Stahlberg have been in the 1st Squadron with the other volunteers?' Sergeant-major's answer: 'It's probably because he's not noble, Sir.' Pause.

Next order: 'Hands up who can ride!' Six hands flew up. 'One step forward, those six. Enter them.'

Of course I could ride, otherwise I would not have enlisted in the cavalry. I had sat horses from childhood, in Kieckow, in Paetzig, I had ridden across the fields with my uncle, had gone on horseback to bathe and to visit the Tresckows in Wartenberg, and with Uncle Ewald von Kleist in Schmenzin, and even in Berlin, when one of my many uncles came visiting and paid for me to ride from Tattersall Beermann by the Zoo Station across the Tiergarten to the Brandenburg Gate.

Then we were divided up, six to a room. The man with thick glasses was in my room. On the top floor was the Quartermaster's store, where we collected the first items of our uniform: underwear, footcloths (instead of socks), breeches and boots, overalls, our coarsely-woven working uniform. Then the young soldier learned to stuff his pallet with fresh straw. Even the things that one's mother did at home had now to be learned by every recruit 'according to regulations', including sewing on a button!

The three hundred metres from our quarters to the stables were not walked but marched, generally singing. The tall fellow had to give the note, he seemed to know something about music.

Then we were assigned to groups, or 'patrols', and acquired our 'instructors', a sergeant and a lance-corporal, known as 'twelvers' because they had signed on for twelve years in the old Reichswehr [predecessor to the Wehrmacht].

There were over a hundred horses in each squadron. I was given a chestnut gelding called 'Heldensang', because he was the biggest horse in the squadron, an eleven-year-old crib-biter, hard-mouthed and still harder-jowled, as would soon appear, but hardest of all in the trot. But he 'covered' me, so that in silhouette my heels were scarcely visible, a veritable 'monument of a horse'!

We learned grooming and everything else connected with our horses' well-being. Even as a country child, I had never dreamed how much pampering is meted out to the horses of the Prussian cavalry. 'Horse first, weapons next, man last!'

The first riding lesson came on the third day. The shining horse, without blanket and saddle, was led out of his stall on the snaffle. More than sixty recruits stood with their horses at the end of the barrack square, ranged in one line down the middle of the jumping ground. The oldest of the lieutenants, who was in command of riding training, ordered: 'The six who can already ride, three paces forward.' Then came the next order: 'Trooper Stahlberg, three more paces forward. Mount!' I had learned that in Kieckow as a child. I leaped and sat on the horse's back. The lieutenant's next order was: 'Scissors about face!' A sergeant stepped up to my horse, to hold him if necessary, but Heldensang knew it all and stood as if rooted to the spot. So now I was sitting backwards on the gelding's shining back, facing his tail, awaiting the lieutenant's next order. When it came it was not what I expected: 'One round of the jumping ground, gallop!' I looked down at my lieutenant in astonishment. 'Didn't you understand the order?'

I heard the sergeant whisper: 'Not possible, let yourself fall on the jump and keep relaxed.' So I gave Heldensang my knees and slapped him on the buttock. The sergeant, with his hand on the snaffle, ran a few paces in the direction of the first obstacle, Heldensang jumped, and I was rolling in the sand. Heldensang, the old trooper, stopped immediately after the jump and waited until I was up again.

The remaining five who, like me, 'could ride already' went through the same manoeuvre. None of us was seriously injured, and yet I found the exercise a macabre joke. Probably it was meant to be like crossing the equator for the cavalry.

Of course a cavalryman's training calls for hardness: in order to have his horse under his control later on in every conceivable situation, the rider must first sit deep in the saddle. But the military saddle in those days was hard and the recruits had to take the saddle they were given, including some that were poorly suited to the shape and size of their own bodies. After a few weeks, the young soldiers were sitting their horses so well that the squadron could go to the parade ground in closed ranks and the instructors could begin with the first simple dressage exercises.

Inevitably, we had to take the oath, to the person of Adolf Hitler, as the regulation stood since the death of Reich President Hindenburg [Field Marshal Paul von Hindenburg, who had commanded the Kaiser's armies, 1916–18] on 2 August of the previous year. The officers had to prepare the recruits over a good many hours of instruction, of which the one before the oath was taken was conducted by the squadron commander himself. I looked forward to it in some suspense.

Today, Lewinski began, he would talk through the words of our oath with us. He said that the oath had been changed the year before, the former Reichswehr had sworn a different oath from the one we would be taking tomorrow. I put up my hand to ask a question: would he recite the former oath to us for purposes of comparison, so that we could hear the difference? The captain looked at me in some astonishment. 'Well, why not,' he said eventually. He still knew it by heart, and we could certainly hear it:

'I swear by God this sacred oath, that I will always serve my people and
    Fatherland
faithfully and honestly and be prepared as a brave soldier to risk my life for
    this oath
at any time.'

The new oath that we were to swear tomorrow ran otherwise:

'I swear by God this sacred oath, that I shall be unreservedly obedient to the
    leader of
the German Reich and people, Adolf Hitler, Supreme Commander of the
    Wehrmacht,
and prepared as a brave soldier to risk my life for this oath at any time.'

The chief then said we need not learn the oath by heart; at the ceremony next day he would recite passages aloud and we had only to repeat them. That was the end of our lesson.

Each evening, as darkness fell, we heard the duty trumpeter at the barrack gates sounding the Last Post. Then the duty sergeant went through the rooms to make sure all the recruits were in bed and to put out the lights. That night I started a conversation in the darkness with my five room-mates – all of them came from Silesia. 'What do you think of the oath we're going to take tomorrow?' I asked. A long silence. 'Are you all asleep?' I asked. All five said 'No' almost simultaneously. But what were we meant to think, said one, orders were orders. Then I heard our little fellow, the one with the thick glasses: 'At home we do something that makes the oath invalid.'

I had never heard of this and asked him to tell us about it. 'Right then, listen,' he began. 'You raise your right hand to swear and at the same time you make a fist with your left. That way you can swear what you like but the oath is no good. Only afterwards – or that's the way we do it – you have to confess it to your minister.'

I thought this was priceless, but all my other room-mates confirmed it. At all events it would be a good idea to put your left hand in your pocket during the procedure. 'All the same,' I began again, 'I wouldn't advise this left fist tomorrow. When we're standing in ranks anyone behind you might see you and denounce you later which would mean you'd be in a lot of trouble.' 'You're right,' he said, 'then I shall clench my fist in my mind and I'll still confess it, my minister will go along with that.' I said: 'So you don't want to swear?' Oh yes, he wanted to swear, but not for Adolf. He liked the old oath the chief had told us about today much better. 'So do I,' I said, 'all the same I'll be swearing for Adolf tomorrow.' Then the questions came from all sides: 'Aha, then you're a Nazi?' 'I'll tell you something,' I said angrily, 'I'm no Nazi, and I don't care for Adolf, but I'll do it for my Fatherland.' Another long pause. Then I heard the little man with the thick glasses again: 'I like that, you know. But it means a lot of care tomorrow.' 'No,' I said, 'you have only to say "Adolf Hitler" and think "Fatherland". That's not difficult, and in any case it will be much easier for you tomorrow than for me because you were conscripted by the army district command and I am a volunteer.'

The next morning the whole cavalry regiment was drawn up in the great barrack square on foot, in square formation. For the first time I heard our excellent body of trumpeters. For the first time I saw and heard our regimental commander, Lieutenant-Colonel Arno von Lenski. As he walked along the lines, I found myself briefly eye to eye with him. He was wearing a monocle; I did not care for him. Then he made a rousing speech, speaking not of the 'Führer, Adolf Hitler', but of 'our beloved Führer Adolf Hitler'. No wonder that, with all these protestations of affection, he later became a member of the notorious National Socialist People's Court. However, it was not until the war was over that he attained his highest military honours, ending his career as a General in the National People's Army in the German Democratic Republic.

Officer tuition generally appeared on the duty roster once a week. One day the subject was 'Orders and Obedience'. Question: 'What orders must the soldier carry out?' Answer: 'Every order must be carried out.' Question: 'What orders should the soldier not carry out?' Pause: laborious thought by the recruits. Lieutenant's answer: 'An order should not be carried out if the soldier realizes that by carrying it out he would be committing a crime.' Question: 'What does the soldier do when he realizes this?' Answer: 'He refuses to carry out the order.' Next question: 'Give examples.' Answer: 'Murder, manslaughter, looting, rape, killing prisoners.'

In another lesson we were taught about The Hague Convention on Land Warfare of 1899 and its regulations on the distinction between civilian personnel and soldiers, the treatment of the wounded and prisoners, peace emissaries, respect for the person, honour and property of inhabitants of foreign countries in the event of war. These matters were treated with great conscientiousness in our regiment.

On another occasion we discussed the law of self-defence, because we young soldiers were now bearing arms. The relevant paragraph of the German Reich Penal Code had to be learned by heart and we were questioned on it a week later. I remember it now:

'No punishable action has been committed if the action was necessitated by self-defence. Self-defence means the defence which is necessary in order to prevent an immediate, illegal attack on oneself or another person.'

After a number of recruits had repeated the paragraph correctly and in full, the rest of the period was spent explaining each individual word, with examples. The greatest value was attached to the words 'illegal attack on . . . another person'.

The case of self-defence became complicated if, for instance, one happened to witness a crime. Was it then better simply to look away, to avoid the disagreeable consequences one might incur by intervening, or should one act to prevent the obvious crime in time? After all, the wording actually was: '. . . defence which is necessary . . .'

'You Share the Responsibility'

Infantry training with machine-gun practice was also on the duty rota. This part of the basic training demanded extreme physical effort from the young soldier. We were young and healthy and did not mind being pushed to the limit of our capacities. I myself thought of it as intensive training in a sport.

But we had the little fellow with the thick glasses in our group. Though he sometimes astonished me with extremely intelligent remarks, I often noticed that immediately afterwards he would be dreaming away absentmindedly. In any case he was physically not up to the riding training and drill. His gait was hunched and he could not straighten himself up. When we had to march singly across the barrack square with our backs straight, he always relapsed into a kind of amble. His left hand swung forward with his left leg, the right did the same. I could not understand why he had been sent to the Army at all when he was called up. He looked to me like a case for the doctor, and probably for the psychiatrist.

Unfortunately, our sergeant and lance-corporal saw things differently and took pleasure in teasing and tormenting him from the very first day. If he was given an order in arms drill he had to think before carrying it out after a fashion. So the two instructors took pleasure in making an unwilling clown of this unfortunate at every opportunity.

Today it was time for machine-gun practice. In groups, close to a toolshed, with the MG13 Dreise in our hands, we obeyed the orders: 'Up, at the double' – 'Down, up, at the double – down' – three, four, five, six times and more. Suddenly our problem child was on the ground, face down. One more pitiful effort, then – so it seemed – he passed out. Our two instructors were already standing beside him, shouting at the prostrate man, but to no effect: there he lay still, face down.

Then something extraordinary happened. The sergeant and the lance-corporal grabbed the man and set him on his feet. At first his legs would not carry him, but finally he was standing up. Then our sergeant shouted, so loudly that we could all hear him: 'We'll make a soldier out of you, like it or not!' They took him between them by the arms and disappeared round the other side of the toolshed.

We stayed where we were, bewildered. After only a few moments loud screams came from behind the shed. I froze, but then I heard quite clearly: 'Help – help!' I ran round the shed.

There he lay, on his back now, our room-mate. Beside him stood the sergeant and lance-corporal, kicking the prostrate man in the stomach again and again. I shouted as loudly as I could: '*Herr Unteroffizier!*' The two spun round, stared at me, and came towards me as if nothing had happened. The three of us rejoined the rest of our group, the sergeant ordered two others to bring him to our room because he was not feeling well. They did not pass us on the way to the barracks but went round the far side of the shed, supporting our fellow soldier like a wounded man.

At lunchtime we found him lying on his bed in our room. He said he was better already and did not need to report sick, he would come out and join in the afternoon duty.

My account of that day is not over. In the afternoon I was called to the telephone in the orderly room: an officer wanted to speak to me. Who could it be? This was my first telephone call as a soldier. 'Trooper Stahlberg speaking,' I announced. 'Henning here,' was the reply. It really was Henning Tresckow, who had just concluded a General Staff tour in Schwedt and told me to get two hours off and meet him at the Café Wieck on the Schlossfreiheit.

The sergeant-major, evidently impressed that one of his recruits should have a General Staff officer for a cousin, agreed but insisted that I should report to him correctly dressed. When I reappeared in his office, having changed my clothes, he stood me to attention and walked round inspecting me. So off I went to town at top speed, my heavy sword at my side.

A number of military vehicles were parked in front of the Café Wieck and I went in to the building in some excitement. Frau Wieck, well known in Schwedt as a 'soldiers' mother', stopped me this time at the entrance. I could not go in, because she had reserved the house for a single party. She was all the more astonished when I said that I had been invited to that very party. 'But then you could at least have put on a decent uniform jacket instead of those fatigues!' she exclaimed. 'We haven't collected them from the stores yet,' I excused myself – but the sergeant-major had approved my dress.

When I opened the door to the restaurant, all I could make out was that the room was filled with officers under a cloud of tobacco smoke, and that they all had a red stripe on their trousers. I closed the door behind me and, as instructed under 'conduct in public', stood to attention to await events.

After a brief moment Henning was before me, shaking me by the hand. The room had fallen silent and I felt all eyes on me. Henning called in a loud voice: 'Gentlemen! May I present my nephew

Stahlberg. He is serving here as a volunteer and I have asked him to come because I want to hear from him how he likes military service.'

I hung my sword, belt and cap on the clothes-stand and found myself sitting at one of the little café tables, hemmed in by General Staff officers. I was relieved that no one started questioning me immediately, so that I could devote myself exclusively to the apple pie and whipped cream that Henning had ordered for me. Only once Henning commented drily: 'Well, how do you feel in the midst of the top brass of the German General Staff?' I have no idea whether and how I answered and I no longer remember which of the Army's later celebrities were sitting round me. What I do remember is that I was astonished how casually and openly they talked.

As soon as my plate was empty, Henning got up to take a short walk with me on the Schlossfreiheit. I could never have guessed how many, many 'short walks' he would take with me in the future.

Outside the door I was no longer nephew, but 'cousin' and now he positively showered me with questions. He wanted to hear my impressions, my views and my criticisms of the training for the first year of general conscription. I was to talk to him quite freely and without regard to the difference in our military rank.

My judgement was predominantly positive, but finally I also told him the nasty story of that morning, including the fact that my shout to the sergeant had put a stop to the sadistic abuse. Henning stood still. 'You undoubtedly did the right thing,' he said, 'but that's not enough. You now share the responsibility of ensuring that nothing like this ever happens again in your squadron!'

'How can I do that, as a recruit?' I asked helplessly. Henning's face was grave. 'Who's your squadron commander?' he asked. I gave him Lewinski's name. 'You will go to him today and report the incident. In all probability you will have to repeat the story, the way you told it to me, before a court martial, under oath. Can you do that? In other words, will you stick to what you said?' I drew a breath and said: 'Yes.'

Then he gave me a friendly clap on the shoulder, said goodbye and told me to give his greetings to Captain von Lewinski.

The rest is quickly told. When I told our sergeant-major after the evening roll-call that I wished to speak to the commander on a personal matter, he wanted to know my reasons. I said that it was personal and I would only tell the captain himself. 'Very well,' he said, 'you are entitled to insist, every soldier has the right to speak to his chief personally.'

Next morning I was standing before the captain. 'I have a report to make, Sir,' I began, 'but before I begin there are two things I have to say. Firstly, I had two hours' leave yesterday to meet my cousin, Major von Tresckow of the General Staff, in the town. He has been a kind of godfather to me for many years, though he is not officially my godfather. He told me yesterday to make the report that I am now making to you, Sir. He also asked me to give you his greetings.

'Secondly, I wish to say that I am very glad to be a member of the 3rd Squadron and would like to remain so.'

Then I described the events of the day before in detail.

Lewinski was chalk-white when I finished. 'That happened in my squadron?' he stammered. Then he asked if none of his officers had been present at machine-gun drill. I said no, and he dismissed me with the words: 'Thank you for your report.' But I felt that he was struggling for self-control.

Some two weeks later I was a witness at a court martial called at regimental headquarters. My examination was brief and I did not have to take the oath because the accused had confessed. When the verdict was pronounced I was already back on stable duty, but the word soon went all round the Regiment that the two accused men had been condemned to a suspended prison sentence and demotion. They had also been transferred to another regiment. Our room-mate with the pebble glasses was released from the Army a few days later as unfit.

Never again was I to witness such an assault in the Army. The service remained hard, but correct. I was certainly 'cut' by some sergeants for months and treated with icy reserve, but I believe there are many advantages in a certain solidarity among NCOs. There were some sergeants in the squadron who 'out of comradeship with the victim of recruit Stahlberg', 'revenged themselves' on me. I took it in good part, and am glad to have spent a year as a cavalry recruit. You have to have ridden with a squadron to understand the fascination exerted by horses en masse, for the horse is a herd animal by nature.

Captain von Lewinski told us one day that squadron drill in closed ranks was unfortunately no longer part of cavalry training, although this was a touchstone for ability and discipline as a rider. If the cavalry had a future now, it lay exclusively in military reconnaissance and scouting in small groups, and yet one day he drew his sword and explained the old command signals with the weapon to the squadron. We tore off across the parade ground, the horses' hooves thundering on the earth. It was like some mighty natural cataclysm. What must it have been like in the days when not only a squadron, but a whole regiment or even several brigades appeared on the horizon?

Even our regimental trumpeters practised on horseback. If a squadron had performed particularly well it might be rewarded by the mounted band waiting outside the town to station itself in front of the squadron, the drummer, with his Kaiser Wilhelm beard and white gauntlets, on his magnificent, large-framed grey, the kettle-drums decorated with yellow saddle-cloths on either side. He guided his horse with his feet, the snaffle-reins ending at his stirrups.

The music started up at the first house on the edge of town and it was like the old nursery rhyme: 'When the soldiers come marching through the town, every maiden puts on her favourite gown . . .' Then the captain, riding at the head of his squadron, drew his sword, and at the command: 'Draw swords!' a hundred gleaming side-arms hissed into the air, to be propped for the parade on the right thigh, so that the blade lay on the shoulder, glittering in the sun.

And so we crossed the town, to the 'Cavalry March of the Great Elector' or even our regimental march, the 'Hohenfriedberger', which had been my father's regimental march as well.

TIM BISHOP

One Young Soldier

Recruit training, Life Guards, pre-war

The rank and file of most European armies have always contained some better-educated recruits, who are known in the British army as 'gentleman rankers'. One such was Tim Bishop, the son of an officer, who, on failing to gain a place at Sandhurst, the British military college for future officers of the infantry and cavalry, enlisted in 1934 as a private soldier (trooper) in the Life Guards. The Life Guards, which forms with the Royal Horse Guards the Household Cavalry, is the senior regiment of cavalry in the army and has a direct relationship with the sovereign.

Ceremonially uniformed in scarlet tunics, with black thigh boots, burnished steel breastplates and tall plumed helmets, and mounted on black horses, they escorted the King on state occasions and mounted a daily guard in Whitehall, the London thoroughfare on which the principal government offices stand. Much of a recruit's day was spent cleaning stables, grooming his horse to a high gloss and cleaning his uniform and accoutrements to a gleaming finish. Life Guards troopers, unlike those in Reiterregiment 6 that Stahlberg joined, were not expected to understand military law or the duty of the soldier to political authority. Such matters would have been thought dangerously abstract by the troopers' sergeants and officers. The object of recruit training in the Life Guards and all other British cavalry regiments before the Second World War was to produce a perfectly turned-out soldier who could ride, care for his horse, perform mounted drills and obey orders without question. Tim Bishop, who later became an officer in the 12th Lancers, perfectly conveys the atmosphere of the narrow world in which such recruits were raised.

*   *   *

Now, however, we were but the rawest recruits, who had not yet even drawn canvas dungarees in which to work in the stables. Smart uniforms, guards, escorts, 'box'-horses [for ceremonial guard] and stick orderlies [senior officers' messengers] were a world away and a hard world at that. A Foot Guardsman joined for three years with the colours and it was said that it took those three years to produce him fully trained. We joined for eight (and four on the Reserve) and our recruits' course alone would take a year. In other words we had let ourselves in for twelve years, but, thanks to the war, I was leading a squadron of 12th Royal Lancers within ten years. However, to my disappointment, it was not a gleaming sword but a much-used broom which I was handed (or had hurled at my back) in the half light of the following dawn. Thereafter I spent much time learning by experience the truth within the local jest: 'Join the Army and see the world. Join the Life Guards and sweep it.' But with nearly four hundred horses on the strength, almost non-stop sweeping was essential to keep the place like a new pin. And like a new pin did we keep it.

Several things took a bit of understanding. First, perhaps, the shouting of the NCOs at Reveille: 'Rise and shine, bed in line etc. etc. The sun is burnin' your bleedin' eye-ball'; and this on a pitch dark and freezing morning in January. Then there was the unofficial but total and dreadful ban on the wearing of pyjamas – that and having to share 'ablution' facilities. In that society it would have been as wrong to wear them as would be a top hat and cut-away coat out cub-hunting. The shirt and underclothes worn during the day were the recognized form of night attire and that was indisputably that. And it is nearly true that you can get used to anything in time. One was usually so tired that it did not matter much anyway. Secondly, the mucking out of stable which followed at 6.30 a.m. using only the bare hands. No stable implements were permitted young soldiers. Nor was so much as a wisp of straw wasted. Bedding that would have been thrown out of a civilian stable was carried, in warm, sticky, ammonia-scented and ton-weight armfuls, to be dumped in the yard (in bad weather under the lean-to sheds) to be spread to dry by 'old' soldiers with forks. It was carried in again, lighter and less pungent, at evening stables. Meanwhile these were swilled out until the last hay-seed was washed away. Breakfast could have been eaten off the floor. And when wisps [for rubbing down horses] were made for use at evening stables you made them from this used bedding, never from hay or clean straw. Thirdly, the method employed for removal of grease from new leather spur-straps (the type used for swan-necked spurs and jackboots). They were simply dropped into the urinal and left to lie there for a matter of days. They were then fished out, washed, dried – and, with considerable 'elbow-grease' – at last worked up to a fine polish. Fourthly, the shaving with cold water. One got used to that, too, but the Royal Navy complained bitterly and not really surprisingly when some of its members stayed with us at Windsor for a royal funeral. I was orderly corporal at the time and in the end just had to arrange for hot water from the cook-house to be taken up to them, 'in case,' as I explained to my indignant comrades-in-arms, 'we have another Invergordon!' [The naval protest against pay reductions in 1931.]

Fifthly, the food. A slab of yellow, soap-like cheese and one large white and pungent slice of raw onion was quite frequently the last meal of the day – unless, of course, you had sufficient private means to buy your own supper in the canteen later on. Very few had. And a breakfast of bacon smothered in fried onions could surely stymie the hungriest. Sixthly, having to wear ankle boots with nailed soles. In order to get about the stables and yard one adopted a sort of ostler's trot, but at first it was like working on ice. It did make life doubly difficult. The boots of, for instance, Corporal of Horse Wheatley had rubber soles. No doubt he had specific permission. But why did not we all? A quiet, hyper-efficient NCO, his approach was not only sure but silent, keeping the soldiers in his troop on their (steel-plated) toes, wide awake and alert. Seventhly, being constantly taken to task in no uncertain manner for something that could not possibly by any stretch of the imagination be your fault (such as being hauled over the coals because your horse's feet had gone away from under him on the tarmac and given you a crashing fall!).

Eighthly, trying to obey the orders of NCOs who might have

been talking Chinese, Russian or Arabic until you got the hang of it. Very early in my Army career I understood the guard commander to tell me to 'nibble the apples for a rhyme in Whaddon Woods'. Rapped out sharply, this strange command had the green listener at a disadvantage, yet to look baffled or to hesitate thereabouts, simply did not do. 'Corporal!' you shouted – and made yourself scarce, finding as quickly as possible a sympathetic interpreter. ('He wants you to nip up the stairs [apples and pears, Cockney rhyming slang] to the canteen for a cheese roll [a rhymer], a bun [a wad] and a packet of Woodbine cigarettes.') Like so many things, easy when you know how. The Life Guards had a language of their own and it was as well to learn it with all dispatch. For instance, the order to 'Get a rift on that chain'. To polish your horse's collar-chain you first dropped it in the gutter of your horse's stall and rubbed it about with your foot. You then picked it up, washed and dried it by rubbing it vigorously on the top of an unopened bale of straw. When it was dry, you put it into a sack which you had previously filled with shredded paper or clean straw. You then shook and tossed the sack until you had got a rift on the chain, a rift meaning, I suppose, the required sparkle.

Again, if someone was described as 'ticking' or 'bobbery' it meant that he was angry. And if an NCO warned you that you would be 'First relief chicks', he meant that your sentry duty would be carried out under the arches either side of the stables' entrance into the Horse Guards. This was officially known as 'Over the Arms', but unofficially as 'The Chicken Run' and was where the least smart members of the guard were posted. Initially, like so much else, it all called desperately for interpretation. A rifle was a 'bandook', 'browned off' was fed up, 'gippo' was gravy and 'duff' was pudding. The fair sex were 'cows'.

'Where is your Tom Clark?' was another question that could leave the raw recruit silent. A Tom Clark was a stable rubber dipped into a bucket of water and then wrung out as dry as possible. Put to immediate use it was a wonderful thing with which to lay the dust on the surface of a horse's coat just before showing him out to an NCO after grooming.

But it did seem to me as time went by that things were made as arduous and uncomfortable as was humanly, or inhumanely, possible. The times for recruits' parades were so arranged that it was impossible, for instance, to be dismissed from one riding school, turn in your horse, change your clothes and present yourself punctually and properly turned out for the next – perhaps the square (meaning drill with or without swords or rifles) – unless, that is, you went without breakfast. By missing this meal, however, we managed to do it and thereby, I suppose, learned over and over again that it is possible to achieve the impossible. A recruit received much the same treatment as does a tennis ball during a prolonged rally. The Inspector-General of Cavalry said of recruits in the 1890s that they 'hardly had time to eat their dinners'. Forty years later, neither had we.

Anyway, I never suffered such discomfort or was so rushed, even during six years of war. Perhaps anyone bothering to read this might think 'What did the fool expect?' Well, the fool expected what he got, though there is a difference between what you expect and the actual experiencing of it. But the fool was comforted in the knowledge, even then, that the happiest regiments are the best disciplined.

But to this day I still consider it pretty mean of the War Department of those times to expect soldiers, who were daily supposed to represent all that was smartest in the British Army and whom tourists from all over the world travelled miles to see, to wash *and shave* in cold water. However, one cannot deny that the general policy was one that worked. If you have a lot of young men who, because they are young men, are fairly pleased with themselves, you have to give them 'a bit of stick' or discipline suffers. For pleased with ourselves we must have been to have wanted to don a uniform that was going to take so much wearing and looking after (we had, for a start, got two different cap badges, let alone swords), and [to have wanted to ride] a horse that also had to be looked after and never left because, always stabled, it was incapable of looking after itself. 'Il faut souffrir pour être beau?' Suffer we assuredly did. Hell was most generously dispensed.

GEORGE ORWELL
Wounded (2)

Like so much else written by George Orwell, whose gift was as much for originality as fine writing, his account of his sensations on being wounded have few parallels in the literature of war. It is entirely characteristic of Orwell that, even in the throes of pain from a severe wound, he was able to collect his impressions and register them for subsequent translation into documentary form.

Orwell, an anti-Fascist by conscience, decided soon after the outbreak of the Spanish Civil War (1936–9) to go to Spain to fight on the Republican side. On arrival, chance rather than choice directed him into the ranks of the POUM militia, the fighting branch of a minority Marxist party, at odds with the increasingly dominant Communist Party and eventually to be brutally persecuted by it. Those events lay in the future when Orwell was hit. He had been serving in the trenches in the mountains of central Spain in the depths of a bitter winter. Because, unlike most of the POUM volunteers, Spanish or foreign, he had a little military training – gained in the cadet corps at Eton – he had been given a junior command appointment. He was relieving sentries when a stray bullet from the lines opposite him found its target.

Orwell's wound may well have saved his life, for he was still convalescing when the Communist Party began its brutal purge of the POUM in which many of the latter's members were murdered.

*   *   *

I had been about ten days at the front when it happened. The whole experience of being hit by a bullet is very interesting and I think it is worth describing in detail.

It was at the corner of the parapet, at five o'clock in the morning. This was always a dangerous time, because we had the dawn at our backs, and if you stuck your head above the parapet it was clearly outlined against the sky. I was talking to the sentries preparatory to changing the guard. Suddenly, in the very middle of saying something, I felt – it is very hard to describe what I felt, though I remember it with the utmost vividness.

Roughly speaking it was the sensation of being at the centre of an explosion. There seemed to be a loud bang and a blinding flash of light all round me, and I felt a tremendous shock – no pain, only a violent shock, such as you get from an electric terminal; with it a sense of utter weakness, a feeling of being stricken and shrivelled up to nothing. The sandbags in front of me receded into immense distance. I fancy you would feel much the same if you were struck by lightning. I knew immediately that I was hit, but because of the seeming bang and flash I thought it was a rifle nearby that had gone off accidentally and shot me. All this happened in a space of time much less than a second. The next moment my knees crumpled up and I was falling, my head hitting the ground with a violent bang which, to my relief, did not hurt. I had a numb, dazed feeling, a consciousness of being very badly hurt, but no pain in the ordinary sense.

The American sentry I had been talking to had started forward. 'Gosh! Are you hit?' People gathered round. There was the usual fuss – 'Lift him up! Where's he hit? Get his shirt open!' etc., etc. The American called for a knife to cut my shirt open. I knew that there was one in my pocket and tried to get it out, but discovered that my right arm was paralysed. Not being in pain, I felt a vague satisfaction. This ought to please my wife, I thought; she had always wanted me to be wounded, which would save me from being killed when the great battle came. It was only now that it occurred to me to wonder where I was hit, and how badly; I could feel nothing, but I was conscious that the bullet had struck me somewhere in the front of my body. When I tried to speak I found that I had no voice, only a faint squeak, but at the second attempt I managed to ask where I was hit. In the throat, they said. Harry Webb, our stretcher-bearer, had brought a bandage and one of the little bottles of alcohol they gave us for field-dressings. As they lifted me up a lot of blood poured out of my mouth, and I heard a Spaniard behind me say that the bullet had gone clear through my neck. I felt the alcohol, which at ordinary times would sting like the devil, splash on to the wound as a pleasant coolness.

They laid me down again while somebody fetched a stretcher. As soon as I knew that the bullet had gone clean through my neck I took it for granted that I was done for. I had never heard of a man or an animal getting a bullet through the middle of the neck and surviving it. The blood was dribbling out of the corner of my mouth. 'The artery's gone,' I thought. I wondered how long you last when your carotid artery is cut; not many minutes, presumably. Everything was very blurry. There must have been about two minutes during which I assumed that I was killed. And that too was interesting – I mean it is interesting to know what your thoughts would be at such a time. My first thought, conventionally enough, was for my wife. My second was a violent resentment at having to leave this world which, when all is said and done, suits me so well. I had time to feel this very vividly. The stupid mischance infuriated me. The meaninglessness of it! To be bumped off, not even in battle, but in this stale corner of the trenches, thanks to a moment's carelessness! I thought, too, of the man who had shot me and wondered what he was like, whether he was a Spaniard or a foreigner, whether he knew he had got me, and so forth. I could not feel any resentment against him. I reflected that as he was a Fascist I would have killed him if I could, but that if he had been taken prisoner and brought before me at this moment I would merely have congratulated him on his good shooting. It may be, though, that if you were really dying your thoughts would be quite different.

They had just got me on to the stretcher when my paralysed right arm came to life and began hurting damnably. At the time I imagined that I must have broken it in falling; but the pain reassured me, for I knew that your sensations do not become more acute when you are dying. I began to feel more normal and to be sorry for the four poor devils who were sweating and slithering with the stretcher on their shoulders. It was a mile and a half to the ambulance, and vile going, over lumpy, slippery tracks. I knew what a sweat it was, having helped to carry a wounded man down a day or two earlier. The leaves of the silver poplars which, in places, fringed our trenches brushed against my face; I thought what a good thing it was to be alive in a world where silver poplars grow. But all the while the pain in my arm was diabolical, making me swear and then try not to swear, because every time I breathed too hard the blood bubbled out of my mouth.

BRUCE SHAND
Previous Engagements

Bruce Shand's light-hearted reminiscence of life in a 'good' regiment during the Phoney War of 1939–40 in France epitomizes the studiedly unserious flavour of pre-war soldiering in the British army. Commissioned from Sandhurst into the 12th Lancers in 1937, he had spent the years of peace riding to hounds, playing polo and making friends, of whom he accumulated a large number.

His insouciance was, however, an affectation. A man of keen intelligence and marked literary gifts, he proved when the real war came both an efficient and a brave soldier, twice winning the Military Cross before he was taken prisoner in the Western Desert. Still thriving, he personifies the type of *beau sabreur* who had officered European armies since the days of Louis XIV and of which a few survive to this day.

*   *   *

The Phoney War in France, 1940 (for fun)

After seeming hours the stocky form of Captain Dowell, RAMC came through the door. 'Well, Bruce, what's the matter with him?' 'Probably clap,' I diagnosed. 'Well, we'd better give him a catheter.'

Once again a heavy sweat broke out on me, but with the doctor's arrival I had averted my eyes from the patient to look out of the window and could see his truck and the orderly now talking to Sergeant Ditton. With a presence of mind that has seldom visited me since, I opened the window and shouted to Ditton to send the orderly up and then removed myself cravenly from this urinary battlefield. Stinker [Dowell, the medical officer] bore me no ill will and later came up to our mess to have a much-needed drink saying that he'd 'given the old bastard a good blow through'.

Later, in the winter, I managed to become stricken with 'flu which developed into mild pleurisy. Stinker was in his element and put in much time listening to my lungs, finally recommending that I needed to recuperate in a warm climate. I had hoped that I might have returned temporarily to England but, before long, I was on the Blue Train en route for Cannes.

Dozy Willis, who would have had friends on Pitcairn Island should a visit have been contemplated there, had long known Lady Trent, widow of the original Jesse Boot, the cash chemist. Part of the year she lived in a handsome villa on the Californie hill, which she and her husband had equipped as an officers' convalescent home in the previous war. She was patriotically prepared to repeat the performance but as events turned out I was to be the only beneficiary of her bounty.

I left Fonquevillers on New Year's Day, lunched with some friends in Paris and boarded the Blue Train in the evening. During dinner I had some desultory conversation with a middle-aged Hungarian lady who spoke tolerable English. I had quite a lot of magazines and papers with me, some of which she asked to borrow, saying that she should return them in the morning. I slept remarkably well until the *wagon-lit* attendant called me the next day, bringing in the journals, about which I had forgotten. He gave the rather surprising information that the lady had been removed in the night at Dijon by the French police. I looked carefully through the *Bystander*, *Horse & Hound* and other unlikely publications for hidden messages, without avail.

Lady Trent's White Russian chauffeur met me at the station in a dark green Rolls-Royce and I moved for ten days into a sybaritic existence. She was a delightful and unaffected person living with her unmarried sister in considerable comfort and served by admirable servants. I was treated far too well, especially as many people seemed to think I had been wounded, despite there being no fighting. Indeed, there was a rather embarrassing moment when a young French naval officer and I, as symbols of allied unity, were made to stand on a dais in the Carlton Hotel, at some charity gala, clasping each other by the hand before both national anthems were played. On another occasion, taken to luncheon by Lady T, in the house where Prince Leopold died of haemophilia, I drew as my neighbour the widow of Marshal Joffre, an interesting link with the past. She was, I think, his second wife, a formidably upholstered lady. Luckily I had been informed that 'Papa' had been a great trencherman, so we cruised fairly easily along gastronomic channels.

A retired doctor of the Indian Medical Service and a very social French physician examined my chest rather perfunctorily before lunch one day and agreed that I was sound. Back I went to Fonquevillers, though strangely I have no recollection of the return journey from Cannes.

No sooner had I returned to 'A' Squadron than there was an alert and we moved up for the second time to the Belgian frontier. We were billeted near Roubaix, with our mess in a turreted villa of unbelievable hideousness. However, at least it was warm, a great blessing in that very cold winter, and we made the best of our circumstances with a battalion of the Coldstream Guards as congenial neighbours. We must have stayed there for a week before the alarm subsided and it was during this period that our regimental chaplain preached a sermon of such power ('Be strong and of a good courage') that a couple of troopers and certainly one young officer were obliged to retire from the congregation.

This interesting cleric, the Revd Godfrey Macmanaway, had joined us when we arrived in France. An Ulsterman, currently Rector of Londonderry and the son of a former Bishop of Clogher, he had been an airman in the First War and proudly wore RFC [Royal Flying Corps] wings on his tunic. Extremely convivial, eloquent in the pulpit (he never prepared his sermons), rather idle and immensely social, he was an asset to any assembly, and there was little he did not appear to have done in his life. At a later date we briefly shared quarters and I observed that he invariably had his servant bring him a glass of warm whisky and water before getting up in the morning. He then smoked a powerful pipe before arising to his ablutions and a hearty breakfast. He claimed never to have had a hangover.

Assiduously political, he resigned his living in 1950 when elected to Westminster as Unionist Member for West Belfast. Confusion then arose as to whether he should be permitted to take his seat, since there was doubt about whether the Church of Ireland was disestablished or not. After considerable excitement – I think his case warranted a leading article in *The Times* – he was disqualified. He returned to Belfast in some chagrin and died the following year of a fall or a stroke when leaving the Ulster Club.

A constant feature of my life was Trooper Smallridge, who had been assigned to me before the war as first servant. A Welshman from a mining family in Treherbert, Glamorganshire, he was roughly my age, perhaps a little older, a remarkably good driver, small and lithe, with a slight stutter. He made friends wherever he went, from my august grandmother (his youngest sister came to work for her as a housemaid) to the most ill-assorted Arabs.

Before the war he cannot have had too many military duties (officers' servants were the bane of sergeant-majors – they could never get their hands on them for their own purposes), as most of his time must have been taken up in attending to the voluminous wardrobe that was needed for even an ordinary subaltern's life. He came away with me all the winter when I was hunting, cleaning my clothes and boots to a degree of perfection that will never be seen again. He still lives in Treherbert and we correspond and telephone periodically, though he would be appalled if he could see the undisciplined attire in which I am writing these words. Our ways only parted when I was wounded and captured. He soldiered on until after the end of the war, finally leaving the regiment at Villach in Austria some time in the summer of 1945. He returned then to me for a brief time while I was still in the Army, bringing with him my

Everyman copy of Lord Chesterfield's *Letters to His Son*, which had cost about 2/6d when I purchased it in Reigate in 1941. It had travelled with me to the Middle East, was with me throughout the desert campaign and was in my jeep when I was taken prisoner. Smallridge hung on to it for the rest of the war and it went with him through the remainder of the North African campaign and then to Italy. I still possess this battered and much travelled volume.

Despite the various alarms, life during that particular hard winter and spring was both monotonous and uncertain. Our training was restricted both by geographical boundaries and by a limitation on the use of petrol. Living in close proximity, with not too much to do, was irksome at times, especially for those with a temperament like Andrew Horsburgh-Porter, who was the epitome of the free-range officer. Someone coming into the small farmhouse parlour that we used for a squadron mess said that it resembled a scene from *Journey's End*, but 'above ground and without shellfire!' Our only view was that of the midden.

WINSTON CHURCHILL
BBC Broadcast, London, 19 May 1940

The false calm of the Phoney War was shattered on 10 May 1940, when the German panzer divisions broke through the defensive lines of the French and British armies in France and Belgium and began a whirlwind advance which, nine days later, would bring their spearheads to the English Channel. By the end of the month, the British Expeditionary Force would begin its seaborne evacuation from Dunkirk. On 21 June the French government would sign an armistice, at a ceremony staged on the same spot, and in the same railway carriage, where the Germans had signed the armistice with the French and their allies in November 1918.

The military crisis precipitated a political crisis in Britain. The House of Commons indicated its loss of confidence in the premiership of Neville Chamberlain. He was succeeded as Prime Minister by Winston Churchill, the First Lord of the Admiralty, on 10 May, the first day of the catastrophe. On 19 May, he made his first broadcast as Prime Minister to the British people. It was to be followed by many others during the war. None, in retrospect, seems more inspiring than this, in which he set out his beliefs in the rightness of the Allied cause – resistance to tyranny, the championship of freedom – and his whole-hearted confidence in eventual victory. What might have sounded hollow words, had Britain succumbed, ring out, even fifty years after the victory he did so much to win, with a magnificent defiance. They were to bring heart to his people then and throughout the five succeeding years of struggle.

*   *   *

I speak to you for the first time as Prime Minister in a solemn hour for the life of our country, of our Empire, of our Allies, and, above all, of the cause of Freedom. A tremendous battle is raging in France and Flanders. The Germans, by a remarkable combination of air bombing and heavily armoured tanks, have broken through the French defences north of the Maginot Line, and strong columns of their armoured vehicles are ravaging the open country, which for the first day or two was without defenders. They have penetrated deeply and spread alarm and confusion in their track. Behind them there are now appearing infantry in lorries, and behind them, again, the large masses are moving forward. The regroupment of the French armies to make head against, and also to strike at, this intruding wedge has been proceeding for several days, largely assisted by the magnificent efforts of the Royal Air Force.

We must not allow ourselves to be intimidated by the presence of these armoured vehicles in unexpected places behind our lines. If they are behind our Front, the French are also at many points fighting actively behind theirs. Both sides are therefore in an extremely dangerous position. And if the French Army, and our own Army, are well handled, as I believe they will be; if the French retain that genius for recovery and counter-attack for which they have so long been famous; and if the British Army shows the dogged endurance and solid fighting power of which there have been so many examples in the past – then a sudden transformation of the scene might spring into being.

It would be foolish, however, to disguise the gravity of the hour. It would be still more foolish to lose heart and courage or to suppose that well-trained, well-equipped armies numbering three or four millions of men can be overcome in the space of a few weeks, or even months, by a scoop, or raid, or mechanized engagement of the masses, which will enable the qualities of the French and British soldiers to be matched squarely against those of their adversaries. For myself, I have invincible confidence in the French Army and its leaders. Only a very small part of that splendid army has yet been heavily engaged; and only a very small part of France has yet been invaded. There is good evidence to show that practically the whole of the specialized and mechanized forces of the enemy have been already thrown into the battle; and we know that very heavy losses have been inflicted upon them. No officer or man, no brigade or division, which grapples at close quarters with the enemy, wherever encountered, can fail to make a worthy contribution to the general result. The Armies must cast away the idea of resisting behind concrete lines or natural obstacles, and must realize that mastery can only be regained by furious and unrelenting assault. And this spirit must not only animate the High Command, but must inspire every fighting man.

In the air – often at serious odds – often at odds hitherto thought overwhelming – we have been clawing down three or four to one of our enemies; and the relative balance of the British and German Air Forces is now considerably more favourable to us than at the beginning of the battle. In cutting down the German bombers, we are fighting our own battle as well as that of France. My confidence in our ability to fight it out to the finish with the German Air Force has been strengthened by the fierce encounters which have taken place and are taking place. At the same time, our heavy bombers are striking nightly at the taproot of German mechanized power, and have already inflicted serious damage upon the oil refineries on which the Nazi effort to dominate the world directly depends.

We must expect that as soon as stability is reached on the Western Front, the bulk of that hideous apparatus of aggression which gashed Holland into ruin and slavery in a few days, will be turned upon us. I am sure I speak for all when I say we are ready to face it; to endure it; and to retaliate against it – to any extent that the unwritten laws of war permit. There will be many men and many women in this Island who when the ordeal comes upon them, as come it will, will feel comfort, and even a pride that they are sharing the perils of our lads at the Front – soldiers, sailors and airmen, God bless them – and are drawing away from them a part at least of the onslaught they have to bear. Is not this the appointed time for all to make the utmost exertions in their power? If the battle is to be won, we must provide our men with ever-increasing quantities of the weapons and ammunition they need. We must have, and have quickly, more aeroplanes, more tanks, more shells, more guns. There is imperious need for these vital munitions. They increase our strength against the powerfully armed enemy. They replace the wastage of the obstinate struggle; and the knowledge that wastage will speedily be replaced enables us to draw more readily upon our reserves and throw them in now that everything counts so much.

Our task is not only to win the battle – but to win the War. After this battle in France abates its force, there will come the battle for our island – for all that Britain is, and all that Britain means. That will be the struggle. In that supreme emergency we shall not hesitate to take every step, even the most drastic, to call forth from our people the last ounce and the last inch of effort of which they are capable. The interests of property, the hours of labour, are nothing compared with the struggle for life and honour, for right and freedom, to which we have vowed ourselves.

I have received from the Chiefs of the French Republic, and in particular from its indomitable Prime Minister, M. Reynaud, the most sacred pledges that whatever happens they will fight to the end, be it bitter or be it glorious. Nay, if we fight to the end, it can only be glorious.

Having received His Majesty's commission, I have found an administration of men and women of every party and of almost every point of view. We have differed and quarrelled in the past; but now one bond unites us all – to wage war until victory is won, and never to surrender ourselves to servitude and shame, whatever the cost and the agony may be. This is one of the most awe-striking periods in the long history of France and Britain. It is also beyond doubt the most sublime. Side by side, unaided except by their kith and kin in the great Dominions and by the wide Empires which rest beneath their shield – side by side, the British and French peoples have advanced to rescue not only Europe but mankind from the foulest and most soul-destroying tyranny which has ever darkened and stained the pages of history. Behind them – behind us – behind the armies and fleets of Britain and France – gather a group of shattered States and bludgeoned races: the Czechs, the Poles, the Norwegians, the Danes, the Dutch, the Belgians – upon all of whom the long night of barbarism will descend, unbroken even by a star of hope, unless we conquer, as conquer we must; as conquer we shall.

Today is Trinity Sunday. Centuries ago words were written to be a call and a spur to the faithful servants of Truth and Justice: 'Arm yourselves, and be ye men of valour, and be in readiness for the conflict; for it is better for us to perish in battle than to look upon the outrage of our nation and our altar. As the Will of God is in Heaven, even so let it be.'

STUART MILNER-BARRY
Codebreakers

One of the chief props to Churchill's confidence during the early and dark days of the war with Germany was the knowledge that the British Government Code and Cipher School (GC&CS), located at Bletchley Park in Bedfordshire, was beginning to read Germany's most secret military ciphers. These, enciphered on a compact, electrical switching machine known as 'Enigma', were believed by the German armed forces to be totally secure, since any single letter might be translated by the machine in any one of 200,000,000 different ways. It was calculated by its users that the effort to decipher an Enigma message would defeat any group of analysts, however large, using mathematical attack to test out the permutations.

In theory, the Germans were correct. Enigma, if strictly used, provided an impenetrable ciphering system. What the Germans did not recognize was that Enigma operators would be prone to taking short cuts, to the repetitive use of formula messages and to other procedural mistakes that would provide the British cryptologists with entry points into the Enigma traffic. Progress at first was slow, despite much help given by the Poles, who were the first to break into Enigma, and some ciphers, used by meticulous German operators, were never penetrated. By the end of 1940, however, the mathematicians, chess champions, crossword solvers, etymologists and other exceptional brains collected at Bletchley had begun to break into several Enigma ciphers in 'real time' – near or at the time that German recipients were decrypting the messages intended for their sole use – and so were able to anticipate and forestall German operational moves. The Enigma secret, codenamed Ultra, was one of the most important of Britain's – and later America's – war-winning techniques.

*   *   *

Hut 6: Early Days

When the war broke out in 1939 I was in the Argentine, playing chess for the British team in the Olympiad. My great friend and rival, C. H. O'D. Alexander, was another member, as was Harry Golombek, later chess correspondent of *The Times*. We returned home immediately, and before long found ourselves at Bletchley Park, where we remained for the duration. We had been recruited by Gordon Welchman, an old friend of mine at Cambridge.

Welchman was a brilliant mathematician, a research fellow at Sidney Sussex. He and I came up to Trinity College, Cambridge, in the same term (October 1925). We became friends, and, as I lived in Cambridge with my mother (who was Cambridge born and bred, the daughter of Dr W. H. Besant, a well-known mathematical fellow of St John's College), we saw a lot of Gordon at our home in Park Terrace. In due course Welchman married and came to live in Cambridge himself.

Alexander was three or four years younger than Welchman and myself, but I had known him ever since, in 1924, he had beaten me at Hastings in the British Boys' Chess Championship, which I had won the previous year at the age of 16. Hugh Alexander was a scholar of King's [College, Cambridge] and also a mathematician, though not, I fancy, in Welchman's class. He came from Birmingham, where my sister (the eldest of our family) was a lecturer in English and Anglo-Saxon at the University. When he left Cambridge he became a schoolmaster and taught mathematics at Winchester College. But before long he married an Australian girl some years older than himself. She was unhappy in the cloistered atmosphere at Winchester, and eventually he gave up teaching at which he was very good, and took up a business career in London for which he was ill adapted; he was far too untidy even to look like a businessman.

Welchman had, I think, been recruited into intelligence work before the war broke out, and he had already done a good deal of research into the mysteries of the Enigma machine. These of course came easily enough to Alexander, but I am (in spite of my ancestry) almost innumerate. I therefore found Gordon Welchman's patient explanations very difficult to follow, and to this day I could not claim that I fully understood how the machine worked, let alone what was involved in the problems of breaking and reading the Enigma cipher. Fortunately this did not matter too much, as I was able to make myself useful in other ways – principally because I had a working knowledge of German.

To return to the beginning of the war: fairly early in 1940 we three found ourselves installed at The Shoulder of Mutton Inn, Old Bletchley, about a mile from the entrance to Bletchley Park. Here Hugh and I were most comfortably looked after by an amiable landlady, Mrs Bowden. As an inn-keeper she did not seem to be unduly burdened by rationing, and we were able (among other privileges) to invite selected colleagues to supper on Sunday nights, which was a great boon. Welchman moved out fairly soon to live with his wife in the town, but Hugh and I happily remained at the Shoulder until the end of the war.

When we first came to GC&CS, Bletchley Park was a tiny organization, probably not more than thirty strong. It consisted of a few old-time professionals who had worked in Room 40 [Naval Intelligence] at the Admiralty in the First World War, such, for example, as Dillwyn Knox, a Fellow of King's who died during the war, and A. G. Denniston, and new recruits such as Welchman and Alan Turing. Knox had, so I understood, been defeated by the Enigma, and the main credit for solving the Enigma and subsequently exploiting its success, should (subject to the Poles) probably go to the other two.

Turing was a strange and ultimately a tragic figure. But as an admirable biography of him has been written [Alan Hodges, *Alan Turing: The Enigma of Intelligence*, 1983], I shall say no more here. Welchman, on the other hand has, I think, never received his just deserts, quite apart from being ridiculously persecuted on security grounds for revealing, some forty years after the event, how the job of breaking the Enigma had been done. Welchman was a visionary, and a very practical visionary at that. In spite of Knox's failure, he always believed the Enigma could be broken. He also realized the enormous importance of the success, and took it for granted that, when the phoney war ended, the Germans would rely principally on the Enigma for their military communications. He foresaw much of what would be involved in the way of expansion of staff, machinery (the bombes [see p. 361]), and all the other necessary substructure. And he had the fire in his belly that enabled him to cajole higher authority into supplying our wants. If Gordon Welchman had not been there, I doubt if Ultra would have played the part that it undoubtedly did in shortening the war.

The first essential, once the initial breakthrough had been made, was the expansion of staff. This was done to begin with on an 'old-boy' basis. Welchman knew a few undergraduate mathematicians whom he had supervised at Sidney Sussex. I lived in Cambridge and had the advantage of a close connection with Newnham College, where my sister Alda had been Vice-Principal until her untimely death in 1938. So I was able to recruit a few girls from both Newnham and Girton Colleges, such as Mary Wilson, Wendy Hinde, Margaret Usborne, and Jane Reynolds (later Jane Monroe), who formed, together with a number of undergraduates from men's colleges, the invaluable nucleus of the original Hut 6. At about the same time I made a very profitable foray to the Scottish Universities.

As a result of these efforts, we were able, even during the period of the phoney war, to put ourselves on to a three-shift basis, and there of course we stayed throughout the war. This was during the winter and spring of 1940, and the decrypts themselves were mostly practice by the Germans – nursery rhymes and the like. But the exercise gave *us* invaluable practice as well as the Germans, and provided us with a battery of cribs, of which we were able to make invaluable use when the war became real.

This happened, as I remember, with the invasion of Norway, about the end of April and beginning of May 1940. The keys [German enciphering systems] we had been breaking were a practice key and the Red (general) key of the German Air Force. The Red key became of vital importance immediately, and remained so all through the war and in all the main theatres of war except Africa. Another key was the Army–Air Force key, the Yellow, used only in Norway.

The Red was the great standby that kept Hut 6 going. I cannot remember any period when we were held up for more than a few days at a time. Indeed, if we had been, it was by no means certain that we would have been able to get started again. The fact that we nearly always broke the Red, often quite early in the day, may give the impression that it was as easy as falling off a log; indeed Hut 3, our intelligence opposite numbers to whom all our decrypts were sent immediately, took it for granted that they would receive their daily ration. But there was in fact no certainty at all about it.

This leads me to stress how extraordinarily lucky we were. The Germans regarded the Enigma as a perfectly secure machine, proof against cryptanalysts however talented and ingenious they might be. Several times during the war, when it became inescapably clear that a great deal of intelligence was finding its way to the Allies, the Germans set up commissions of inquiry into the security of the Enigma. They always came to the conclusion that the machine was invulnerable and that the leakage must be due to secret agents. And the fact is that, had it not been for human error, compounded by a single design quirk, the Enigma was intrinsically a perfectly secure machine. If the wireless operators who enciphered the messages had followed strictly the procedures laid down, the messages might have been unbreakable. But of course that was the whole trouble. It is easy enough in principle to observe the relatively simple rules, such as, for example, avoiding enciphering the address and title of the recipient in the same way at the beginning and the end of the message. But if you are a harassed machine operator, it is so much easier to do what you have always done than consciously to force yourself to put your 'Xs' in different positions, or not to put 'Xs' in at all for punctuation, that it is very difficult to blame those concerned.

In claiming credit, as I have done, for Turing and Welchman for the initial breakthrough in the Enigma, I should stress the original and vital contribution made by the Poles. The Poles had always been brilliant cryptographers, and had been breaking and reading some Enigma ciphers since the early 1930s. When the war broke out, however, the Germans made a major change in the machine which put the Poles out of business; but shortly before the war and in anticipation of it, the French Intelligence Chief (Colonel Bertrand)

and ourselves had got together with the Poles, who not only gave us an Enigma machine, but shared with us all their knowledge and experience. They had a group of distinguished mathematicians, including one, Marian Rejewski, and another, Colonel Langer, whom I met after the war. It was always a mystery to me that the Polish contingent was not incorporated at Bletchley during the war, where they would no doubt have made an invaluable contribution; but in fact they were sidetracked in France and had to be evacuated when the Germans overran the whole of the country. I can only assume there were security doubts, and I believe the Poles continued to operate their own organization; but I feel there must have been a sad waste of resources somewhere.

So much has been written on the technical side, including, of course, Gordon Welchman's own *The Hut Six Story* [1982], and many erudite productions by Polish cryptanalysts, that it would be pointless to add to it. But for the non-technical layman who would like to have some idea of what was going on, it might be of interest if, as a non-technical layman myself, I were to try to describe things as I saw them.

The Enigma, an electric machine, looked like a typewriter with a typewriter keyboard. But it could be set up in millions of different ways, so that if you typed out an ordinary sentence on it you would get nothing but a jumble of nonsense groups. In order to turn these nonsense groups back into the original German you therefore had to have a machine set up exactly as the original operator had it. Merely having a machine did you no good at all, and since the Germans changed their machine settings every twenty-four hours, you were helpless unless you knew in advance, or could in some way work out, what the new setting for the next day was to be, out of the millions available.

Since we did not know, we had to start from scratch every day; and the basis of breaking was known as the 'crib'. Take, for example, a routine message like *keine besonderen Ereignisse* ('no special developments'). This might appear in the enciphered version as:

Text: KEINE BESON DEREN EREIG NISSE
Cipher: ACDOU LMNRS TDOPS FCIMN RSTDO

This jumble would form the core of the message. But, of course, most messages had to have an address and a signatory; thus there would have to be two or three groups beforehand and two or three at the end, so that, even if you knew that somewhere in a short message (e.g. twenty groups of five) there lurked the five groups which represented *keine besonderen Ereignisse*, you did not know which five consecutive groups they were.

Here we were greatly assisted by an idiosyncrasy of the Enigma machine. This was that a letter 'could never go to itself': in other words, if you tapped A on your machine, you might get any letter appearing except A.

As you will see, this was a great help in eliminating possible candidates. If you look back at *keine besonderen Ereignisse*, you will see that in the enciphered version beginning ACDOU there is nowhere a clash of letters between the German and the enciphered groups. But if, for example, you were looking for the most likely position in your twenty groups for the *keine besonderen Ereignisse* sequence, and you noticed that under the B in *besonderen* was another B, you would know for a certainty that that could not be the right position for the core of the message, and would have to look elsewhere.

Let us assume, however, that you have guessed the right position for the five *keine besonderen Ereignisse* groups. The problem is now to set up your own machine in an identical position in all respects to the German machine. If you could do that, you would have broken the Enigma Red key for that day; and then you could decipher all the messages sent over the air by the Germans for that day in that particular key. But how was this problem to be resolved?

In theory, of course, it could be done by running through all possible settings of the machine until, as by a miracle, the tapping of ACDOU LMNRS etc. produced KEINE BESONDEREN EREIGNISSE. But in practice this was completely impossible. It might have taken years to run through all the possible permutations and combinations, and the war might well have been over before we had succeeded with even one key for one day. However, given that we were confident that somewhere in a jumble of consecutive five-letter groups was concealed a text of intelligible German –

twenty-five letters such as in *keine besonderen Ereignisse* should normally be quite sufficient – that would put us well on the road to success. So long a stretch would enormously cut down the number of machine settings that would have to be tried; and modern technology had, in fact, provided a machine (known as the bombe) which would normally provide the correct setting of the Enigma within a few hours or so.

In fact, the Germans used a number of keys every day. Some we were able to break and some not. As I explained earlier, the Red was the great standby, but at times others were of even greater operational importance. There was, for example, the Brown. The Brown was the key used by Kampfgruppe [Luftwaffe battle group] 100, which was responsible for the so-called Baedeker raids on key targets such as Bristol, Birmingham, Coventry, and so on [in retaliation for RAF raids on equivalently historic German towns]. My recollection is that the Brown was a relatively easy key to break, and that we often broke it fairly early in the day. Obviously, the earlier the better, so that counter-measures could be put in hand: sometimes even squadrons in the air, or at least such anti-aircraft ground measures as were possible. Unfortunately, the Germans did not tell us what the targets were. They simply referred to them by numbers (*Ziel* [target] so-and-so), and we might or might not be able to guess what town was intended. There was an obvious risk that, if we took counter-measures, we should give away that we knew from Enigma that X was the target that day. There was a story, which can still be heard from time to time, that it was known that Coventry was to be the target for a devastating raid, and that Churchill forbade any special precautions for fear of giving away the all important secret that we were reading the Enigma. What Churchill would have done had he been confronted with this terrible dilemma I have, of course, no idea. But it so happens that we did not know the target for that night until too late; in fact, as I remember, we were expecting a very heavy raid on London.

While Brown was, I suppose, the most dramatic of the various keys with which we were concerned, it was when Rommel invaded Egypt that Hut 6 really came into its own. The German expeditionary force had its own keys, which we named after particular birds, and the most important bird was Chaffinch. I think we read Chaffinch pretty consistently after the spring of 1942. And since the Germans had no other regular means of communication with their forces abroad, we were given a pretty complete picture of their strength and disposition, as well as their intentions; and, as a by-product, a lot of information about their U-boat movements in the Mediterranean. (This was quite distinct from the U-boat warfare in the Atlantic, which was looked after by Hut 8 under Alexander and which dealt in the U-boats' own naval Enigma key). I remember I made Chaffinch my own particular concern, if that day's key had not been broken by the time I came on deck.

Hut 6 was the cryptanalytical section; Hut 3 the intelligence section. We broke the keys, the messages were deciphered as they came in by our deciphering section, and the contents were passed immediately to Hut 3. It was Hut 3's responsibility to decide what they meant, to decide the degree of importance and urgency, and to report accordingly to London. Naturally, it was essential that we should keep in the closest touch with Hut 3, which was organized in watches just as we were; and the heads of our respective watches were in continual discussion about the priorities for breaking and deciphering. In the early days, before the production of bombes overtook the requirement, this sometimes produced very difficult discussions between Hut 6 and Hut 8, because we shared the same bombes. But, fortunately, Alexander and I were such close friends as well as colleagues, that I do not remember any serious dispute ever arising. In any case, I took it for granted that ultimately breaking the U-boat cipher must take priority, for failure to read it could lose us the war. But naturally we took guidance from Hut 3 about which of our Hut 6 ciphers took priority in our sphere.

Once again, we were very fortunate. Hut 3 was staffed rather differently from Hut 6; whereas Hut 6 had a high proportion of undergraduates (even one or two straight from school), those in Hut 3 were for the most part older. They were, naturally, German scholars and not cryptanalysts, many of them schoolmasters or young dons of their University (predominantly Cambridge as it happened). But they were presided over by Group-Captain Jones. Jones was not a scholar or an academic; I suppose he must have had some knowledge of German, but primarily he was a businessman coming, I think, from Lancashire. I do not know what brought him to Bletchley, but it proved a brilliant choice. He was a genuinely modest man who regarded himself as having little to contribute compared with the boffins with whom he was surrounded; in fact he was a first-rate administrator who was liked and trusted by everyone. After the war he stayed on in the service, and ended up as the head of GCHQ [Government Communications Headquarters] at Cheltenham.

I do not imagine that any war since classical times, if ever, has been fought in which one side read consistently the main military and naval intelligence of the other. Of course, the Germans read a fair amount of our own codes, but in nothing like the comprehensive and all-embracing manner in which we read theirs. It was rather like a game of bridge in which we were shown the opponents' cards before the hand was played. One might well have expected that, with the cards so stacked on one side, so big an advantage would be reaped by the reading side that the result of the contest could hardly be in doubt. But, although the intelligence balance was enormously in our favour, its only really decisive effect was in the Battle of the Atlantic. It can, I think, be pretty confidently asserted that, had we not at the most crucial times and for long periods read the U-boat ciphers, we should have lost the war. I have seen it argued that, in the long run, it would have made no difference because the Americans would have more than made up any conceivable losses in our – and their – shipping; but this makes big assumptions about how the United Kingdom was to be kept from starving in the meantime. Certainly, when, as did happen in 1942, Hut 8 was held up for months on end on the U-boat ciphers, the losses in the Battle of the Atlantic rose in the most alarming way. This was particularly so after the entry of the United States into the war, before the Americans had had time to organize a convoy system on the American littoral.

So far as the Army and Air Force were concerned, the advantage conferred on the Allies must have meant a very great saving of lives and, I imagine, a considerable shortening of the war. I would not suggest that it made the difference between victory and defeat. Indeed, in the early days the disparity in overall strength and readiness between the Germans and the Allies was such that no amount of knowing their intentions could prevent them carrying them out; only later, when the strengths became more equal, did the intelligence advantage become a match-winning factor, and remained so until the last days of Hitler.

I felt at the time – and looking back on it after more than half a century, I feel exactly the same – that I was extraordinarily lucky to have found myself in that particular job at that time; and I suspect that almost everybody concerned in the Enigma operation at every level felt the same. We were slightly ashamed, or at any rate unhappy, that we should be leading such relatively safe and comfortable lives when a large proportion of the population, whether civilian or in uniform, was in so much danger. But at the same time, we could not help realizing that our work, individually and collectively, was of enormous importance in the conduct of the war. For that reason alone there was a spirit of camaraderie which I think never failed us, even when the war generally seemed to be going badly, or when we found ourselves temporarily delayed in delivering the goods. There was a perpetual excitement about each day's breaks, at whatever time of day or night they might come. To the chess player, it was rather like a long-running tournament with several rounds being played every day, and never any certainty that the luck would continue to hold. The friendships that were formed during the war have in many cases survived fifty years later, and that applies equally to some of the small but very high-powered American contingent, headed on the cryptanalytical side by Captain (as he then was) William Bundy. No more admirable representative of our great new ally could possibly have been selected. Many of us feared, I think, that the Americans, with their enormous superiority in manpower and resources of every kind, would quickly reduce Hut 6 – and no doubt Hut 8 as well – to a very subordinate and minor role. Nothing of the kind happened. Bundy and his colleagues were anxious only to learn and to help. They learnt very quickly and their help was invaluable. Above all, able and high-powered as they were, they were modesty itself in their demeanour. There could have been no happier partnership.

It goes without saying that, since we were frequently reading the enemy's communications as soon as their intended recipients, some pretty dramatic messages were brought to me in Hut 6. I remember especially a message describing in considerable detail the German plans for the invasion of Crete in some fortnight's time [May 1941], and also thinking to myself (quite wrongly, alas!), 'Now, we really have got them this time.' And also the desperate message from the German commander in the Battle of Normandy in 1944, which heralded the collapse of the German resistance in the Cotentin peninsula and enabled the American tanks to break for Paris. This kind of message, shown to us maybe in the middle of the night, gave one an extraordinary sensation of living with history. But of course it was not individual messages of this kind that mattered most, though the Prime Minister took the liveliest interest in them and No. 10 [Downing Street] had to be kept constantly in touch. It was the complete build-up which Hut 3 was able to construct of the whole German side of the fence that counted, and that could have been achieved in no other way.

The Cretan episode was, from the Hut 6 point of view, the greatest disappointment of the war. It seemed a near certainty that, with General Freyberg warned that the crucial point of the invasion was to be the airborne attack on the Maleme airport, and the time and every detail of the operation spelt out for us in advance, and given the appalling difficulty and danger of any airborne invasion in the best of circumstances, the attack would be ignominiously thrown back; and we awaited the operation with anxiety but also with a considerable degree of confidence.

In the event, it was a 'damned close-run thing'. The Germans took Crete from the air, and we lost a great deal of shipping in trying to save it. The best that could be said about it, from the Allied point of view, was that, though the conquest of Crete was ultimately achieved, it was so enormously expensive that the Germans never attempted an airborne invasion on that scale again. We fully expected, I believe, that Crete would be followed by Malta; but it never came. So, since the loss of Malta would have been an appalling catastrophe, we can at least be assured that the defenders of Crete lost their lives in a good cause.

PETER CREMER
U-333

The intelligence derived from breaking Enigma transmissions – the product was known as Ultra in Britain, because it had the highest of all security classifications – was of vital importance in fighting the Battle of the Atlantic. The identification of U-boat movements through Ultra intercepts allowed convoys to be directed away from their patrol lines, aircraft to be vectored to their positions and escort groups of destroyers and frigates to be hurried to their concentration areas. The cipher war in the Battle of the Atlantic swayed one way and another. British naval codes, particularly the Long Naval Code No. 3, were read by the Germans while the British were reading Enigma. At the end of 1942 and well into 1943, the British lost the U-boat key altogether, with calamitous results for convoy sailings. Overall, however, Ultra intelligence was a crucial factor in the winning of the Battle of the Atlantic against the U-boats.

This description of the effects of escort attack on a U-boat during a convoy battle graphically conveys the horror of a successful depth-charging. By November 1943, to which this episode dates, the Battle of the Atlantic was largely won. A determined U-boat could still inflict damage unless relentlessly attacked and kept down, as U-333 was by HMS *Exe*, a River-class sloop, one of the hundreds of small ships that were the mainstay of the Allied effort in the Atlantic battle. Kapitän-

leutnant Cremer survived this, his second, patrol, against Convoy SL (Sierra Leone) 139, but was sunk on his fifth, in July
1944. A majority of U-boats were sunk on their first patrols.
The U-boat force lost 70 per cent of its manpower during the
war, the highest proportion of casualties suffered by any arm
of service in any combatant country.

*   *   *

On 13 November the sailing was reported of a convoy with ships
from Gibraltar and North African ports which next day joined up
with a Sierra Leone convoy about 100 miles south of Cape St Vincent
and now consisted of 66 freighters: it was code named SL 139/
MKS 30. To start with it was accompanied by 40th Escort Group,
but during the passage to Britain 7th and 5th Escort Groups were
fetched from other convoys and the 4th from Belfast, so that the
66 merchant ships were gradually surrounded by 28 escort vessels:
frigates, corvettes, destroyers and the Canadian anti-aircraft cruiser
HMCS *Prince Robert* which had come from Plymouth – the Luftwaffe
now had only a fitful presence over the ocean, but reconnaissance
aircraft and even bomber formations were occasionally spotted.
RAF Wellingtons from 171 Squadron in Gibraltar provided air
protection, reinforced by Mosquitos and Beaufighters off Cape
Ortegal. Direct convoy protection was then to be assumed by aircraft
from Cornwall (RCAF [Royal Canadian Air Force] 422 Squadron
and RAF Liberators from 453 Squadron).

Against this doubly and trebly screened convoy we had the Schill
Groups I, II and III in three stop lines about a day's sailing apart.
Air Command Atlantic was involved with 25 long-distance bombers
of type HE [Heinkel] 177. This time the convoy battle promised to
be both vast and varied.

On 16 November the convoy was first sighted by a German
aircraft, and thereafter there were several sightings. In the morning
hours of 18 November the ships sailing up from the south came
into action against the U-boats on watch in the north. The scene
was roughly midway between the Azores and the Portuguese coast.
This was a grey November morning after a clear moonlit night. The

sickle stayed for a long time in the sky before it was properly light. Small clouds came up. A slight wind was blowing from the north-east, and when the sun shone in the course of the morning it showed a sea stirred into a slight swell by the winds of recent days. U-333 was moving underwater and when I occasionally raised the periscope spray splashed against the lens. The empty horizon was a sharp dividing line which in the rhythm of the sea rose up, then disappeared behind the wave-tops. The clock showed 11.30.

It was relatively quiet in the boat. Amid the gurgle and wash of the water came other sounds, weak at first, then growing stronger. The operators signalled ships' propellers from the south. I let things start gently and hung on the periscope. After a while the silhouette of a great many freighters appeared. Fourteen rows of ships were coming in their full width with foaming bow waves directly towards me. It was a unique sight: the expected convoy SL 139/MKS 30. Chance would have it that U-333 was the first boat to intercept the enemy.

In front sailed two escort vessels, clearly destroyers. U-333 lay roughly in the middle between them, in an attack position which would probably not recur. I only needed to let myself drop into the convoy and attack like a pike in a carp pond. All tubes were ready for an underwater shot, flooded and with bow caps open. All I had to do was lie still, let the enemy draw closer and then: at him with a roar! But everything turned out quite differently.

The destroyers were signalling to one another. The right-hand one was zig-zagging continually while the left kept to a straight course. I had my periscope up again, hoping it would not be noticeable in the light motion of the sea, when suddenly I saw an aircraft flit past my lens barely 30 metres above the water. At the same moment the locating signals of the enemy Asdic [sonar – underwater sound-ranging system] struck the U-boat's side with their horrible ping-ping-ping-ping, so loud I might have sent them out myself. We had been discovered. Involuntarily everyone held his breath.

The left escort, it turned out to be the frigate *Exe*, was already turning towards us and in a moment was so close I could distinguish details on deck. I intended to fire a spreading salvo of three into the convoy and had already ordered 'salvo ready', meaning after previous misses to be on the safe side and let the ships come closer, despite the menacing frigate whose sailors I could now see running to and fro. I was still staring obstinately through the periscope when a pattern of ten depth charges exploded with a deafening roar round the boat. We had got into the middle of a carpet.

The effect was terrible and is hard to describe. Suddenly everything went black and everything stopped, even the motors. In the whirl of the shock waves the rudderless boat was seized like a cork and thrust upwards. There was a cracking and creaking noise, the world seemed to have come to an end, then crashes and thuds as the boat was thrown on to its side and everything loose came adrift. I managed to grab the steel strop on the periscope, then my legs were pulled from under me. We had collided with the frigate's bottom which was now thrusting away above us, steel against steel. Certainly the British were no less shaken than we, seeing that [according to a British after-action report] 'just before the first charge exploded, the ratings on watch in the boiler room heard the periscope scrape down the side' . . .

Seconds later, the periscope broke off. The swaying boat reared up, struck the hull of the *Exe* with its conning tower and the control room and listening compartment immediately flooded. The water quickly rose above the floor plates. The light of a torch lying on the chart table showed a picture of devastation. All the indicator gear was hanging loose, the glass was splintered, light bulbs had burst. Cable ends spread in bundles through the control room, the emergency lighting accumulators [batteries] had torn free. Before I even got to the depth-keeping controls the boat was again shaken by the heaviest depth charges. Like a stone we slipped backwards towards the ocean bed which here lay 5,000 metres below.

From the engine room the hydrostatic external pressure, indicating depth, was passed on from mouth to mouth and the fall of the boat stopped by blowing the tanks with compressed air. It rose slowly, then faster and faster until it had to be flooded again so as not to shoot out of the water like an arrow. Beams from the torches flicked over the walls glistening with moisture. It trickled and poured. As none of the pumps was working the water that had come in was transferred in buckets from hand to hand from the lower-lying stern – where it was already above the coaming of the alleyway hatch – into the central bilge. Gradually the boat swung back from the slanting position to the horizontal.

Fortunately the switchboard was still dry, there was no short circuit. We could put back the knife switches which had fallen out and in feverish haste got the electric motors working. Though the noise of the port propeller shaft showed that damage had been caused, its turning again was music in our ears.

My log says: 'Decide to hold the boat by all possible means and slowly go deeper. Damage very great and cannot yet be assessed.'

From the British viewpoint we were 'in the centre of the (first) pattern when it exploded' and 'the first attack must certainly have damaged the U-boat so severely that it was unable to surface.' When an oil patch was sighted at 11.56 and a sample was collected it was believed we had sunk.

But in fact we had let ourselves drop from 60 to 140 metres. Meanwhile the entire convoy in its whole length went thumping past us overhead. In such a situation that is about the safest place for a stricken U-boat, particularly as any hydrophone [acoustic] contact is lost in propeller wash. Nothing can touch one, unless perhaps a ship is torpedoed and falls on one's head.

But hardly was the mass of the ships past than we were overwhelmed with a drumfire such as I had never yet experienced. And that is saying a lot. It began at midday and went on till 20.55 like a continuous thunderstorm, now close, now further away, the heavy-sounding depth charges and the lighter Hedgehogs [multi-barrelled mortar bombs]. And each time we thought, 'Now there'll be a direct hit,' but in fact the explosions detonated further away, we had to wipe the cold sweat from our faces. So-called heroism has not much to do with it. And when finally the torture ended and the great silence began we refused to believe it, but stood there wide-eyed, gasping and struggling for breath, waiting for the next series.

Luckily we had little time for reflection. There was too much to do. The worst damage had to be repaired. Damage to instruments (speed and trim indicators, water and pressure gauges, depth recorder) belonged to the lesser evils. Broken telephones and radio could be accepted. Even a destroyed fire-control panel lost significance for survival, particularly as the heads of all the torpedoes in the tubes had been dented, not to mention the plastic cap of my acoustic torpedo which I had thought so important . . .

But the starboard diesel had been thrust sideways and fallen from its base, and this was more than problematical. Now by the sweat of our brow we had to wedge and support it with beams. And hardly less serious was the port propeller shaft, which had been bent and was hammering loudly. To complete our misfortune, the radio installation was so badly damaged that despite trying three times I was only able to send a short mutilated signal.

Air was running out. Our bodily exertions had used it up quicker than usual and it had to be improved with potash cartridges and oxygen. There was a stink of battery gas – the ventilation lines had been broken – and of watery oil. Eventually the fug became chokingly thick, and compressed air was getting short. After nine hours of depth charging which had thoroughly shaken the boat and had necessitated our repeatedly blowing the tanks to maintain station, there was hardly any compressed air left. The boat was tending to lose depth again and could be kept trimmed only with difficulty. I had to go up regardless, and so, one hour after the last charge had exploded and the great, perhaps deceptive silence had fallen, I brought U-333 to the surface.

Up above it was dark. Sea state 5 to 6 with heavy swell, the slim outline of the boat hiding in the troughs. Somewhere there was a destroyer, but she spotted nothing. The watch came up and we inspected the scope of the exterior damage. The forward net-cutter was broken, the bridge cowling bent forward. Both periscopes were useless, the attack periscope bent, the night-sighting periscope broken. Radio direction-finder and anti-aircraft guns had gone, as though shaven off. The lurching boat had a list. It was only just floating above water, and I had another shock, in so far as one was capable of more, to see air bubbles surging from both sides. Apparently all the ballast tanks had cracks. U-333 could float only to a limited extent.

The diesels would not start. Despite repeated blowing, the boat would not stay on the surface but slowly sank downwards by the stern. We had to submerge again so as not to drown on the bridge. Last man down as always, I shut the conning tower hatch – or tried to. This time it stuck. Meanwhile the boat was submerging completely and streams of water poured through the opening. I clung to the wheel until I fell into the control room. We blew the last of the compressed air into the tanks and the hatch slowly emerged from the swirling water. I had swallowed a great deal, was soaking wet and numbed, but with the help of the 2WO, was clearheaded enough to find the curious cause of the defect: part of a blade knocked off from the propeller of the frigate *Exe* had slipped and blocked the hatch.

Things were against us – as though a renewed attempt was being made to do away with us. After many experiments we finally got the port diesel going – it began to function 'slow ahead' – and eventually the starboard diesel started as well. The ballast tanks were no longer airtight, but with the exhaust gases we blew into them we could roughly keep a balance with the water coming in. But the boat was still more or less unstable and threatened to drop away under our feet. And somewhere water was continually dribbling inboard where we struggled with the damaged pumps. We had no alternative but to move 'dynamically' and gently get away.

Next day already on the way home I wrote in the log: 'Surfacing goes better', and the day after, 'considerably better, which means I have got accustomed to the condition of slowly sinking.' Though that speaks volumes, it says nothing about our wet feet. Yes indeed, we had been given a pasting. My people patched everything possible with what lay to hand. That was very little. And so we went hobbling home with a wreck. And with all the iron around us, navigated with the magnetic needle alone, for the gyro compass too was broken.

HUGH DUNDAS

Flying Start

Hugh Dundas was one of Britain's leading fighter aces of the Second World War. In this reconstruction of pilots' banter and operational chatter, he catches the atmosphere of a Fighter Command squadron in Southern England in the summer of 1941. The Battle of Britain, the Royal Air Force's historic defeat of the Luftwaffe, was over but the pressure to keep the Luftwaffe at bay was still strong, the air battle over and beyond the English Channel intense, and pilots' lives short. Douglas Bader, the legless survivor of the 1940 battles, was already celebrated for his longevity as well as unparalleled courage.

*   *   *

On some faded sheets of paper there has survived a description which I wrote, at that time, of an afternoon and evening which could have been one of many. It is incomplete, a fragment, not part of a diary or larger chronicle. I cannot remember writing it; I do not know exactly when I wrote it, for it is undated. Nor do I know whether it ever had an ending or whether perhaps I just got tired of writing and went to bed. But such as it is, it brings back sharply the feel and taste of those far-off days when I was very young and just discovering life and death stretched out its hand to touch me every day. I quote it, just as it was written then:

It was hot in the garden, lying face down on the lawn, a pot of iced shandy by my hand, Robin (my golden retriever) huffing and puffing and panting at the ants. Odd to be lying there peacefully, listening to the click of croquet balls, the blur of voices, the gramophone. The shandy sharp, cold, stimulating.

'Hullo, Cocky.'

'Hullo, Johnnie.'

'Get a squirt this morning, Cocky?'

'Yes, Johnnie, I got a squirt. Missed the bastard as usual, though.'

'Another show this afternoon, Cocky. Take off 15.30.'

'Yes, I know; take off 15.30.' Three hours ago, over Lille. It happened yesterday, and last week, and last month. It will happen again in exactly two and a half hours, and tomorrow, and next month.

The grass smelt sweet in the garden, and the shandy was good, and Robin's panting, and the gramophone playing 'Momma may I go out dancing – yes, my darling daughter.'

It was hot at dispersal and the grass, what was left of it, brown and oil-stained. The Spitfires creaked and twanged in the sun.

'Everything under control, Hally?' (Flying Officer Hall was the squadron engineer officer.)

'Yes, Cocky, everything under control. DB's not ready yet, but it will be.' (DB were the identification letters of Bader's plane.)

'Well, for Christ's sake see that it is, or there'll be some laughing-off to do.'

'It will be ready, Cocky.'

'OK, Hally.'

Inside is as hot as outside. The pilots, dressed almost as they like, lie about sweating.

'Chalk please, Durham.'

They all watch as I chalk initials under the diagram of twelve aircraft in three sections of four. Nobody moves much until I have finished and written the time of take-off.

'Smith, you'll be with DB. Nip, you and I on his right. Johnnie, you with the CO and two of B Flight. OK?'

'OK, Cocky.'

Here comes DB.

'Why the bloody hell isn't my aircraft ready? Cocky, my bloody aeroplane's not ready. We take off in 20 minutes. Where's that prick Hally?'

'It's OK, DB, it'll be ready. I've seen Hally.'

'Well, look at the bloody thing. They haven't even got the cowlings on yet. Oi, Hally, come here!'

Christ, I wish we could get going.

'Chewing gum, Johnnie, please. Thanks pal.'

'OK, DB?'

'Yes, Cocky, it's going to be OK.'

We walk together again, as far as the road.

'Well, good luck Cocky. And watch my tail, you old bastard.'

'I'll do that DB. Good luck.'

Just time for two or three more puffs before climbing into A for Apple.

'Everything OK, Goodlad?' (the fitter who looked after my plane).

'OK, sir.'

'Good show. Bloody hot.'

Climbing in, the hottest thing of all. The old girl shimmers like an oven, twangs and creaks.

'Good luck, sir.'

'Thanks.'

Up the line DB's motor starts. 610 [Squadron] have formed up and are beginning to move off across the airfield as we taxi out – DB, myself, Smithie, Nip, then two composite sections from both flights.

Straggle over the grandstand at Goodwood in a right-hand turn and set course east in a steady climb, Ken's twelve [formation of twelve fighters] a little above and behind to the left, Stan's out to the right. Ten thousand feet over Shoreham. The old familiar, nostalgic taste in the mouth. Brighton – Maxim's last Saturday night; dancing with Diana in the Norfolk. Beachy, once a soft summer playground, now a gaunt buttress sticking its chin bluntly out towards our enemies. Spread out now into wide semi-independent fours. Glint of perspex way out and above to the south shows Stan and his boys nicely placed between us and the sun. Dungeness slides slowly past to port and we still climb steadily, straight on, way out in front.

Twenty-five thousand.

'Levelling out.'

Puffs of black ten thousand feet below show where the bombers [they are escorting] are crossing between Boulogne and Le Touquet.

Six big cigars with tiers of protective fighters milling above them.

'Hello, Douglas, Woody calling. There are fifty plus gaining height to the east.'

'OK Woody.'

'Put your corks in, boys.' Stan.

Over the coast at Hardelot we nose ahead without altering course.

'DB, there's some stuff at three o'clock, climbing round to the south-west.'

'OK, I see it. Stan, you deal with them if necessary.'

'OK, OK. Don't get excited.'

Usual remarks. Usual shouts of warning. Usual bad language. Usual bloody Huns climbing round the usual bloody way.

St Omer on the left. We fly on, straight and steady in our fours, towards Lille. Stan's voice:

'They're behind us, Walker squadron. Stand by to break.'

Then: 'Look out, Walker. Breaking starboard.'

Looking over my shoulder to the right and above I see the specks and glints which are Stan's planes break up into the fight, a quick impression of machines diving, climbing, gyrating. Stan, Fan, Tony, Derek and the rest of them are fighting for their lives up there.

Close to the target area now. More black puffs below show where the bombers are running in through the flak.

'Billy here, DB. There's a lot of stuff coming round at three o'clock, slightly above.'

Quick look to the right. Where the hell? Christ, yes! There they are, the sods. A typical long, fast, climbing straggle of [Messerschmitt] 109s.

'More below, DB, to port.'

'OK, going down. Ken, watch those buggers behind.'

'OK, DB.'

'Come on, Cocky.'

Down after DB. The Huns are climbing fast to the south. Have to get in quick before those sods up above get at us. Turn right, open up slightly. We are diving to two or three hundred feet below their level. DB goes for the one on the left. Nipple is on my right. Johnnie slides across beyond him. Getting in range now. Wait for it, wait for DB and open up all together. 250 yards ... 200 ... wish to Christ I felt safer behind ... 150. DB opens up. I pull my nose up slightly to put the dot a little ahead of his orange spinner. Hold it and squeeze, cannon and machine guns together ... correct slightly ... you're hitting the bastard ... wisp of smoke.

'BREAK, Rusty squadron, for Christ's sake BREAK!'

Stick hard over and back into tummy, peak revs and haul her round. Tracers curl past ... orange nose impression not forty yards off ... slacken turn for a second ... hell of a mêlée ... better keep turning, keep turning, keep turning.

There's a chance, now. Ease off, nose up, give her two lengths' lead and fire. Now break, don't hang around, break! Tracers again ... a huge orange spinner and three little tongues of flame spitting at me for a second in a semi-head-on attack. Round, round, so that she judders and nearly spins. Then they're all gone, gone as usual as suddenly as they came.

'Cocky, where the hell are you? Are you with me, Cocky?'

There he is, I think. Lucky to find him after that shambles.

'OK, DB, coming up on your starboard now.'

'Right behind you, Cocky.' That's Johnnie calling.

'OK Johnnie, I see you.'

Good show; the old firm's still together.

It was cooler, on the lawn, and still. The shadows from the tall trees stretched out to the east. Robin lay beside me pressing his muzzle into the grass, huffing at insects. The pint pot of Pimms was cool in my hands and the ice clinked when I moved. The cucumber out of the drink was good and cold and sharp when I sucked it.

'Hullo Cocky.'

'What-ho Johnnie.'

'Tough about Derek.'

'Yes, Johnnie; and Mab.'

The croquet balls sounded loud to my ear, pressed in the grass. The distant gramophone started again on 'Momma, may I go out dancing'.

'Come on, you old bastard, let's drink up and get out of here.'

The tide washed up the creek to Bosham and splashed against

the balcony of the Old Ship. We sat and sipped our good, warm, heartening brandy and watched the red sun dip through the western haze, watched the stars light one by one, watched the two swans gliding past like ghost ships.

'Cocky.'

'Yes, Johnnie.'

'Readiness at four a.m.'

'OK let's go.'

That was the way of it at Tangmere in high summer 1941. That, word for word, is how I wrote it down, in some moment of self-release on eleven sheets of pale blue writing paper which then lay unregarded among other old papers for twenty years.

GEORGE KENNARD
Loopy

George ('Loopy') Kennard, like Bruce Shand, was a pre-war regular officer in the British cavalry. He shared Shand's light-hearted attitude to life, displayed at its best in desperate circumstances. Circumstances were desperate at the end of the British campaign of 1941 in Greece, to which Winston Churchill had sent a large contingent of the Western Desert Force (later Eighth Army) to assist the Greeks in their resistance to the German invasion in early April. Kennard's regiment, the 4th Queen's Own Hussars, took part in the fighting withdrawal down the east coast of Greece but was one of those overwhelmed by the weight of the German offensive.

*   *   *

We reached the Peloponnese, where we hoped we might be allowed to stand and fight. By now the Squadron had a seventy-mile front from Corinth to Patras, four tanks, a handful of trucks and a hundred men. Here there was a respite as the enemy, too, had its resupply troubles. All went quiet apart from the plane that machine-gunned Billy and me as we clung to our motor-cycle on a mountain pass. What a sleep we had that night in the olive groves, but at dawn bigger aircraft appeared and the parachutists dropped like autumn leaves as the Germans tried to seize the bridge over the canal. We took a heavy toll of them but they got on to the bridge and our Sappers tried to blow it. Nothing happened and on they came, accompanied by a newsreel truck filming for the benefit of Berlin cinema audiences. Then up it went, bridge, trucks, troops and camera. Whoopee!

Close to that bridge, the Colonel and the Adjutant, returning from Headquarters with fresh orders, ran into an ambush; for them the war was over. Clem took command and on we went, what was left of us, to Kalamata, Greece's most southerly port. There we were to be embarked to fight another day. As we set off, Billy Hornby was captured. Later he was to be put into a POW cage in Salonika where he saw his chance to make a run for it. They spotted him; a few pfennigs' worth of German powder and that bundle of gaiety and fun lay dying – dead. Ja, he was shot while trying to escape.

A few miles from Kalamata we paused while Clem went on to try and discover what the powers that be, if there were any, wanted us to do. We were out on our feet and, one by one, we collapsed by the side of the road in sleep, some thirty men and one anti-tank rifle, grimly clutched by Trooper Small. I awoke to the sound of machine-gun fire and of John de Moraville exhorting Small to fire at a German armoured car. He did his best but the bolt jammed. I ran to a garden shed inside which were more of our men; I told them to beat it quickly to the south. I badly wanted to go with them but couldn't leave John and Small. The volume of firing increased as I did my best to retain my dignity; as I arrived John was trying to kick the gun into action; suddenly he spun round and said: 'The buggers shot me,' and so they had. In his neck was a neat entry hole and no exit; it didn't seem to affect him much, except to increase the volume of oaths he was directing at the enemy as the gun still refused to fire. I was trying to plug the wound with a handkerchief when a polite, Teutonic voice, speaking in a modulated English accent informed me that we were now prisoners of war. It was my turn to spin round. I gaped: 'Good God, Otto, wie gehts? What on earth are you doing here?' It was Otto Hertzog, who had often come to stay at the von Mitzlaffs [with whom Kennard had stayed before the war] and was their cousin.

Otto's men were busy rounding up the remainder of our party, including those who had bolted from the garden shed; three had been killed in the initial burst of fire. We were all shepherded together on the road and I tried to introduce Otto to John. The latter threw his field-glasses on to the ground and said: 'For God's sake, Loopy, stop talking to that bloody German.' I suddenly realized that I might appear to be some sort of fifth-columnist and contented myself with trying to find out what would happen to us now. Otto told me that we were to be taken to Kalamata by vehicle and from there sent to a prison camp. We would be well treated and John would be taken to hospital. He added that he would take me out to dinner when he had some leave. All this seemed good news; Kalamata was, as far as we knew, full of British troops still fighting, and it should be relatively easy to make a break when we arrived there. We were put into an open truck and driven off singing 'Roll Out the Barrel' at the tops of our voices; this seemed to annoy the German escort considerably. We redoubled our vocal efforts.

The Germans appeared to have no clue how many British and Allied troops there were in Kalamata, or perhaps because they thought that most by this time were unarmed, they didn't care. We drove on into the outskirts of the town in the gathering dusk and were fired on. We were bundled into a cellar and I decided that this was my moment. As I gathered up my courage, I heard shouted German orders for the deployment of a large field gun which was to begin firing shells on to the beaches now packed with largely unarmed troops waiting for the naval evacuation. Our guard dropped dead, shot by one side or the other and I hopped it. There was some erratic shooting from both sides as I ran down the street and eventually I found myself among some New Zealanders. I explained that I knew the whereabouts of the gun that was even now causing havoc and death on the beaches and a Sergeant said: 'Right, Sir, hop into this truck and we'll go get him.' The truck was a small one and Sergeant Hinton had only a revolver; I was unarmed. I had just started to mutter, 'Not bloody likely . . .' and other excuses when I was elbowed aside by a huge Australian. 'Come on then Taffy,' said Sergeant Hinton, and off they charged, straight at the gun. But it wasn't the only gun, it was supported from both flanks by automatic weapons. There was a vast commotion, then silence – the gun lying on its side and the truck on its roof. The German crews were dead and Hinton was prone in the road seemingly oblivious to the fact that he had been shot in the stomach. Taffy was away in the distance chasing more Germans in the darkness.

Sergeant Hinton won a Victoria Cross, presented to him in a POW camp. None could have been more deserved, but as far as I know Taffy, whose name I never discovered, got nothing.

On my own once again, I thought I would go back to the cellar to find John and the others. I passed a badly injured German officer lying by his defunct gun and gave him some cigarettes. The cellar was empty but a little further on I came across a dozen or so German prisoners being led away with their hands up. Among them was Otto. I took them over and headed for the beach. Otto was nervous; he obviously thought that I had shot his sentry earlier and that he was likely to be similarly treated. I told him that if the Navy came he would be taken to Egypt and treated well as a POW. I would, I said, take him out to dinner.

On the beach I gave up my prisoners to the New Zealanders, and tried without success to find the remnants of my Regiment. I wandered back into the town and fell in with a New Zealander carrying a Bren gun. We had barely introduced ourselves when we heard the now familiar pop, pop, popping of a German motorcycle combination. We rushed into the nearest house and took up position at a first-floor window. More Germans appeared, a target too good to miss. I grabbed the gun and emptied the magazine. At that moment, the last few weeks and months overwhelmed me. Night after night, bomb after bomb, no regiment left, Cecilia and the baby, no mail, a fuck-up on the beaches, Billy gone and me here, with a Bren gun, and out there the enemy. The New Zealander went out to look at the dead and the dying. I did not and remember only how much nicer they looked without their steel helmets which had rolled off as they fell.

We went back to the beach and heard that the Navy would take off only the sick and the wounded, believing that the German Divisions had arrived at the coast in force. We knew better, but there was nothing more to do. I was taken to the Brigade Commander who, knowing that I spoke German, had a job for me. As the Navy had refused to evacuate us he was forced to surrender some 10,000 men at dawn. I was to convey this message to the German Head-quarters so that the packed beaches would not be bombed at first light. I was to take my German Officer friend with me as a guide. I told Otto, who addressed his fellow captives before we left. They let out a tremendous cheer, quickly stifled under threats from the infuriated Australians.

Otto and I set off once more. No training manual that either of us had read covered our mission. We decided that I would go first shouting: 'I am a British officer with a German,' on our side of the town, after which he would lead and make similar noises in what we imagined might be the German sector.

We wandered round and round and through the square, the scene of my brainstorm. I told him what I had done. He shrugged. He said it was a pity; they were good men; one had just won an Iron Cross, First Class. I muttered, 'Fuck him' under my breath.

By now the poisonous thorn in my foot, my only wound, had made walking almost impossible and it was with much relief that we eventually found Otto's headquarters. He left me outside and went in. The sentry gave me a cigarette. After about half an hour I was ushered in. It was just like a film set. Around a candlelit table sat half-a-dozen officers. All got up, saluted me and gave me food and schnapps. I told them of the beaches, the coming surrender and the lack of food and water. I asked them, please, to stop the bombers. They said they would and mentioned that they had lost nearly 75 per cent of their force. I went to sleep on a mattress, still wearing my revolver, for which I had never had any ammunition. 'Loopy,' said Otto, 'tomorrow I shall have to take your pistol away.' And tomorrow they did, but not until after I had motored with the German commander to the Brigadier on the beach and he, faced with no alternative, handed over to captivity 10,000 men.

When soldiers are taken prisoner their first reaction is to blame someone – anyone. Many on that beach blamed the Brigadier. Many more blamed the Navy. Some blamed Wavell, some the Government and some the Greeks. For my part, I blame a combination of circumstances which, once under way, led to an unstoppable chaos. Recrimination is fruitless.

The Germans struck a medal and gave their highest award to Oberleutenant Otto Hertzog, who as a prisoner, 'frightened the British High Command into surrendering'.

STUDS TERKEL
The Good War

The tide of the Second World War turned against Hitler when Japan, Germany's ally under the terms of the Tripartite Pact (the third member being Italy), decided to attack the United States by a surprise raid on the American Pacific Fleet at Pearl Harbor on 7 December 1941. The Pact bound Germany to come to Japan's assistance only in the case of hostile action by a third power, exactly the opposite of the circumstances the Pearl Harbor raid had produced. Hitler, against the protestations of his Foreign Minister, Ribbentrop, nevertheless decided to make common cause with Japan in its war against America and himself declared war on the United States on 11 December.

It was the most disastrous decision of his leadership, condemning his country to a defeat that was inevitable once the power of the American economy, the largest in the world, was engaged against him. Studs Terkel, the pioneer of oral history, records, through an interview with Paul Edwards, how American economic power was mobilized for the Second World War. The world depression of the 1930s had hit the United States hard, throwing 12 million Americans out of work at its depth. Between 1941 and 1945, unemployment disappeared, the nation's gross product doubled and the American economy was, at the war's end, equal in size to that of the rest of the world combined. The Second World War made the United States the most powerful country on the globe, a status it retains to the present day.

*   *   *

Paul Edwards

I was living in Winner, a little South Dakota cowtown, a rootin', tootin' cowtown (laughs) west of the Missouri River. We'd just spent Saturday night in a pretty rugged fashion, drinking and carousing a bit. So I got up late Sunday. I had a headache. A friend came in and said, 'Turn on the radio. The Japanese have bombed Pearl Harbor.'

That fall I had been named head of the Junior Red Cross for the county. On December 8, I went down to St Louis and signed up as a field director for the American Red Cross. I was immediately sent to Fort Riley, Kansas, with the assimilated rank of lieutenant. The Eighth and Ninth Cavalry were stationed there, totally black. The officers were all white. They had been for years a showpiece for parades. They were great horsemen. They were great drill. They were a lot of old-time sergeants with hash marks [service stripes] from here to there, up their arms. They had built a morale of their own. They had pride. Boy, you put those old sergeants out in front of a troop of guys on black horses or bays, you put them on parade grounds, it gave you a thrill. It gave them a thrill.

They had been drafted out of the poorest families of the Midwest. Most of them sent fifteen of their twenty-one dollars a month to their families back home on welfare. They used to pay twenty-one silver dollars. There was this big husky black leanin' up against the wall. He'd drop the silver dollars like a gambler and they'd go clink, clink, clink. He said, 'Man, I'm well off. Hear that money clinkin' in my pocket?' He never had that much money.

Most of the draft boards were composed of men from Main Street across America. They were quite punitive towards the welfare crowd. Volunteerism was not yet in effect. This was before December 7. Most of these kids expected to be let out after a year of the draft. A lot of 'em had plans. All of a sudden they were thrown up against the reality of total war. We had a problem with suicides. Guards were put out at night. They would not issue ammunition. I watched the transition from peace to war.

The role of the Red Cross was peculiar. You were neither man, beast, nor fowl. You were with the army but not of it. You were under military control, but you didn't enjoy the privileges of being an officer among officers. Your assignment often put you up against the army with regard to the individual man. The command didn't like a challenge to their control. You bunked with 'em, you ate with 'em, so this tension made the job difficult.

Things moved fast with the expansion of the army. I wound up at Fort Meade in Sturgis, South Dakota, near home. That's the old Seventh Cavalry headquarters. Custer's. Thousands and thousands of horses were raised at these remount stations, Nebraska, elsewhere. All of a sudden, they weren't needed: the transition to the jeep, the scout cars. The romanticism of the cavalry was still very strong. Officers hated giving it up. I remember old Colonel Hooker at the Nebraska remount station. He pounded the table when the Japanese had sunk two British warships [the battleships *Repulse* and *Prince of Wales*] off the coast of Singapore: 'I know that country down there. Goddamn it, they'll never take a square foot of it until they get our men down there on horses and donkeys.' (Laughs.) There was a lot of resentment as we moved from one era into another, from horse to motor.

All of a sudden, I was sent down to Camp Barkley, Texas. The needs of the soldiers were rooted in the Depression. You jerk a young man out of his family. Next thing you know his father dies. Who'll support the family? He's just been married, his wife dies, he's left with a child. You had the business of AWOL and family circumstances. A black kid at Fort Riley came to me with a telegram. It said, 'Daddy died last night. Come home at once. Bring an overcoat.' The simple need: bring an overcoat.

It was segregated army at that time. There were hundreds of labor battalions that were totally black, still under white officers. They were an abysmal shame to the nation, if the facts were known. They were held to a work schedule, a seven-day week. They were almost imprisoned. It was cold, it was mud, it rained. Here they were pouring concrete bases for the Eighth Air Force. It was hell to pour concrete runways in the downpours. And no leave, no nothing. At that time, there wasn't one black field director of the Red Cross.

Camp Barkley was just outside Abilene. It was a righteous Baptist town, still had prohibition. We had forty thousand troops around this little town of thirty thousand. That made for prejudice against the soldiers. They were patriotic towns, but my God, look at our restaurants, we don't get in. Look at our girls, they're bein' insulted. It was hard to get quarters off the base. Once I rented a converted chicken coop. Yet the town drowned in money.

There was an alliance between the Baptists and the taxi drivers that kept the town dry. The hotel keepers wanted to open it up, but you'd find a lot of Baptist money and cabdrivers boostin' prohibition. 'Cause the drivers were sellin' us booze at five dollars a pint. (Laughs.) The righteous and the unclean. (Laughs.) Of course, they weren't crazy about blacks. I remember a little black singer who came out to entertain the troops. Here were about ten thousand guys, and this little girl worked that show like you wouldn't believe. She ended gettin' them all to clap and sing in a state of ecstasy, almost. She couldn't get a room in a hotel that night. We had to find her quarters.

Out of nowhere, I get a call to Dallas. They need a supervisor in Great Britain immediately. To a country boy from South Dakota, that's reachin' for the moon.

I left for the UK on December 7, 1942, the anniversary of Pearl Harbor. We were all rounded up and put on the *Queen Mary*. A whole division. There must've been fifteen thousand people on that ship. This was a time of high secrecy. How the hell you keep the movements of the *Queen Mary* secret is a good trick. (Laughs.) But you still went around and played spook.

We had a violent crossing. The *Queen Mary* tipped to forty-two degrees and she was supposed only capable of leaning to forty degrees. A big wave hit us and brought us back up. We hotbedded, half down and half up, during the night and day. We had two meals a day, largely of boiled English fish. Everybody was so sick, it didn't make any difference.

We had a remarkable storm one awful night. The portholes were smashed and the sea was coming in on us. There was almost a riot as people rushed for the gangway and the MPs [military police] stood there with pistols: 'Get back down! Get back down!' The lifeboats had broken loose from their moorings and they came in against the side of the boat. The crash had broken the portholes. We made it to Glasgow.

The Red Cross has a split personality. When the war broke out, the leadership was extremely conservative. It depended on rich donors. The orientation was very upper-class, snobbish. There was a marked difference between these volunteers and the staff. Eventually the pro staff took over because of the dimensions of our chores.

You'll hear those who damn and those who praise the Red Cross. The leadership was loaded with prejudice – to blacks, to Jews. When I first went in, there were hardly any Jews. A gentlemen's agreement. You didn't talk about it. I remember when the first Jewish field directors arrived. Odd, when you think we were fighting Hitler.

Until we were well into the war, they segregated the blood plasma of the blacks from the whites. I must say the Red Cross was the mover and shaker and changer, because not only was there no scientific sense to it, the economics was bad. There was always that double dilemma. There were no black field directors. Many towns were off-limits to black troops. They wouldn't let whites go into others. We had black and white towns. Many racial incidents developed and the Red Cross found itself in the middle. Historically, we came down on the right side.

I'll never forget the first black guy I got as a field director in England. He's an All-American end [national football player], about six foot four, a prince of a man. I went down to set him in command. And I just caught Billy-all-hell from the commanding officers. Whaddaya mean sending this nigger down here? I said, 'You'll take him or none.' They took him. In six weeks, the man was a hero. Even these officers – he just did a job, a heroic one, with the morale of the black soldiers.

We developed these on-site clubs for the Eighth Air Force. I like to think I had a hand in it. When a soldier got a pass in the city, his recreation problem was pretty well solved. But it was while he was on the base, and his buddies were getting shot down, and the mud and the cold, and you're with those guys and the next night they're on a mission – and the next day they're gone. It would tear you up.

I arrived there on the fourteenth of December, 1942, and I stayed until March of '44.

When the first contingent was moved to North Africa, they left behind a lot of pregnant girls, commitments to marry. Remember, the Americans were dashing and daring and had money in their pockets. The factory girls came from little provincial towns. I tried to arrange marriages by proxy. We did it over the telephone. I tried to get the Church of England – I went down to see the old bishop. Couldn't we get an exemption so we could have these proxy marriages? The old man was about ninety with a secretary about eighty-five. He had one of these old-fashioned trumpets. I yelled into the thing, but he turned me down.

I went to see Churchill's [younger] brother [Jack], who lived in the south. I told him about this disappointment. He said, 'Nonsense, son, don't worry about it. All those soldiers from your country are in good physical condition, aren't they? They're all inspected and examined?' I said yes. He said, 'Why, we'll just tell the girls to go ahead and have the babies and we'll adopt them and call them the King's children and raise them. They'll be good for the blood and bone of the country. We lost lots of our best men in this war.' Sounded like a stud farm in a way (laughs), but it was quite sensible. He was less bombastic than Churchill, but a real character.

I was called one day to the Command in Grosvenor Square [Eisenhower's London headquarters]. An officer of General Jake

Devers' staff cornered me: 'I want you to see this light [lieutenant-] colonel. He has a problem. We cannot deal with him.' The guy in trouble was an oceanographer. He studied the tides and coasts for a landing. A terribly important guy. I went to his appartment, knocked a number of times. Finally, a guy came to the door. He looked haggard, terrible, messy: 'I don't want to talk to you.' I said, 'I was interested in your name, because when I was in high school I played football against a small town in Oregon and there was a guy playin' there –' He started to cry. He was the guy. We played tackle against each other at a football game back in the twenties.

Well, the story is he had an affair with an Englishwoman, who suddenly turns up and tells him she's pregnant. She's threatening to write his wife. He has two nice kids and is scared to death. I said, 'Let me talk to the woman.' She was demanding a thousand pounds, clear out of his reach. She was a cockney and kind of garrulous. I thought, This doesn't quite ring true. She said, 'Make it five hundred and I'll go down to Bournemouth and have the baby.' I made an appointment with her, but also with our chief nurse and a policewoman. In the meantime, I had her investigated by Scotland Yard. Turns out she's a well-known hustler. She comes to see me and I have our nurse ready and ask for a physical inspection. And there she was with a padded blanket inside her clothes. So we blew that out of the water.

I go tell the light colonel the facts. He thanks me profusely, he weeps and says, God, he'll never forget me. Two weeks later, I run into him down at the Grosvenor; he looks right at me and looks away and never speaks a word. I never felt too bad about it. The guy had to forget.

In spring of '44, we were staffing for the Normandy invasion. What were our responsibilities? Should we have side arms? Do we go in with the landing? Yeah, we lost four men in the landings. We had very high losses later.

When you were hired by the Red Cross, you were draft-exempt. They took men who were overage or had infirmities. I tried many things to get in the service, but I had this football injury. We had teachers, coaches, this, that, and the other. Most of us were family men. We had a two-thousand dollar insurance policy, that's all.

Meantime, I'd been sent back to the US to make a cross-country speaking tour. Fund raising, that sort of thing. That's when I heard about the four guys we lost. In Minnesota and Iowa, I ran into the wives of two of these guys. Each had been a teacher. Each left a wife and two kids and a policy of two thousand dollars. I went back to Washington and asked: What about raising the insurance? They go in with the army, why can't we put them under the GI insurance provision, ten thousand dollars?

I got Senator Chan Gurney to introduce the bill. It would have gone through Congress like that. (Snaps fingers.) I thought I'd cut a fat hog. I went back to headquarters and reported to my bosses. And, by golly, they turned it down. It just made me madder than hell. There was no reason for it. We were back to that snob approach. They said, 'Well, we were afraid we would lose too much control of our people to the army.' That was b.s. [bullshit], because you can't control a dead man, right?

I was so angry – I was still in uniform – I got my cap and said to my wife, 'I'm leavin' this.' I drove out to South Dakota and went hunting – with Bob Feller, Rollie Hemsley, old baseball players. And I did too much drinking. I got a phone call to come back. I did and ran into the same attitude. So when I got a job with UNRRA, I quit.

My old boss said, 'Are you going with that worldwide WPA [Works Progress Administration – social services agency]?' We were back to that contemptuous attitude toward welfare. The word that's been used to beat poor people over the head ever since Roosevelt's time.

UNRRA – the United Nations Relief and Rehabilitation Administration – was chartered in unknown circumstances to help the war-torn areas of the world. To people used to established ways, this kind of venture is a threat to their values. It was chartered in a conference in Atlantic City in 1943 by nations fighting fascism. Its purpose was to rehabilitate nations devastated by war.

I was one of the early employees, about fifty, sixty of us. My first assignment was to go to the Middle East and run the refugee camps.

There were thousands of Yugoslavs, fifty thousand, sixty thousand. We brought in something like fifty thousand Greeks, who'd come off the islands through Turkey. I had a Greek camp at Gaza [in what was then Palestine]. That was my base. My territory ran clear north to the Turkish border. The Turkish army would take them over, knock out their gold teeth, jerk the gold rings out of their ears, and push 'em over the border into Syria. We had camps along the Suez Canal, camps in Egypt. Palestine was full of refugees.

In Palestine we had Royalist Yugoslavs, and down in the Suez we had thirty-eight thousand Red Star Yugoslavs. Tito's crowd. We had to separate them. They were deadly enemies.

We had areas for tuberculars. We had typhoid, we had typhus, we had scarlatina. They were in terrible condition, starved, dying. Every evening we would have a mass burial in a big ditch.

I had great admiration for the Jugs [Yugoslavs]. They were tough, resilient. One time an airplane crashed nearby. They took the scrap metal and made cooking utensils. They took the tires and made rubber-soled shoes.

I sometimes worked in cooperation with the Joint Distribution Committee, a Jewish agency. When the Germans captured Greece, they shipped a lot of Jews to one of the extermination camps. The JDC and others intervened with Franco, who intervened with Hitler, and they were released on the promise that Americans would take them off his hands. They brought them down in boxcars to Spain. The Americans picked them up and took them to Casablanca. We brought them over by boat to Alexandria, put 'em on a train, and brought them up to my camp, six miles south of Gaza.

We had a ninety-year-old man there. His wife was eighty-eight. Their daughter and son-in-law came in, from up near Tel Aviv, in a little car. They had escaped out through Rumania and Turkey, right? Would I consider releasing the old couple to their care in their home? As soon as they could get some space. Space was very dear at the time. We got an exemption to bring them to [British-controlled] Palestine. The English had barred the advent of any Jews there, right? But I said yes. I never had much regard for stupid regulations. So they left, elated. Would you believe that six miles north of Gaza, there was a sudden storm and a wall of water swept that young couple to the sea? Tell me about fate, friend. How do you break this news to the old couple?

UNRRA soon came under attack in Congress. Because the Soviet Union was part of it. Herbert Lehman was my boss. He had no more personality than a musk ox, but he was always on the side of the angels. It was a fight, always. I sent my last group back home to Greece in the fall of '45.

I was sent to Czechoslovakia as deputy director for UNRRA, under a Russian chief of mission, General Peter Alexander Alexeyev's industrial rehabilitation. We brought the first cotton up the Danube. We got textiles in so their mills could get back to work. We got repairs for the machinery, which was exhausted 'cause the Germans had worked it to death. We helped start the forest industries again. And agriculture.

I was director of Slovakia. It was odd. Slovakia had been collaborationist under Father Tiso. I attended his hanging. God, I hated Hitler and all his flunkies from day one. I started in South Dakota. There was strong sympathy for him in the Lutheran Church. They weren't bad people but the German culture was strong. The president of my college, just come back from Germany, spoke at our assembly: 'I've seen a nation that had such poverty and hardship pull off an economic miracle. Everybody has a job. And this at a time when millions in our country don't.' Oh boy. As for Russia, the streak of hostility was always just beneath the surface. The Russian alliance was never a thing of the heart. It was a calculated stratagem to defeat Hitler, with the help of American technology. I've had Russian generals tell me what won the war for them was that old Studebaker six-wheel truck. That thing was a genius of transport. It was tough, it was wiry. When the evacuation of Austria came about in '45, '46, I saw a stream of these American trucks coming out of there, older than hell, rusted and spoutin' steam and smoke. But they were still movin'.

In my experience, dealing with the Russians was like when I was a kid ridin' on freight trains. Once in a while you get in a boxcar that had a flat wheel. The wheel goes around in perfect circles, it's fine. All of a sudden – bump. When I left Czechoslovakia to go to Germany to direct the DP [displaced persons] operation, the Czechs gave a big dinner. Alexeyev came up and put his arms around me – (Suddenly cries angrily) Goddamn it, we missed something. This is one of the tragedies of the death of Roosevelt. There was a blind spot. We're on the threshold of destruction because it didn't work.

I'm not nor never was a communist. Matter of fact, I lost my standing in the international community by helping people escape from the communists after they took over Czechoslovakia. I was there when they took over. Czechoslovakia needed a communist revolution like it needed a hole in the head. Several of my friends were put under house arrest and persecuted. So I took my passports and got seventeen people over the border, including two cabinet members. I had a system: a false stamp which was like a Czech approval for going in and out. I lost my wife's passport in no man's land between the Czech and German borders. It was picked up by a German farmer and sent to the American embassy. They called me in. They said I abused my American passport. I had to resign, and I left in October '48, I went home in disgrace. I was helping people escape what they supposedly hated. It was one of the ironies of our time. Later on, in the McCarthy era, it came to haunt my professional life. I was marked unreliable.

This is just another small fallout of the Cold War. I feel to this day it didn't have to be this way. I've been to Russia a number of times. They're so bloody fearful of us, you can't believe it. To talk about Russian superiority is to be totally unaware of the dysfunctioning of Russian machinery, shortage of skills, inefficiency. Twenty million people killed in World War Two. If you know these things, you get furious.

When I come back to the US, here was [the leading American journalist] Walter Winchell, the dean of boobality. He had the American people in a state of total fright. I listened to the stuff, I couldn't believe it. I was living in Czechoslovakia at the time. I remember driving from Prague to Berlin up through the Russian zone. I saw a lieutenant-colonel in a buggy behind a horse and a cow. The scarcity of supplies, the thinness – it sickens me to remember the distortions on which this Russophobia is built.

We ran into it in Greece, when we started to repatriate these people. Churchill insisted on putting King George back on the throne, and the Greeks didn't want the son of a bitch. These were just people who resented what was being done to them. The English forced him back there, and they created a communist revolution. The more you fed it, the more it became like a fire.

Fiorello LaGuardia succeeded Lehman as UNRRA boss. One of his men asked me to go to Germany to run the DP operation in the American zone. There were about six million DPs there at the time. I really didn't want to go. You had responsibility without authority. The army ran things. A command decision could just wipe you out. And they did.

This was December '46. We still had about four-and-a-half million in the American zone: Poles, Russians, Ukrainians, Lithuanians, Estonians, Latvians. And some Jews; there only were about 280,000 of them left. While I was still in Czechoslovakia, we moved quite a few Jews out of Russia. They came across Poland and down the border across Slovakia to Vienna and then up into the US zone . . . We diverted food and medicine, what have you, from the Czech mission. It was too pitiable.

I went underground for three weeks before I began to run the office. I'd get in my car and stop in at camps, see who was running them, what spirit prevailed, who was running whom. And I came back to Heidelberg, our headquarters. It was an undestroyed town. We had a staff of fourteen thousand. It involved millions of people and everything from care of infants, to food, to shelter, to clothing, to transport, to death. Some of the social workers tried to work from textbooks. It didn't work. I went to the basics: shelter, food, sickness. That was it, right?

By this time, the warriors had gone home. The whole attitude toward the Germans had changed. The Cold War had set in. General Lucius Clay was made head of the US zone. It was his mission – as he told me once – 'to get those damn Jews out of here because my job is to rehabilitate the German economy and these people are eatin' up our groceries.' The thing was to build up Germany as a counter-force to the Russians, and the DPs were a drag on the economy.

Of course, you had an anti-Russian factor at work among the DPs. A lot of Latvians, Lithuanians, and Estonians had elected to join the Third Reich. And don't think they were coerced. They wanted to go. They killed their own Jews. They didn't need any help. They were a bunch of bastards.

The Estonians and the Latvians, especially, are a beautiful people, and the Germans loved 'em. Blue eyes, blond hair. Almost every officer, if he had a rank of colonel, had one of these women in his bed. They were choice women, right? They were the dancers, the entertainers, what have you. And they had infiltrated our command. We had a G-5 section that had to do with the displaced persons and prisoners of war. It was wildly anti-Russian. It had largely to do with who you're sleeping with. I can't tell you what the influence of the bedroom is on military and political policy, my friend. When I went to Prague, the Pankratz Prison was run by an SS group from Latvia.

They told their horror stories, which were probably true. The Soviets weren't pasties, believe me. They were ruthless, especially when they saw them as German collaborators. I know the Russians were trying to repatriate people. They had persuasion teams. We did no forced returning. The army did, but we didn't. About three million out of the four-and-a-half million went home within a year and a half.

[*Terkel*:] Were the DPs screened by us?

One of the first things I did was put through a *Fragebogen*. It was a questionnaire. Where were you on such a such a date? What work were you doing? I wanted to screen 'em out. The army objected. They raised Billy-all-hell. They called me up to Frankfurt and read the riot act to me. I had a press conference. It hit the old Paris [i.e. the *International*] *Herald-Trib* and the *New York Times*. From then on, I was anathema to the army. It was war between us now.

By this time, the combat troops had gone home and the second echelon had moved in. They were all in bed with the Germans and they particularly turned on the Jewish DPs like you wouldn't believe. It fell just short of persecution.

[*Terkel*:] Who's they?

The US Army. You had the sycophants, most of whom had been collaborators, right? The Jew came out of the ovens and he said, Screw you, Jack. He's lost his fear of death, life, hell and fire and damnation, because he's been there, right? All of a sudden the war was over and he thought he was on the winning side. I remember incidents that would kill you.

We were in this camp and a little redheaded guy – a pock-marked, tough little wiry Jew, who's survived the ovens – they were asking him, 'Do you think you want to go to Israel?' He said, 'I don't think, I'm goin' there.' He used good uncouth GI language he'd picked up on the way. They said, 'Suppose we don't let you?' He said, 'You keep me from goin'? How the hell you gonna keep me from goin'?' He broke down and wept when we came outside.

His story: he'd come out of the ovens weighing somethin' like sixty-five pounds. He'd built himself back, he'd married, had a baby. He and another guy had been walking through the camp, through the streets of the town [Landsberg] where Hitler wrote *Mein Kampf* [in prison, 1923–4]. Two GIs, half loaded, with some German girls, went by, and one of the girls called him *Judensau* – you Jewish pig – 'cause they didn't get off the sidewalk. The GIs started pushin' 'em around and these little guys gave it back at 'em and there was a fight and the MPs came and they arrested 'em and they sent this guy to prison for a year. So he was full of this anger when I talked to him.

He joined the Haganah [Zionist military organization in Palestine], moved there down through Italy. His wife and son were left in camp. War broke out with the Arabs. I didn't know he'd gone. One morning at Heidelberg, at seven in the morning, a woman came to see me. This guy's wife. She had a letter. Her husband was killed down by Gaza. I put her on the first legal movement, to go to Israel. We moved people illegally by the thousands down through France and Italy.

What more is there to say? After I was canned for my Czechoslo-vakian adventure, I ran a ski lodge in Vermont. I was called back as director of information for the islands and possessions of the United States. In those days, Hawaii, Alaska, Puerto Rico, and the Canal Zone, right? One morning I found everything off my desk. The security department called me in and told me, No problem, Mr Edwards. We just decided we didn't want you. Questions had arisen. It took me four years to find out what the charges were. It was cleared up, and I came back to the United Nations with top clearance. It's a chronicle of a well-used life. And how nutty the Cold War makes us.

To many people, the war brought about a realization that there ain't no hidin' place down here. That the world is unified in pain as well as opportunity. We had twenty, twenty-five years of greatness in our country, when we reached out to the rest of the world with help. Some of it was foolish, some of it was misspent, some was in error. Many follies. But we had a great reaching out. We took fifteen, eighteen million men overseas. For the first time they saw pain and poverty in dimensions they had never known before. At heart, Americans had a period of unbelievable generosity toward the rest of the world, of which they knew little. It was an act of such faith.

Now we're being pinched back into the meanness of the soul that had grabbed a new middle class that came out of poverty as did I. We squeezed our soul dry of pity. If it were just pity, that'd be one thing. But reason itself denies this. You can't repeal the speed of sound. You can't repeal the speed of communication. You can't repeal the interlinking of social orders around the world. It's impossible.

While the rest of the world came out bruised and scarred and nearly destroyed, we came out with the most unbelievable machinery, tools, manpower, money. The war was fun for America – if you'll pardon my bitterness. I'm not talking about the poor souls who lost sons and daughters in the war. But for the rest of us, the war was a hell of a good time. Farmers in South Dakota that I administered relief to, and gave 'em bully beef and four dollars a week to feed their families, when I came home were worth a quarter-million dollars, right? What was true there was true all over America. New gratifications they'd never known in their lives. Mass travel, mass vacations, everything else came out of it. And the rest of the world was bleeding and in pain. But it's forgotten now.

World War Two? It's a war I still would go to.

OBITUARY
David Stirling

Special forces have become the most feared elements of the world's armies in the second half of the twentieth century. Their role has acquired a particular importance with the rise of international terrorism, which confronts governments with the need to forestall acts of sabotage and hostage-taking or to intervene swiftly and decisively while such activities are in progress. Special forces also play a key role in operations behind enemy lines, particularly, as in the Gulf War of 1990–1, in the hunt for hidden weapon systems and in missions designed to paralyse enemy command centres. Small in size, such units, when composed of select personnel trained to the highest standards of efficiency, achieve results out of all proportion to their size.

The pioneer of the special forces idea was David Stirling, who recognized that, at a time when Britain's armed forces were heavily outnumbered by those of Germany, the balance could be partly redressed by deploying quality against quantity. The value of his idea was proved in action, and his Special Air Service Regiment (SAS) has become the model for all other special units in every country. It retains its reputation as the most effective of all of them.

*   *   *

Colonel Sir David Stirling, who has died aged 74, was the creator of the Special Air Service, which subsequently became an elite regiment of the British Army and won the admiration of many foreign countries which tried to imitate it.

Stirling won a DSO in 1942 and was appointed OBE in 1946. Maj-Gen Robert Laycock, head of Combined Operations, said he was one of the most under-decorated soldiers of the Second World War. This was probably because there was no senior officer or other eyewitness of his exploits to recommend him for just reward.

In 1941 Stirling was nicknamed the 'Phantom Major' by the Germans for his remarkable exploits far behind their lines in the Western Desert. In the 15 months before he was captured, he and his desert raiders destroyed aircraft, mined roads, derailed trains, fired petrol dumps, blew up ammunition depots, hijacked lorries and killed many times their own number. Rommel admitted that Stirling's men caused more damage than any other British unit of equal strength.

In 1942 the SAS was given the status of a full regiment. Montgomery said of its creator: 'The boy Stirling is quite mad. However, in war there is a place for mad people.' Nevertheless Montgomery refused to allow Stirling to pick recruits at will from his army.

Stirling himself designed the Regiment's cap badge, bearing the words 'Who Dares Wins'. The motto summed up his philosophy.

The Egyptian appearance of the SAS wings was due to the fact that they were modelled on a fresco in Shepheards Hotel, Cairo, where there was a symbolical ibis with outstretched wings. The ibis was removed and a parachute substituted. The 'winged dagger' badge was meant to resemble Excalibur – the sword of freedom.

Archibald David Stirling was born on November 15 1915, the son of Brig-Gen Archibald Stirling of Keir and his wife, Margaret, fourth daughter of the 13th Lord Lovat. David's brother, William Stirling, commanded the 2nd SAS Regiment.

Young David was educated at Ampleforth and Trinity College, Cambridge, but he was sent down after a year and began to study painting. On the outbreak of the Second World War he was in the Rocky Mountains practising climbing with the ultimate object of attempting Everest.

He served with the Scots Guards (the family regiment) for the first six months of the war, and then transferred to No. 3 Commando and went to the Middle East as a member of Bob Laycock's 'Layforce', which planned to capture Rhodes. When 'Layforce' was disbanded, Stirling and a few of his Commando friends decided to teach themselves parachuting with a view to landing behind German lines in the desert and destroying aircraft on the ground.

They 'acquired' parachutes and the use of a dangerously unsuitable old Valentia [bomber] aircraft. Inevitably Stirling was injured, but, in June 1941, while still on crutches, he managed to gatecrash GHQ, Middle East, and gain the approval of the C-in-C, Gen Auchinleck, to enrol 66 of his colleagues for his new enterprise.

In the early days the SAS was known as 'L Detachment' of the Special Air Service Brigade, although the latter did not exist. Training was extremely arduous. On one occasion two parachutists were killed owing to faulty static-line clips. Stirling identified the fault, made new clips and tested them himself at dawn the next day.

'Were you scared?' he was asked later. 'Terrified,' he replied, 'but what else could I do?'

The first venture by parachute, on November 17 1941, was a total disaster, because of a sudden sandstorm with winds of 90 mph. Of the 66 who set out, only 22 survived.

Undeterred, Stirling continued with his plans, now using trucks and the expertise of the Long Range Desert Group to navigate in the desert. His revised plan for the unit's employment was to travel deep into the desert by truck or jeep, walk several miles to the target airfield, arrive by night, and plant specially timed bombs to explode when all the dispersed German aircraft had been visited.

The bombs were fused by special time-pencils, invented by his colleague, J. S. Lewes, an Australian and former Oxford rowing blue who, with R. B. Mayne (later to win four DSOs), helped to create the unit's think-tank. Stirling had a genius for recruiting suitable people and, among others, John Verney, Fitzroy Maclean, Randolph Churchill and Roy Farran joined him.

The vital achievement of the SAS was that it destroyed on the ground the latest German aircraft, such as Messerschmitt 109Fs (armed with cannon) which in the sky totally outclassed the ageing Hurricanes and Gloster Gauntlets of the scanty Desert Air Force. One of its most spectacular exploits was the raid on Sidi Haneish airfield, when 18 jeeps, each carrying four Vickers K machine-guns, drove straight down the central runway, destroying Junkers, Heinkels, Messerschmitts and Stukas. They completed their work by driving around the perimeter, destroying no fewer than 40 aircraft.

Soon the SAS was raiding far and wide, taking pressure off Malta by destroying the airfields from which German bombers took off; it also raided Crete several times. In 1943, while his regiment was operating in the restricted area of northern Tunisia, Stirling was captured as 500 Germans surrounded the cave in which he was sleeping. He soon escaped, but he was recaptured.

After being flown to Italy he escaped four more times, but each time his height – 6ft 5in – gave him away. Eventually the Germans interned him in Colditz [the prison camp for escapees].

By the time of his final capture, the SAS – which was prepared to reach its targets by parachute, canoe, jeep, submarine or on foot over vast distances – had destroyed 350 German aircraft and numerous hangars, supply dumps, bridges, roads, and vehicles. It had inflicted many casualties and also drawn off many Germans to try to guard their airfields.

While Stirling was a prisoner of war, the regiment and its sub-unit, the Special Boat Squadron, were also ranging from Italy and the Adriatic to the islands in the Aegean and surrounding seas. They played a leading part in disrupting German communications in France.

On his release Stirling went to live in Rhodesia and Kenya, where he founded the Capricorn Africa Society with the objective of promoting racial equality, tolerance and understanding. He was the society's president for 12 years and made more friends among the black than the white community.

In 1959, when he returned to England, he became involved with the syndication of television programmes, and won the franchise for operating Hong Kong's television service. This became Television International Enterprises, of which he was chairman.

Stirling was always careful not to interfere in any way with the SAS which, having been disbanded, was reconstituted to fight in the Malayan Emergency [1948–60]. His military expertise, however, and wish to be concerned with projects beneficial to Britain drew him into advising units countering terrorism and subversion in countries where Britain had interests.

In 1967 Stirling and his friends created the Watchguard Organization, which, based in Guernsey, employed ex-SAS soldiers to provide bodyguards for Middle Eastern rulers and others. Occasionally, as in Kenya and Dhofar, he was overruled by Whitehall which sent the SAS, with its larger resources, instead.

By 1972 Stirling thought there were too many groups providing similar services, and for reasons of profit rather than patriotism, and he resigned from Watchguard. By this time the highly reputable Control Risks International was operating to frustrate kidnappers, prevent hijacks, and negotiate releases.

During the 1970s there were occasional antagonistic probings by the Press as to how far Stirling was involved with mercenary or secret organizations. There were attempts to link his name with Gen. Sir Walter Walker's vigilante organization for civil defence and to imply that he might harbour hostile intentions towards the government. This was pure slander, as both Walker and Stirling had often made it clear that they were working for the British government of whatever political party, and not against it.

In 1979 Stirling won substantial damages and costs in settlement of a High Court libel action against the magazine *Time Out*, which had published an article implying that Stirling's capture in North Africa in 1943 showed that he was a coward. Since Stirling was a man of legendary bravery, it was difficult to see why such an absurd accusation could have been made.

Extremely courteous, soft-spoken and self-effacing, David Stirling was worshipped by the men of the unit he had created and many more outside it. For a man of his size, he could move extremely swiftly and silently; in his younger days he had been able to stalk a stag and kill it with a knife.

He regarded killing the enemy as an unfortunate necessity. His most memorable characteristics were his creative vision, his cultured outlook, leadership, patience and iron determination – he was always adamant that the SAS soldier must be governed by self-discipline and never appear to be a heroic figure.

David Stirling had a wide and varied circle of friends, took a lively interest in all games of chance, and was a considerable bon vivant.

He was knighted in the New Year Honours list of 1990. He never married.

6 November 1990

KEITH DOUGLAS
(1920–44)
Aristocrats

'I think I am becoming a God'

The noble horse with courage in his eye,
clean in the bone, looks up at a shellburst:
away fly the images of the shires
but he puts the pipe back in his mouth.

Peter was unfortunately killed by an 88:
it took his leg away, he died in the ambulance.
I saw him crawling on the sand; he said
It's most unfair, they've shot my foot off.

How can I live among this gentle
obsolescent breed of heroes, and not weep?
Unicorns, almost,
for they are falling into two legends
in which their stupidity and chivalry
are celebrated. Each, fool and hero, will be an immortal.

The plains were their cricket pitch
and in the mountains the tremendous drop fences

brought down some of the runners. Here then
under the stones and earth they dispose themselves,
I think with their famous unconcern.
It is not gunfire I hear but a hunting horn.

Enfidaville, Tunisia, 1943

Note: The quotation heading the verse was, according to Suetonius, a remark
made by the Emperor Vespasian while he lay dying.

ALEXANDER STAHLBERG
Bounden Duty (2)

Alexander Stahlberg, who had begun his military life as a
cavalry volunteer, became an officer and in 1942 was appointed
aide-de-camp (ADC) to Field Marshal Erich von Manstein,
commander of Army Group South on Germany's front
in Russia. Manstein had already established his reputation as
the leading practitioner of mobile operations on the Eastern
Front.

At the time of Stahlberg's appointment, the Red Army had
just broken through the German line north and south of
Stalingrad, at points held by contingents from Germany's allies,
Italians, Romanians and Hungarians, and had encircled the
German Sixth Army inside the city. In December, Hitler would
order Manstein to mount a counter-offensive (Operation Winter Storm) with the object of breaking through the Russian
encirclement and restoring contact with Sixth Army. As Hitler
refused to contemplate allowing Sixth Army to break out
towards Manstein, the operation would fail and the defenders
of Stalingrad would eventually succumb to cold and starvation. Field Marshal Paulus, their commander, surrendered

the Stalingrad pocket on 1 February 1943, against Hitler's orders expressly forbidding capitulation.

* * *

The next day, 18 November 1942, I reached Vitebsk and the headquarters of the Eleventh Army shortly before dark, in a car belonging to the headquarters. I had myself announced to the Commander-in-Chief with some trepidation. When I entered Manstein's study he rose from his armchair, laid aside a book he had been reading, tapped the ash from a fat cigar and accepted my posting orders. Then he shook hands and asked me to sit down opposite him.

He was wearing a white linen uniform jacket of the kind we liked to wear off duty in peacetime. For the first time in my life, I saw the golden epaulettes with the crossed marshal's batons close at hand.

He began the conversation in a very friendly, not to say charming manner. [Stahlberg's cousin, Henning] Tresckow had told him something about me, including the fact that his [Manstein's] brother-in-law, Conrad von Loesch, his wife's brother, a casualty of the Polish campaign, had been married to a cousin of mine. So there would be plenty for us to talk about.

Then he asked me about myself, and especially about my military career. I began by telling him that I had joined the 6th (Prussian) Cavalry Regiment as a volunteer in 1935 because the NSDAP [the Nazi Party] in Stettin had tried to force me to become a Party Member. Without going into this point, he exclaimed with obvious pleasure that in that case I must have been serving alongside his previous aide, Specht. That was how I learned that 'Pepo', as we had called him in Schwedt, had been killed. The Field Marshal wanted to know if I had known him well and I had to say no, as I was older than he. During Reserve exercises, as an NCO, I had had to give the officer cadets riding instruction once or twice. Specht had been remarkable for sitting his horse perfectly from the first day, and had been better at all the cavalry exercises than many a veteran trooper – and certainly better than I was. The Field Marshal enjoyed talking about him and I could see how deeply Specht's death had affected him.

I also referred to the death of his eldest son, Gero – Tresckow had told me of it – and he talked to me about him. Then he suddenly reached behind him and gave me a letter from his desk to read. It came from the editorial office of the *Völkischer Beobachter*, the official party newspaper of the NSDAP. In words of contrived courtesy it informed the Field Marshal that the paper was prepared to print the notice of his son Gero's death only if the reference to the biblical text (Acts 8:39) which was to appear above the notice was removed. The verse ended with the words: '. . . he went on his way rejoicing.' Through this – undoubtedly perfectly inoffensive – text his wife and he had wanted to express two things: firstly, that he and his family were Christians, and secondly, that Gero had been a particularly happy human being. He would therefore tell the *Völkischer Beobachter* that the notice was to appear unchanged and with the biblical text. After all, he had given his rank when he signed it. He looked at me questioningly.

After a pause, I asked: 'Does it have to be the *Völkischer Beobachter*, sir? The *Deutsche Allgemeine Zeitung* seems to me a more suitable place for this notice. It too is conformist, of course, but I have always noticed that "our families" prefer that paper.' He answered quickly that the *Deutsche Allgemeine* had received the notice in the same post as the *Völkischer Beobachter* and had not refused to print it. That was why he was going to insist that the *Völkischer Beobachter* should also print it in full.

At that point Manstein was obviously unaware that the *Deutsche Allgemeine* had already carried the death announcement on 7 November, having deleted the biblical text without reference to the family. The newspaper had acted independently, in accordance with the rules of the Ministry of Propaganda for the obituaries of the fallen, whereas the Party newspaper had at least written. When writing this account, I sought out Gero von Manstein's obituary in the *Völkischer Beobachter* at the Press Institute of the Free University of Berlin. It was published on 22 November 1942, without the biblical text. So that was how the Nazi Party treated even Field Marshals.

All this lies in the past now, but it epitomizes National Socialist propaganda policy, with its curious nuances, and it reveals the inhumanity and rigour of the demands it made on people. This first conversation with Manstein covered other subjects as well, and I was well aware that I was being tested. He made me report in detail on the fighting at Tikhvin and above all the battle south of Lake Ladoga, when he had led us. Quite suddenly he interrupted me. 'I'll make you an offer,' he said. 'If you like, we'll give each other a try.' I accepted at once, because I was impressed by his personality and his approach to a much younger man. I felt I could work well for him.

Then he outlined my job, of which I had only a vague idea, summing up: 'You will be my constant companion, you will be present at all my conversations, you will take brief minutes of our daily doings, in so far as they are important. You will listen to my telephone conversations, write for me and keep my files, both the military and some of the private ones.'

I interjected that I was not clear to what extent a lieutenant could share the official and other life of a Field Marshal; above all, I could not imagine that there were not some matters which a Field Marshal had to deal with privately and in the strictest of confidence. He rejected this at once, emphatically: in wartime it did not apply, at least not for him. I persisted: after all, he might be talking to the Führer. 'Then you will be there, unless accompanying officers are excluded on his orders.' Then I objected that although I could type a little, I had not learned shorthand. 'That is actually a good thing,' he said. 'I do not care for shorthand in the work of the General Staff, because the shorthand-writer records everything he hears, including trivia. I expect you to recognize what is important and what is not. The less you write, the better. I hope,' he continued, 'that I shall soon be able to dictate to you. And I expect you – though not overnight – to be able to transfer the sense of my dictation reliably to paper from your notes.'

I felt quite dizzy at what was in store for me. After a pause, during which he presumably watched my reactions, he began again, stressing something of extreme importance to him: he had worked on the General Staff in the First World War, so he knew the military hierarchy well enough to know that with each promotion a senior officer ran more risk of isolation. On the so-called path of duty, each upward step was necessarily a kind of additional filter and it was up to me to tell him important things which other people thought should be kept from him. Obviously he did not expect me to burden him with gossip, but he attached great importance to knowing what, in my view, he ought to know. What he meant, in a word, was this: 'Whatever you know I too must know, if you consider it necessary!'

I was delighted. This was a superior entirely to my taste. I was to work for a man of real consequence, who had honoured me with the greatest trust.

I then learned that the entire operations staff of the Eleventh Army (the new Army Group Don) was to move into the waiting command train that evening, to be ready to cross Russia southward towards Stalingrad, where the Red Army had launched a massive offensive a few days before in an attempt to pose a serious threat to the entire German Southern Front, including our Italian, Hungarian and Romanian allies. Hitler had made Manstein Commander-in-Chief of this distant section in order, as the General Staff Officer says: 'to restore the situation', or in plain language, to save what could be saved.

Whilst we waited for the orders from OKH [Army High Command], I made my most important preliminary visits round the headquarters, in precise order of seniority, according to custom.

General Friedrich Schulz was Manstein's Army Chief of Staff and his closest colleague since the Crimean campaign. He welcomed me with great warmth. Word had already gone round among the staff that Tresckow had recommended me, and I had a clear sense of the reputation my cousin enjoyed. Not only that evening, but countless times in the years that followed, people would speak to me of Henning, so that I became aware of the responsibility I shouldered as 'Tresckow's man'.

I at once felt both trust and liking for General Schulz and no less so for his aide-de-camp, Lieutenant Otto Feil, who kept the Army's War Diary. Feil was soon one of my closest friends on that large staff.

Things were quite different with the 'Ia', the First General Staff Officer, Colonel Theodor Busse. When I went to his office, he offered me a chair facing him, turned the light of a standard lamp on me and questioned me about everything he was interested in without my being able to see his face. So there was a wall between us from the start.

The headquarters had as its mobile command post a special railway train of about ten express-train carriages. It was equipped with everything an operational department needed in war, with radio and telegraph equipment as well as its own armament. The Field Marshal had a former saloon carriage which had seen better days and which had been converted into a study and briefing room, with two large sleeping compartments next to it for himself and me, as his aide-de-camp, as well as compartments for the batmen. It was a wonderful old carriage, more elegant than any I had seen before. It was said to have belonged to the Queen of Yugoslavia. Precious art-nouveau marquetry decorated the panelled walls and heavy silk curtains contrasted oddly with its current use.

Two days later, whilst we were visiting two of the corps still under the Field Marshal's command, the expected order arrived from OKH confirming the creation of a new Army Group Headquarters from Headquarters Eleventh Army under the title Army Group Don. The new formation was to move immediately to the South to relieve the Sixth Army, threatened by encirclement at Stalingrad, and to 'reconstruct' the defensive front on both sides of the city.

On that day (20 November), I was present at a strategic discussion of great significance. Only a very defective situation map of Southern Russia was available for the discussion, yet the sparse entries were enough to lead Manstein and his staff to the conclusion that there could be no question of a 'restoration of the former front line'.

I now realized what the 'leadership' of the Supreme Command – in other words, Hitler – really meant: it was neither flexible nor dynamic, but unimaginative, uninventive and, above all, static. I now held the key to my own terrible experiences and the destruction of the 12th Panzer Division in more than a year of battles between Leningrad and the Volkhov.

GEYR VON SCHWEPPENBURG

On the Other Side of the Hill

While the situation on Germany's Eastern Front went from bad to worse in 1943–4, in the west the Wehrmacht enjoyed the sensation of unnatural calm. It was, however, the calm before the storm. Supreme Commander West, Field Marshal Gerd von Rundstedt, knew that an amphibious invasion by the British and Americans was imminent. What he could not tell was when it would come or where it would fall.

General Geyr von Schweppenburg commanded the main armoured counter-attack force on Rundstedt's front. The positioning of its tanks was a source of fierce contention in the months before the invasion. Rundstedt wished to keep the tanks in reserve and commit them to battle only when the location of the landings had been firmly identified, the orthodox military solution. Rommel, his direct subordinate and in operational control of the anti-invasion forces, argued that the tanks must be deployed as near to the beaches as possible, on the grounds that Allied air power would prevent the tanks from moving, once the invasion was under way. Hitler adjudicated, to neither party's satisfaction. While some tanks were left under Rundstedt's control and others allotted to Rommel, a third portion was reserved to Hitler himself to deploy. In the event, the armoured divisions arrived at the beaches too late to hinder the landings.

Schweppenburg's account of the misunderstanding and mistrust prevailing between the different German headquarters in the west in the invasion summer accurately summarizes how far the Wehrmacht's power had declined since the days of effortless conquest in 1939–41. His own headquarters, identified by Ultra intelligence soon after D-Day, was to be completely destroyed by air attack on 10 June.

*   *   *

In a peaceful room in St Germain in the early summer of 1944 I was sitting with the Chief of Staff of von Rundstedt's Western Army Group. Once again our conversation turned to the question of the invasion.

'Our intelligence service has just received a warning from one of the London embassies that the invasion is now imminent,' he remarked. 'But I am not at all sure that this whole business of invasion is not an enormous bluff.'

Indeed, knowing the British intelligence methods, one could seldom tell what was the truth and what was a hoax; but this time I was sure. 'No,' I replied, 'the British, from the Prime Minister down, are prepared to use anything as a hoax, with one exception. That exception is the Crown. The King has seen the troops off. Believe me, the invasion is coming.'

The danger of opinions such as that expressed by the Chief of Staff seemed to me so great that I asked if I might put my views personally to von Rundstedt. The Field Marshal listened in silence. Such was the lethargy in the Army Group Headquarters that I also warned the staff as energetically as I could of the dangers of air attack, and in particular of airborne landings. I knew General Browning [commanding British Airborne Corps] from my time in London. I knew his job at that time was training the airborne troops. [Air Marshal] Slessor I had not then known personally, but I was well aware of his important and dangerous military teaching.

By 1944, the Hitler regime had seriously undermined both the spirit and the principles of the German command system. In the old army, teaching on the subject of command depended upon a cold and sober assessment of every situation, and relied upon the competence of every subordinate to carry out his task in whatever manner seemed best to him. This was replaced by 'intuition' from Berchtesgaden [Hitler's holiday retreat], and by a strict control of every smallest detail from the top. Contrary opinions were not entertained.

Hitler hated the General Staff. He succeeded in splitting it, and reducing it to a monstrous Saints and Sinners Club. Such few Saints as remained by 1944 he hanged after July 20 [the date of the failed attempt by German officers on Hitler's life].

No Allied soldier can understand what the atmosphere of Hitler's madhouse was like. His influence affected everyone – the leading military men as well as those of weaker character. To understand it one must have experienced it oneself. People who have had no experience of the destruction of complete families and the concentration camp cannot fully understand.

The German soldiers who fought and died bravely behind the fictitious Atlantic Wall made a sad picture. The infantry divisions could scarcely be called even third-rate. It was not their loyalty or their courage that was in question, but their physical condition and their equipment. Again and again these formations had been combed out to provide replacements for casualties on the Eastern Front. Almost one-third of the strength of the infantry divisions of the Seventh Army in Normandy was made up of Russians; they lacked all mobility, and their equipment simply did not compare with that of their enemy. At no time could the so-called Atlantic Wall ever seriously have impeded an Allied landing on the Continent.

The panzer divisions, on the other hand, were well trained. The commanders had for the most part had a longer command training than even those of the Waffen SS. It was reckoned later on, by a number of experienced officers, that at this time the panzer divisions were still at least one-third as powerful as they had been in September 1939.

The whole basis of their training lay in recognizing the enemy air superiority over the battlefield. They depended upon fast movement by night and in the twilight, on really accurate shooting, and on the quick and reliable passing of orders. Enemy airborne or parachute landings had to be dealt with immediately, even if they turned out to be dummy landings. Where infantry tactics were concerned, I had ordered a demonstration battalion to be formed in 21st Panzer Division, to show British infantry tactics. These were watched by all the panzer troops.

The fine troops were overwhelmed by weight of numbers. The words of Marshal Timoshenko – 'The steel of the German Army must be melted in the holocaust of the Russian onslaught' – applied once more.

Old Field Marshal von Rundstedt in St Germain, though highly respected, was ailing. He had become lethargic. He had no command over either the Luftwaffe or the German Navy, and armoured tactics were not his strong point. Difficult decisions were avoided. His Chief of Staff tried to iron out differences of opinion by negotiation, even when no compromise was possible and only a tough decision would do.

Before the arrival of Rommel, Field Marshal von Rundstedt had laid down clearly the role of the panzer forces in defence of the Western Front. They were his only really battleworthy formations. Rundstedt had followed the recommendation of his responsible adviser on the subject – in this case myself. The main force of the panzer divisions was to remain well back from the coast, but north of the River Loire. South of the river were three panzer divisions which I had newly formed into the 58th Panzer Corps. It was impossible to tell at this stage whether the invasion would come first from the Mediterranean or from the Channel, but with this deployment, with the SS Panzergrenadier Division Götz von Berlichingen on the Loire, an immediate reaction either northwards or southwards could be achieved.

Rommel, however, brought a new idea. He required the invaders to be pushed back on to the coast. His experience of the very powerful British Air Force in Africa had convinced him (wrongly) that the movement of large mechanized formations was no longer a practical proposition. For months there had been a protracted controversy between him and myself as the adviser on this subject to the Commander-in-Chief. General Guderian, Hitler's highly experienced but often unheeded armoured adviser, supported my opinion without question.

Rommel's ideas, being clearly defensive in concept, were undoubtedly out of keeping with proper armoured tactics. He did not see the difference between the open expanses of the desert and the thickly covered countryside of Normandy. On the other hand, by adopting a more mobile concept of operations, a concentrated counter-attack could at the very least have achieved a temporary victory against the Allies. The insistence on the defence of the coastline threw away from the start the mobility of the panzer formations; worse, they were committed in the impossible hedgerow country, hemmed in by minefields and marshes.

The endeavours of the leading armoured experts to hold a proper force in reserve had two main purposes. The first was to retain some possibility of manoeuvre, and the second was to avoid, in the event of an airborne landing, the breakdown of our whole supply system by cutting us off from our petrol. Due to constant casualties to our supply columns caused by air attacks, our supply system never worked well from the start of the campaign.

Perhaps the last word in the story of this great argument about the employment of the armour comes from one of Rommel's personal orderly officers whom I met briefly (much later) on Stuttgart railway station. He told me that shortly before Rommel's injury on July 17, 1944 [he was severely wounded when a British aircraft strafed his car], the Field Marshal had remarked to him: 'It would perhaps have been better after all to have held the panzer divisions back.'

At the time, the ultimate solution to this lengthy argument came in the form of a directive from Berchtesgaden. This solution was the worst possible: it was neither one thing nor the other. Half the force was to be moved at Rommel's disposal, to the coast; the remainder would stay inland 'for the time being'.

Remember the wise Moltke's maxim: 'Mistakes in preliminary deployment are difficult to correct.' [Moltke the Elder directed Prussia's victories over Austria and France, 1866 and 1870–1.]

As to where the Allied landings would take place, Rommel, Jodl and von Salmuth were all in agreement; the main landing would be in the Pas de Calais. Rommel clung firmly to this belief, even after the battle on the Cotentin peninsula had been fought for some weeks.

Early on June 6 I heard from my Chief of Staff that the invasion had begun. The army group had ordered the panzer divisions which had been held back north of the Loire to move to the coast. I had not even been asked.

I requested immediately that the Panzer Lehr Division should not move before nightfall. My request was refused; or rather, I did not even receive an answer. The armoured grenadier battalion of this division, which was the only one that was properly equipped, was heavily attacked from the air during this daylight move.

On the morning of the invasion a young German pilot had succeeded in carrying out a particularly courageous operation and had flown over the Allied invasion fleet without being detected or engaged. Unfortunately, *his report never reached the High Command.* Had it done so, it would have removed any doubt that this landing in Normandy was in fact the real thing. On June 7 I was ordered from the Headquarters in St Germain to take command of the Caen sector under the direct command of the Seventh Army and indirectly under Rommel's command. (Whatever one may say of Rommel's tactical ability, he was a brave and tough soldier, and his constant appearance at the front line demanded respect. This made it possible for the more senior officers to accept his orders.)

The area around Caen and the neighbouring coastline were well known to me. I had had the job of planning the operations of 24th Infantry Army Corps as the follow-up formation in Operation Sealion (which was the intended invasion of England in 1940).

The boldness of the enemy airborne operations achieved full success – as boldness always does. The landing of the 6th British Airborne Division near Caen was most successful.

A request was sent at once to Rommel's headquarters to allow the 21st Panzer Division to operate against this landing. Unfortunately Rommel was away in Germany at the time, and he had given strict orders that the 21st Panzer Division should not be moved without his own specific order. He had estimated that there was no danger of invasion at this time because of the state of the tides. It was not until eight o'clock on the following morning that his Chief of Staff gave permission for the division to move. By this time it was too late.

When I visited the courageous infantry divisions on the coast north of Caen I found that they had been practically wiped out by the Allied bombardment. On June 8 and 9 I visited the three panzer divisions which had by then been committed. They were fighting a very hard battle. I had received clear orders from the Seventh Army on June 8 in the best traditions of the old German Army; however, shortly afterwards I received conflicting orders from Rommel, from the Army Group Headquarters in St Germain, and from Berchtesgaden. What a state our army had reached!

On the morning of June 10 I visited a regimental headquarters near an abbey on the top of the hill just north of Caen. From there I saw a panzer regiment in action against the Canadians. It was hellish, but in this case the hell came from the sky. The British and Canadian troops were magnificent. This was not surprising when one knew the type of man, and when one knew that a large part of them had been trained by that outstanding soldier (and old friend of mine) General Sir Bernard Paget.

However, after a while I began to think, rightly or wrongly, that the command of these superb troops was not making the best use of them. The command seemed slow and rather pedestrian. It seemed that the Allied intention was to wear down their enemy with their enormous material superiority. It will never be known whether Montgomery had received a private instruction from his Government to avoid for the British troops another bloodbath such as they had suffered in the First World War on the Somme and at Passchendaele. However, it seemed to me that the command of the British and Canadians failed to make the best use of these magnificent troops.

One serious problem which faced us was the dropping of agents by parachute. As soon as they landed they were swallowed up by the local population. Many years later [the New Zealand war correspondent and historian] Chester Wilmot told me that from the moment that my headquarters had left Paris it had been continually shadowed; I could well believe him. The subsequent destruction of my whole headquarters in the late afternoon of June 10 was no doubt the result of the highly organized intelligence service run by the British. Only a few moments before an air attack wiped out my operational headquarters, Rommel and I had left the very command vehicle which was destroyed. I was lucky in that, being 'off-side', I was only slightly wounded.

The command of the Caen sector was then taken over by 1st SS Panzer Corps under General Sepp Dietrich on the orders of Seventh Army. Dietrich had been a Bavarian cavalry sergeant. He was brave and friendly and to me he was always most loyal. However, in a fast-moving and attacking battle he was not up to his job.

The counter-attack which I had planned to take place on June 10 had to be cancelled. Then the British 7th Armoured Division made its appearance. The Desert Rats first started to make life unpleasant for us in the area of Bayeux.

In accordance with my orders I went off to Paris to reorganize my new staff. By about June 23 I had taken over my new command which consisted of three panzer corps and one infantry corps. I was under the direct command of Rommel. As soon as I took over command I called together the four chiefs of staff of these formations. They were all well-trained and experienced soldiers and I had no reason to think that any of them were suicidal maniacs. I then put to them a question to which I required the answer Yes or No. The question was: 'Do you think it possible to push the Allied invasion back into the sea?' Nobody would answer for fear of the results. For this, although it was Hitler's dream, was in fact nothing short of cloud-cuckoo-land.

I felt it my duty to send an honest report about the situation. Fortunately I took the precaution of keeping a photostat copy of this report. Both Rommel and Rundstedt agreed with my opinions. My neighbour, General Hausser, the Commander-in-Chief of the Seventh Army, also supported me.

Rundstedt and I were as a result relieved of our commands, and Rommel, who was good enough to try to support me, told me that he expected to be the next to go.

The most clear-thinking soldier on the Western Front, General Dollmann, latterly the Commander-in-Chief of Seventh Army, soon disappeared from the picture. Luckily for him his sudden death saved him from a subsequent court martial. In any case he, as a practising Catholic, had always been suspect, as was I. By seniority I should have been his successor, but I was superseded. In a sworn statement made subsequently, Rommel's Chief of Staff said that Hitler had always mistrusted me. He was right.

In ending these notes on the invasion, I must stress that what I have said is far from the whole story. No landing or lodgement attempted by the Allies could ever have been defeated by us without an air force, and this we utterly lacked. The command organization set up by Rommel and von Rundstedt in the West was not unlike the Roman system of changing command daily between the two consuls. The result, of course, was Cannae [Hannibal's crushing victory over the Romans in 216 BC].

The differences of opinion throughout the German High Command at the time of the invasion put one in mind of a story told by Field Marshal Lord Ironside [Chief of the – British – Imperial General Staff in 1940, at the time of the German conquest of France] in his memoirs; these had been written, it must be remembered, four years before by a member of the other side! He tells how General Gamelin had warned the French politicians in the hour of emergency that endlessly sitting round a table, thousands of miles from the scene of the battle, would never produce any better answer than the strategists of the Café de Commerce. The German café strategists were at this time sitting in Berchtesgaden.

The effect of Hitler's and Rommel's coastal defence policy on the German troops echoes the bitter remark made to Lord Ironside by the French Commander-in-Chief on the Western Front in 1940, General Billotte: '*Nous crevons derrière des obstacles.*' ('We are rotting behind our defences.')

To think that these were the tactics chosen by the successors of von Schlieffen! [Field Marshal Count von Schlieffen, architect of the famous plan for Germany's defeat of France and Russia which, disastrously altered, was employed in August 1914.] They do not deserve the title of 'strategy', for true strategy is always bold. But one thing is certain, no matter how bold our strategy had been, the final result in the West would never have been changed. The war had already been lost in Stalingrad, in Africa, with the destruction of the production lines for the air force, and, indeed, with the destruction of the Luftwaffe itself. The invasion was the final act in the tragedy of the Third Reich. The remainder of the war after the success in Normandy was only a prolonged epilogue, in which the events of July 20 were no more than the final spasm before the old Germany breathed her last.

5 June 1964

MARIE-LOUISE OSMONT
Normandy Diary

Marie-Louise Osmont, the wife of a Norman doctor and landed gentleman, spent the war in the Château of Périers-sur-le-Dan, a tiny village just inland of one of the beaches (Sword Beach) chosen by the British for the D-Day landings, 6 June 1944.

Her house had been requisitioned by the Germans, who occupied most of it, together with the outbuildings and park. She, however, retained a few rooms and therefore saw the Germans at close hand. Though not unfriendly to them as individuals, she regarded them as interlopers and made no pretence of accepting their presence.

Périers and its château were the scene of fierce fighting on D-Day. The village stands on a ridge which, in the late afternoon, was attacked by the German 21st Panzer Division, trying to break over the crest, reach Sword Beach and drive the British into the sea. British tanks and anti-tank gunners on the ridge succeeded in hitting the panzers and turning them back. Périers was as close as any German counter-attack got to the landing beaches on D-Day or afterwards. Marie-Louise's account is therefore of the keenest interest. It is, indeed, a unique record of the invasion days, a full journal of the time written by a civilian and a woman.

Marie-Louise was wounded in the fighting for Périers by a shell splinter in the back but recovered, remained at the château and continued her journal until the Battle of Normandy had been won.

*　　*　　*

June 6, 1944
Landing!!

During the night of the fifth to the sixth, I am awakened by a considerable rumbling of airplanes and by cannon fire, prolonged but fairly far away. Then noises in the garden and in the house: talking, loading ammunition boxes, nailing. I get up, go to the window. I see the big fifteen-ton truck arriving, coming from the drive and pulling up in front of the stoop, and another truck backing up to the dining-room window. I gather that a departure has begun, and I envision the unit moving to a new camp, and in the middle of the night as always. I'm annoyed at the idea of changing troops. I stay up, wondering. I watch through the keyhole of my door, which faces that of the *Kommandantur* [CO's office]. In the light, I catch sight of shadows moving in the office. They're dragging sacks, boxes; they come up and go down. I recognize Mr George, the bookkeeper, the *Spiess* [sergeant-major]. They don't look happy. I stay by the window. The airplanes fly over in tight formations, round and round continuously. I envision German airplanes overflying the departure. I'm surprised. But the cannon fire gets closer, intensifies, pounds methodically; what's going on? Great turmoil in the garden. The men have shouted, 'Alarm', from man to man, but the siren hasn't sounded. The *Spiess* fidgets, plays with the dog, with his usual air of a man playing at being important . . . nothing more.

Little by little the gray dawn comes up, but this time around, from the intensity of the aircraft and the cannon an idea springs to mind: landing! I get dressed hurriedly. I cross the garden, the men recognize me. In one of the foxholes in front of the house, I recognize one of the young men from the office; he has headphones on his ears, the telephone having been moved there. Airplanes, cannon right on the coast, almost on us. I cross the road, run to the farm, come across Maltemps. 'Well!' I say, 'Is this it, this time?' 'Yes,' he says, 'I think so, and I'm really afraid we're in a sector that's being attacked; that's going to be something!' We're deafened by the airplanes, which make a never-ending round, very low; obviously what I thought were German airplanes are quite simply English ones, protecting the landing. Coming from the sea, a dense artificial cloud; it's ominous and begins to be alarming; the first shells hiss over our heads. I feel cold; I'm agitated. I go back home, dress more warmly, close the doors; I go get Bernice to get into the trench, a quick bowl of milk, and we run – just in time! The shells hiss and explode continually.

In the trench in the farmyard (the one that was dug in 1940) we find three or four Germans: Leo the cook, his helper, and two others, crouching, not proud (except for Leo, who stays outside to watch). We ask them, 'Tommy come?' They say yes, with conviction. Morning in the trench, with overhead the hisses and whines that make you bend even lower. For fun Leo fires a rifle shot at a low-flying airplane, but the *Spiess* appears and chews him out horribly; this is not the time to attract attention. Shells are exploding everywhere, and not far away, with short moments of calm; we take advantage of these to run and deal with the animals, and we return with hearts pounding to burrow into the trench. Each time a shell hisses by too low, I cling to the back of the cook's helper; it makes me feel a little more secure, and he turns around with a vague smile. The fact is that we're all afraid.

Around eleven o'clock, during a lull, I go back to the farm. They show me fifteen or so impact points in the back pasture, impact points on the side toward the church as well, broken branches. We pick up pieces of the missiles, big shell fragments. Mr George and the replacement *Spiess* come to talk with us; we watch the heavy artificial cloud over the sea being driven by the wind, and especially some balloons in the shape of dirigibles, whose purpose we don't understand. We look at them through field glasses. The Germans talk about the landing, but without nervousness, as if speaking of something that hasn't reached us yet; they look at the collected fragments with interest and smile at us as usual. When someone says to them, 'You leave?' they answer, 'No, not leave.' We don't know what to think; probably they don't either.

One of their men is wounded in the knee. He was in one of the trucks parked on the road across from the church; two comrades lend support, and they take him away in a small car.

Around noon a bit of a lull. We leave to try to have lunch; I busy myself with the fire, Bernice with the soup and potatoes; it's cooking. We start to seat ourselves around the table, two mouthfuls of soup, and then everything changes with tremendous speed. Someone – a Frenchman on the road, the soldiers at the gate – someone said: 'The Tommies!' We watch the soldiers. They hide on both sides of the gate, watching in the distance in panic, confusion painted on their faces. And suddenly we hear these words: 'The tanks!' A first burst of tracer bullets, very red, sweeps the gate; the men crouch down. Bernice and I hide in a corner of the room. There's banging in every direction. We're going to have to go somewhere else. Standing in our corner, we gulp a plate of soup, while the *Spiess*, who has been shouting orders, comes with revolver in hand to see whether men are hiding with us. Everything starts happening. Evidently, they're going to try to leave with their trucks. A German tank arrives and takes the *Spiess* away. The shells bang.

The mean *Spiess* had guts. He came and went heedless of the shells, attended to everything – and he probably took responsibility for having the trucks leave; orders from higher up didn't seem to come.

Impossible to stay in this house at the edge of the road with such thin walls. We cut two slices of bread, the same amount of cold meat, and hugging the walls of the outbuildings, we make it to the trench, we fall in, and just in time! There's hissing and banging everywhere, our stomachs are churning, we feel suffocated, there's a smell of gunpowder. We stretch out completely, lying on the straw at the far end.

The afternoon is endless. At one point the sound of footsteps makes us jump up and look toward the opening, expecting anything. Consternation: it's the replacement *Spiess* (the nice dark-haired one), who, with a revolver in his hand, his submachine-gun under his other arm, and followed by a soldier carrying two boxes of ammunition on his shoulder, has come to see whether there are any stragglers still in the holes. He seems exhausted. He's wounded near the ear and there's a trickle of blood; he sits down for a few moments on the edge of the trench, looking at us with sympathy and as if feeling sorry for us. A few words about his wound, a few words about 'the Tommies here', and he leaves. We continue to wait.

The first English soldiers appear in the pasture behind the farm at two o'clock. They come down, submachine-guns and machine-guns under their arms, walking steadily, not trying to hide at all.

Around six o'clock a lull. We get out and go toward the house to care for the animals and get things to spend the night underground.

And then we see the first damage. Branches of the big walnut broken, roof on the outbuildings heavily damaged, a big hole all the way up, a heap of broken roof tiles on the ground, a few windowpanes at my place – hundreds of slates blown off the château, walls cracked, first-floor shutters won't close – but at Bernice's it's worse. An airplane or tank shell (?) has exploded on the paving in her kitchen at the corner of the stairs, and the whole interior of the room is devastated: the big clock, dishes, cooking equipment, walls, everything is riddled with holes, the dishes in broken pieces, as are almost all the windowpanes. The dog Frick that I had shut up in the next room so he wouldn't get killed on the road, is all right and sleeping on a seat. But we realize that if we had stayed there, we would both have been killed. In the face of this certainty, Bernice takes the disaster very well; we try to straighten up the unspeakable mess a little. Out of the question to eat the soup and mashed potatoes that have been prepared; everything is black with dust and full of shards of glass. Someone gives us some soup from the farm. We talk with them for a short while and note that the Germans haven't taken away all the trucks from the drive; there are also a lot of vehicles still in the park.

At the farm, even worse damage. Roof heavily damaged on the hayloft. Dean's room, on the second floor, pierced through by a shell, a beam broken, shell fragments have pulverized his armoire, everything is covered with enough plaster rubble to kill an entire family. Bernadette's room also penetrated, but less so. In the cattle barn a shell has made a very big hole, killed a superb calf, a ewe, and wounded another calf!

In the bedrooms of the farm all the suits, all the clothes are riddled by fragments and unwearable, the hats burned.

In the drive leading to the village, broken trees blocking the way, cut down by the shells, wide breaches in the wall from the passage of the tanks, pillars at the entrance of the old farm on the ground in bits. A cow killed in the pasture.

The English tanks are silhouetted from time to time on the road above Périers. Grand impassioned exchanges on the road with the people from the farm; we are all stupefied by the suddenness of events. I take a few steps down the drive, toward the Deveraux house, and suddenly I see the replacement *Spiess* and his comrade hugging the wall of the pasture. I tell him that he must still have comrades at the guns, since we can still hear the battery firing. You feel that these two men are lost, disoriented, sad. Later, almost night, I see them again, their faces deliberately blackened with charcoal, crossing the park. What will be their fate? How many of them are still in the area, hiding and watching?

Night in the trench, lying on the straw, Bernice and I. It isn't cold. The shells hiss continually over our heads, red streaks in the sky. A few hours of disturbed sleep. Stiff all over.

ERNIE PYLE
Battle and Breakout in Normandy

Ernie Pyle became the most famous American war correspondent of the Second World War, 'the G I journalist'. His matter-of-fact style and his fearlessness in accompanying soldiers into action made the US army feel that he was one of their own. This report of a small action in western Normandy in the third week of the invasion battle epitomizes his methods, both of collecting material and of recording what he saw. On 1 July 1944 the battle for Normandy still hovered in the balance. The break-out from the bridgehead remained a month distant.

Ernie Pyle was killed in action in the Pacific in 1945.

*   *   *

Battle and Breakout in Normandy
By Ernie Pyle

In Normandy – (by wireless) – Lieut. Orion Shockley came over with a map and explained to us just what his company was going to do.

There was a German strong point of pillboxes and machine-gun nests about half a mile down the street ahead of us.

Our troops had made wedges into the city on both sides of us, but nobody had yet been up this street where we were going. The street, they thought, was almost certainly under rifle fire.

'This is how we'll do it,' the lieutenant said. 'A rifle platoon goes first. Right behind them will go part of a heavy-weapons platoon, with machine-guns to cover the first platoon.

'Then comes another rifle platoon. Then a small section with mortars, in case they run into something pretty heavy. Then another rifle platoon. And bringing up the rear, the rest of the heavy-weapons outfit to protect us from behind.

'We don't know what we'll run into, and I don't want to stick you right out in front, so why don't you come along with me? We'll go in the middle of the company.'

I said, 'Okay.' By this time I wasn't scared. You seldom are once you're into something. Anticipation is the worst. Fortunately this little foray came up so suddenly there wasn't time for much anticipation.

The rain kept on coming down, and you could sense that it had set in for the afternoon. None of us had raincoats, and by evening there wasn't a dry thread on any of us. I could go back to a tent for the night, but the soldiers would have to sleep the way they were.

We were just ready to start when all of a sudden bullets came whipping savagely right above our heads.

'It's those damn 20-millimeters again,' the lieutenant said. 'Better hold it up a minute.'

The soldiers all crouched lower behind the wall. The vicious little shells whanged into a grassy hillside just beyond us. A French suburban farmer was hitching up his horses in a barnyard on the hillside. He ran into the house. Shells struck all around it.

Two dead Germans and a dead American still lay in his driveway. We could see them when we moved up a few feet.

The shells stopped, and finally the order to start was given. As we left the protection of the high wall we had to cross a little culvert right out in the open and then make a turn in the road.

The men went forward one at a time. They crouched and ran, apelike, across this dangerous space. Then, beyond the culvert, they filtered to either side of the road, stopping and squatting down every now and then to wait a few moments.

The lieutenant kept yelling at them as they started:

'Spread it out now. Do you want to draw fire on yourselves? Don't bunch up like that. Keep five yards apart. Spread it out, dammit.'

There is an almost irresistible pull to get close to somebody when you are in danger. In spite of themselves, the men would run up close to the fellow ahead for company.

The other lieutenant now called out:

'Now you on the right watch the left side of the street for snipers, and you on the left watch the right side. Cover each other that way.'

. . . The men didn't talk any. They just went. They weren't heroic figures as they moved forward one at a time, a few seconds apart. You think of attackers as being savage and bold. These men were hesitant and cautious. They were really the hunters, but they looked like the hunted. There was a confused excitement and a grim anxiety in their faces.

They seemed terribly pathetic to me. They weren't warriors. They were American boys who by mere chance of fate had wound up with guns in their hands sneaking up a deathladen street in a strange and shattered city in a faraway country in a driving rain. They were afraid, but it was beyond their power to quit. They had no choice.

They were good boys. I talked with them all afternoon as we sneaked slowly forward along the mysterious and rubbled street, and I know they were good boys.

And even though they aren't warriors born to the kill, they win their battles. That's the point.

Scripps-Howard wire copy, 1 July 1944

DAVID SMILEY
Albanian Assignment

The Balkans, between 1941 and 1945, were to be a main centre
of British, and later American, special operations against the
Germans, and, until the armistice of September 1943, the
Italians also. Special Operations Executive (SOE), the organization established by Winston Churchill in 1940 to 'set Europe
ablaze', made early contact with resistance forces in Yugoslavia,
Greece and Albania, supplied them with arms and later infiltrated liaison teams into their 'liberated areas' by submarine,
boat, aircraft and parachute.

One of the main difficulties encountered by SOE agents,
most of whom were serving military officers, was mediating
between the competing resistance factions. Some remained loyal
to the governments-in-exile, others, the more ruthless and unified, were controlled by the local Communist parties, whose
primary allegiance was to the Soviet Union and the cause of a
post-war social revolution. In Yugoslavia the split was between
the royalist Cetniks, led by Draža Mihailović, and Tito's Communist Partisans; in Greece between the Communist ELA and the
nationalist EDES; in Albania, which had been annexed in 1939
by Mussolini, between the Communist LNC and the nationalist
Balli Kombëtar. One of the most successful SOE operatives in
Albania was David Smiley, an officer of the Royal Horse Guards.
His account of the frustrations of organizing a simple ambush
on a road used by German troops in August 1944, just before
Hitler ordered a complete withdrawal from the Balkans, exemplifies the nature of the guerrilla war in southern Europe.

*   *   *

On 10 August I left Shtyllë to carry out an ambush with a Ballist
çeta [Serbian *četa*, troop, hence Cetniks]. This had been arranged by

McLean [a fellow officer] with Safet Butka to test, among other things, the willingness of the Balli Kombettar to fight the Italians and Germans. Fred Nosi and our attached partisans were strongly opposed to this move, and took steps to thwart it. The date fixed for the ambush was 12 August; McLean and I were keen grouse shots, and agreed that this would be an appropriate day to open the season.

Leaving Shtyllë with two guides and the faithful Yugoslavs carrying the Breda [Italian-built machine-gun], I arrived after some hours' march at the village of Kurtes, where I found the whole çeta, some 200 men strong, drawn up for my inspection under Captain Qemal Burimi. He and I both made speeches, then spent the rest of the day in making our plan. Unlike the partisans, who always thought they knew best, Captain Burimi was prepared to listen to my advice and even take it. Next morning we moved nearer the road, not far from the burnt-out village of Barmash, where we made a close reconnaissance in daylight. On our way back to the village we had reached the top of some steep cliffs when we met a çeta of about thirty partisans sitting on top who had clearly been watching us. I recognized the leader, Petrit Dume.

'What are you doing here?' I asked. 'We are going to ambush the road tonight,' he replied. 'Where?' I asked. 'Down there' – he pointed to the corner of the road where Burimi and I had decided to lay the ambush; he had obviously seen us there.

It seemed clear to me that he had been put up to this by the LNC, probably on information from Fred Nosi; the object, without doubt, was to prevent the Balli Kombettar from carrying out an ambush that would give them credit in the eyes of the British and give the lie to the LNC allegations that they were collaborating with the Germans and Italians.

I grew angry at the thought that our plans might come to nought and said to Dume, 'You cannot lay an ambush here tonight as I am doing one with Captain Burimi and his çeta.' Dume replied that he had been given his orders and intended to carry them out. 'If you do an ambush here tonight,' I told him, 'I promise you that the LNC will receive no more arms or money from the British Mission, and we will give all we get to the Balli Kombettar.' I had no authority to make this threat, but it was all I could think up on the spur of the moment. However it seemed to work for, muttering oaths (probably directed at me), Dume led his çeta off down a track in the direction of Korçë. I wondered what his next move would be; I felt sure he had orders to wreck our plans, and would attempt to do so. In the event he did nothing and we saw no more of him.

We had found an ideal spot for the ambush, where the road had been cut out of the side of a very steep mountain and had numerous bends; the mountain itself dropped sheer below the road into a river, and a steep cliff above the road prevented anyone on it from escaping or taking cover. The hills rose again the other side of the river, and I planned to deploy the çeta on these hills; they would be about two hundred yards from the road, with a gully between them and the enemy.

During the night I took a party of ten men with sixteen mines down on to the road, intending to lay the mines in two groups of eight about 250 yards apart but out of sight of each other on either side of a bend in the road. When we had laid the first four mines a Halifax [bomber] flew over, one of the Albanians shouted 'A lorry! a lorry!' and the entire party ran away, never to reappear. I therefore had to dig the holes and lay the remaining mines myself, which took over four hours, luckily without any disturbance, and finished just before daybreak at about five in the morning. I climbed the hill to the çeta positions, exhausted and in a foul temper; there I found a very apologetic Captain Burimi. 'My men have not had much training,' he explained. I agreed. We took up our positions and a short while later my temper was cooled by the fine sight of a big German half-tracked troop-carrier approaching from the direction of Korçë; better still, on nearing I saw it was towing an 88-mm gun. Some of the çeta, now back in their positions near me, shouted 'A tank! a tank!' and ran away, but the less timid of their companions stayed with me; otherwise we were all ready and in position, and I had both my camera and the Breda gun trained on the spot where I had laid the mines. As the carrier drew closer every one of us held his breath; then it went up on the mines with a flash of orange flame followed by a cloud of smoke, and the sound of the explosion echoed through the hills. I had taken a photograph as the mines exploded; by the force of the explosion I estimated that all eight mines must have detonated at once. Once the smoke had cleared everyone opened fire on the troop carrier; I exchanged my camera for the 20-mm Breda, and was delighted to see several of my shots score direct hits. A few Germans jumped out of the carrier and tried to run back down the road but all were shot, and the others tried to take cover behind the carrier. In time the shooting stopped and a silence followed only broken by the groans of some of the wounded. I asked Captain Burimi to send some of his men down to the road to get identifications and to try to push the gun off the road into the river bed below. A few men then cautiously approached the road, some shots rang out – presumably the wounded being finished off – and the gun was then unhitched and pushed over the side, but the half-track proved impossible to move because most of its front was blown away. After about half an hour the men returned from the road very pleased with themselves; I was happy too because they had collected the identifications I had wanted and various pieces of loot, and they told me eighteen Germans had been killed. Twelve were dead in the troop carrier, probably killed by the exploding mines; the six lying on the road behind the carrier had been shot. While they were excitedly telling their story I heard engines and then saw in the distance a large convoy of lorries coming from the direction of Leskovik; this was excellent, for they would blow up on the second group of mines before they could see the wrecked troop carrier. As they approached I counted twenty-three lorries but on checking that all the çeta had moved into their positions was saddened to see that all but about six men, and the two Yugoslavs who appeared to be enjoying themselves immensely, had vanished.

The first lorry went over the mines with a terrific explosion, the whole of the front and the cab disintegrating; I took another photograph. The rest of the convoy immediately halted; two Germans jumped out of the cab of each lorry and ran back down the road while the çeta shot at them. Through my binoculars I could see the bodies of five Germans lying by the first lorry, so I proceeded to shoot at the second with my Breda. I hit it with my third shot and it burst into flames, and I did the same to the third lorry. Had the full çeta remained in position with me I was convinced the entire convoy could have been destroyed, but after about ten minutes the Germans had reorganized themselves to shoot back, to which we replied; when I spotted a small party starting to climb our hill away on a flank, I considered it time to go, so with my attendant caddies, as I called them, we picked up the Breda and walked away.

Although my opinion of Albanian fighting quality was somewhat low I was very satisfied with the results, for we had killed twenty-three Germans without a casualty to ourselves, taken identifications of the First Alpine Division, destroyed a large troop carrier, one 88-mm gun, and three lorries. With better-disciplined and properly trained troops we certainly could have destroyed the lot.

FRANKLIN LINDSAY
Beacons in the Night

The American equivalent of SOE was OSS (Office of Strategic Services), the lineal forebear of the Central Intelligence Agency (CIA). Franklin Lindsay was one of its early operatives, who in 1944 was attached to SOE for duty in Yugoslavia. A mining engineer by training, Frank Lindsay proved a natural special-forces agent, brave, intelligent and of independent mind. He was an aggressive leader of partisans, in so far as they allowed themselves to be led by a foreigner who was not a Communist. He, for his part, suffered all the frustration felt by SOE and OSS officers in their dealings with partisans who were generally more concerned to establish their political dominance over the regions in which they fought, than to engage with the occupying enemy.

In the action recounted in the following extract from his magnificent account of the partisan war, *Beacons in the Night*, some of the partisans' opponents were German, some local

home guards enlisted to the German side. Theirs was the
unhappier lot. After the war Lindsay became head of the
CIA's Eastern Mediterranean division, charged with mounting
subversive operations against the Communist regimes which
had succeeded to power in the aftermath of the partisan campaigns. He was unenthusiastic about his mission, having 'seen
firsthand what a resistance movement entails: horrifying loss
of life, suffering which racks every level of society, wholesale
destruction, the murder of hostages, starvation. No population
that had gone through such an ordeal could attempt it again
without a generation of forgetfulness.' (Thomas Powers, *The
Men Who Kept the Secrets*, New York, 1987.) A generation after
Lindsay served in OSS, the population of what was formerly
Yugoslavia had re-created the ordeal again.

*   *   *

Liberation of a Mountain Valley

Until August 1, 1944, no liberated territory existed in the Partisan
Fourth Zone within the wartime German Reich, the area north of the
Sava River. A few remote mountainous areas were called 'liberated'
simply because of the absence of German troops. Until then the
Partisans had operated only in the mountains. Every village had a
German garrison and Partisan movements were primarily at night.
The Partisans were dispersed in small units except when brought
together for an attack on a German-held target. During June and
July we remained with Fourth Zone headquarters and moved with
them almost every night.

Our security lay not in secrecy, which was impossible to maintain,
but in constant movement. We were dependent on an invisible
screen of civilians – old men, women, and children – to warn us of
approaching German patrols, or of ambushes laid for us on mountain
trails. Although for seven months we were never further than five
miles from the nearest German post, and often much closer, we
were never caught in ambush, and were seldom surprised. In good
part this was luck; many others were ambushed by SS troops or the

White Guard [militia of the pro-German Croat State]. The plain fact was that we could not have survived for even a week if the civilian population had been hostile.

In late July the Fourth Zone headquarters launched a daring offensive of their own. The Partisans assembled their fighting brigades to attack the German garrisons in the Gornja Savinjska Dolina – the narrow valley of the upper Savinja River and its tributary, the Dreta. Descending from the upper slopes of the Karawanken Alps and curving to the east, the valley was surrounded by high mountains. The terrain spread and opened, downstream, only at its eastern junction with another valley descending from the north. There was a reasonable chance it could remain in Partisan hands, but only so long as the Germans did not move in reinforcements from outside Stajerska.

The operation was well planned. All five brigades in Stajerska were employed. The Sixth 'Slander' Brigade was ordered to attack the German garrisons at the villages of Luce and Ljubno. The Eleventh 'Zidansek' Brigade was to mount a diversionary attack on Gornji Grad, and to set up ambushes on the roads leading to the three villages from the German-held towns to the west, north, and east. The three brigades of the Fourteenth Division would form a second protective barrier further to the east by laying ambushes, blockading and mining roads, bridges, and railroad sections, and launching diversionary attacks on other garrisons that might come to the relief of those at Luce, Ljubno, and Gornji Grad. In addition, the Koroski Odred [partisan force] high in the Karawanken Alps would block any German approaches over the high passes to the north. And the odred on the Sava would launch still another diversionary attack on the Litija garrison twenty miles to the south.

The operations began on the night of July 30. Ljubno was attacked from the north by one battalion and from the south by the second battalion. The third was held in reserve. Twelve pillboxes around the perimeter were successively eliminated with grenades lobbed into them. Under covering fire explosive charges were placed against the walls of three strongpoints and detonated. Prisoners were taken in two while the third went up in flames, with no survivors. The remaining strongpoint held out through the night and into the next day.

Gordon Bush and I observed the fighting from a grassy slope on the flank of the mountain overlooking the village. It was a brilliant Alpine summer day. The grass was emerald green, and the panorama of the valley with its fields, drying racks, and houses was breath-takingly beautiful. German soldiers were in the Ljubno church steeple, and we were in plain view a couple of hundred yards away. Suddenly, they began firing on us. Quite irrationally, I felt outraged. We were not shooting at them! But we prudently moved a few yards away where we were out of their sight.

The attack continued as Partisans moved behind walls and houses toward the center of the village where the Germans were holed up in the remaining strongpoint – a heavy stone building. Soon an Italian 75-mm gun appeared on the road into the village. The gun had been dismantled in Dolenjska and man-packed piece by piece across the border, the Sava, and the mountains. It had been reassembled in the woods for the attack and brought to the battle not by a truck, or even by artillery horses, but by six strong Partisans. Together they lifted the trail of the gun, crouching behind the shield as best they could and pushing it into firing position. Not more than 200 yards from the garrison wall they stopped, rammed home a round, closed the breech, laid the gun on the wall by squinting along the barrel, and fired. The wall was breached and a second round fired, which exploded inside the building. The Partisans then stormed through the breach with Sten guns and grenades. There was not much fight left in the defenders. They filed out covered with dust, hands in the air. Several of the garrison had fled to nearby houses and continued to fight. Those houses were destroyed with explosive charges, or by setting them afire, to complete the capture of the village.

At the same time another battalion of the 'Slander' Brigade attacked the garrison at Luce, seven kilometers upstream. The main German force there was holed up in the church rectory. Because of heavy enemy fire the Partisans did not take the strongpoint, but kept it under fire into the next day. In the afternoon they demanded and got the surrender of the garrison.

While this was going on the 'Zidansek' Brigade maintained its attack on Gornji Grad. This was originally planned as only a feint. However, because the attacks against Luce and Ljubno were going well the brigade was ordered to complete the attack. The principal strongpoint was inside a medieval castle surrounded by a heavy stone wall. Rather than attacking it the Partisans demanded surrender, threatening to bring up their artillery – the one piece that had just been used at Ljubno to reduce the strongpoint there. The morning of the second day the remaining garrison surrendered.

Two attempts were made by the Germans to relieve the three garrisons, one from the west over the mountains, and one from the lower valley to the east. But the forces available to them were inadequate, both in number and fighting quality. Neither was able to break through the Partisan blockades.

At the end of the day at Ljubno, as I left the center of the village I suddenly came to an open field behind a peasant house. Several men, some in bedraggled German uniforms and some in civilian clothes, were lined up facing a group of Partisans. Two Partisans had raised their submachine-guns and at that instant both opened fire on the Germans, who crumpled to the ground. Afterwards I pressed Commissar Borstnar to tell me why these men had been executed. I didn't get a straightforward answer. All he said was that they had been guilty of atrocities against the Partisans. Nor could I get an answer on whether there had been any proceedings beforehand. It turned out that of the burgermeisters of the three villages, two had been shot and one spared. The two had been 'bad mayors' and one had been a 'good mayor', he said. I suspected that he had earned his way by secretly aiding the Partisans in the months before.

Two years after the war I was watching a movie in New York when something on the screen reminded me of those executions. I nearly blacked out. Yet at the time in Ljubno I had not been particularly troubled by what I had seen. Psychological protective defenses were obviously at work within me at the time; two years later they had disappeared.

ALEXANDER STAHLBERG
Bounden Duty (3)

Alexander Stahlberg's last reflections on the course of the Second World War, as seen from the level of high command on the German side, are taken from the moment of Nazi Germany's defeat. Most of Germany has been occupied, by the British and Americans from the west, by the Russians from the east. The Battle of Berlin is raging in the ruins of the city, as the Red Army fights street by street against its last-ditch defenders towards the Reich Chancellery. The senior officers and officials of the regime are in flight to Schleswig-Holstein, where the provisional government of Grand Admiral Dönitz, named second and last Führer of the Third Reich by Hitler, would find a brief refuge. Any German who could was seeking to surrender to the Western Allies or to find refuge in their zones of occupation, rightfully fearful of the vengeance the Red Army would take for the Nazi rape of its homeland. The last music Stahlberg hears on the German radio is Bruckner's Seventh Symphony, played at Hitler's order to mourn the surrender of Sixth Army in the ruins of Stalingrad twenty-two months earlier. It is a fitting epilogue to his regime of nihilism and to his own death by suicide in the Reich Chancellery bunker.

*   *   *

April 1945

In retrospect, events in Germany at that time seem unreal.

The British and American Armies have long since crossed the Rhine and penetrated deep into our country. In many places our troops are no longer offering any resistance. Elsewhere German units are still fighting the Allies with self-sacrificial devotion. It is almost incomprehensible that many German divisions in the West should be resisting as fiercely as others in the East. Again and again, vital bridges and strategically important viaducts are being blown up in the West to delay the Allies. Surely the most important thing now is who reaches Berlin first? After all, what matters now is the future of Germany after Hitler.

While the Western Allies have already reached the Weser, an entire Army Group under the command of Field Marshal Model has been surrounded in the Ruhr. On 10 April, in Achterberg, when we hear that the American advance guard has already reached Hanover, Field Marshal von Manstein decides to leave Achterberg. He has no desire to present himself to the enemy troops as a 'prize' until an armistice has been declared.

Our first goal is a village near Bad Oldesloe in Holstein. The drive there in our two cars on 11 April is arduous. On the roads we are met by horse-drawn refugee columns from Pomerania and from East Prussia. The Manstein family, its escort and the luggage are distributed between two cars. I drive the BMW. As I make way for the refugees, I am often forced to take the 'summer road'. Many of the draught horses have lost nails from their shoes and consequently several punctures follow, one after the other. With rolled-up sleeves, I change wheels and mend inner tubes.

On 13 April, in Oldesloe, we hear of the death of the American President, Franklin D. Roosevelt, but none of us feels that his death will alter the course of events in Germany now. On 16 April, we hear that the Soviets too have now launched their assault across the Oder. On 18 April, Army Group Model capitulates in the Ruhr pocket. Model, of whom Hitler had said to Manstein that he 'whizzes all over the place with the troops', which was more valuable than simply continuing to 'operate', shoots himself. But nowhere do we hear that anyone in Berlin is considering that the time may perhaps be ripe to put a stop to the fighting and the senseless bloodletting.

The Field Marshal asks me to find out by telephone where the Army Group headquarters responsible for North Germany is now quartered. On the morning of 19 April, he drives with me to Hamburg. In the Wohltorf district, near Bergedorf, we find the Headquarters of Army Group North-West, in a big old house. For a few days now it has been commanded by Field Marshal Busch, the same Busch who broke down in front of Breitenbuch and me last summer when I had to announce the destruction of his Army Group Centre. He has since been rehabilitated by Hitler, who has no doubt been told with what loyalty and devotion Busch had spoken by Schmundt's coffin in the Tannenberg Monument. [Schmundt was Hitler's adjutant, killed in the 20 July assassination attempt at Führer HQ; the Tannenberg Monument contained Field Marshal Hindenburg's grave.] His new job is the equivalent of being appointed a gravedigger.

Two Field Marshals now stand before the situation map. Their opponent here is the British Field Marshal Bernard Montgomery. It is clear to see that Montgomery will not make a frontal attack on Hamburg, but will circle to the south of the city, cross the Elbe and probably work towards the Baltic near Lübeck. For Manstein, this means that he will have to move from Oldesloe yet again, to avoid being taken prisoner before the end of the fighting.

The two Field Marshals discuss the hopeless situation of the Army Group. Suddenly I hear Busch asking if the Kiel Canal on either side of Rendsburg should be prepared to defend the Southern Front. It takes an effort of will to listen.

I take the opportunity for a private word with Busch's new ADC, whom I know personally. He has taken the position with Busch that Eberhard Breitenbuch used to occupy. I ask him if he can advise me where to find a new refuge for the Manstein family further to the North, in the province of Schleswig-Holstein. He does not need to think for long: the widow of SS *Obergruppenführer* Reinhard Heydrich, murdered in Bohemia [the brutal Deputy-Protector of Bohemia and Moravia and, earlier, architect of the 'final solution to the Jewish question', had been assassinated in Prague in 1942 by Czech nationals parachuted in from Britain], lives on the Baltic island of Fehmarn, in a very beautiful and sufficiently large house with guest rooms, and the lady of the house is charming. One could certainly stay with her 'until the military crisis is over'. I could mention his name to Frau Heydrich. My response could only be silence.

Suddenly the door opens and the Reich Minister for Armaments and Munitions, Albert Speer, stands before us. He has come from Berlin, he says, and spoken to the Führer 'for the last time'. We sit down and listen tensely.

He had flown to Hamburg to persuade the Gauleiter [provincial governor] Karl Kaufmann there, against the Führer's orders, not to have the bridges across the Elbe blown up. The two Field Marshals listen, dumbfounded. 'Against the Führer's orders . . . ?' Speer confirms: 'Yes, against!' The Field Marshal and Commander-in-Chief of the Army Group – the only person with military responsibility here – hears the news after the Gauleiter of Hamburg! So chaos has already spread through the chain of command.

Speer describes his last visit to the underground bunker at the Berlin Reich Chancellery: in the office a trembling, wasted wreck of a sick man sits under the portrait of Frederick the Great, scarcely listening to his visitor. He is clutching the bundle of writing and drawing implements from the desk tray in front of him in one hand and driving them incessantly into the table top, until the points are broken and the table top deeply punctured. Beside him lies an issue of the *Völkischer Beobachter*, now only a few pages long. It lies open at the final part in the series of 'Personal reports by Frederick the Great from the Seven Years War'. Only one book, a volume of Thomas Carlyle's *The History of Friedrich II of Prussia, Called Frederick the Great*, lies near by.

Speer speaks of the two army leaders on whom the Führer's last hopes rest: Generals Wenck and Busse.

We sit up: Wenck? – Busse? – Is it pure chance that two General Staff Officers from the school of Field Marshal Manstein should be the ones to rise to the rank of Army Commanders-in-Chief at the eleventh hour?

Wenck, on whom all eyes turned in November 1942, when he succeeded by a trick in 'capturing' two Romanian armies fleeing westward from the Stalingrad area, a coup that had brought him promotion from Colonel on the General Staff to Major-General. And Busse, Manstein's former Ia [First General Staff Officer] and later – in preference to Tresckow – Manstein's last Chief of Staff: had not General Fellgiebel and our Army Group Communications

Chief Major-General Ernst Mueller warned me in 1943 to be careful in Busse's presence, because he telephoned his brother-in-law, General Burgdorf, at the Führer's headquarters almost every night? Now Burgdorf was Hitler's Chief Adjutant and Busse had been chosen to defend his home town of Frankfurt-on-Oder as Commander-in-Chief of the Ninth Army, and when that was lost, the capital itself.

Speer speaks of the situation around Berlin: they are now expecting the city to be encircled in a matter of days. He asks the Field Marshals if they think it possible that Wenck could push through from the south-west as far as the capital. No answer.

Speer reports frankly that the Führer has ordered him to ensure that all major factories in the German Reich are destroyed before the arrival of the enemy. Under these auspices, he has been travelling from factory to factory for weeks now, urging the directors not to carry out the order. The two Field Marshals remain silent, shaking their heads.

Our programme for this 19 April is not yet finished. Field Marshal von Manstein takes me on one side and asks if I know of a first-class restaurant in Hamburg where one could at least have one more good lunch. I suggest the 'Ehmke' in Gänsemarkt, which looked to me as if it was still open when I drove by. Ehmke also had elegant private rooms on the first floor. He asks me to reserve a private room and to let Frau von Manstein know.

In fact the Ehmke restaurant is still standing and the food is even now like an excursion into 'the good old days'. On that April day, too, Ehmke justifies its fine reputation. While we are celebrating the gastronomic arts of old, we speak freely, as soon as the frock-coated waiters have left us, about Albert Speer and his unexpected disclosures. Scarcely anyone had ever before spoken so freely and openly before two of Hitler's Field Marshals at once.

We are unaware that at this moment Speer is already on his way back to Berlin – after all, he has only just told us that he has been with Hitler 'for the last time'. He certainly had the courage to speak up unequivocally against Hitler in front of two Field Marshals, but he had not dared to say that he wanted to revisit Hitler that very day, to 'see him once again'. So Albert Speer is not yet free of his lord and master, who sits in his underground bunker, surrounding himself to the last with 'Frederick the Great', revealing that the tyrant has never understood, but has only 'used' and hence abused, Prussia and her great king.

From Oldesloe I drive northward on my own at random to find new accommodation. My old friend Irmgard Georgius, Germany's best woman equestrian competitor in the 1930s, gives me some tips in Waldhof. Most of the estates are already overcrowded with refugees from the Eastern provinces and in one of the manor houses the door is shut in my face. 'And now a Field Marshal on top of everything! Wouldn't think of it!'

On the roadside two naval officers wave to me, asking to hitch a ride. One, a U-boat captain, is wearing the Knight's Cross. They want to go to Plön and I am happy to have someone to talk to, though I find it difficult to believe that a U-boat captain, who can steer his submarine through minefields to the east coast of America, is incapable of finding his way on the map through Schleswig-Holstein. However, the other proves to be an extremely useful aerial observer, as British fighter aircraft sweep along the roads, hunting for prey.

For three days I drive from place to place without finding anything suitable. In the twelve years of the 'Third Reich' one has developed a nose for whether the house one is entering is a 'Nazi house'. I do not wish to spend the end of the war in a house with a National Socialist bias.

And then, after all, I find the goal of my desires. In Weissenhaus, on the Baltic, not far from Oldenburg, I meet the owner and his wife: Graf [Count] Clemens Platen invites me in at once for tea, so that we can talk things over together.

We walk through the lovely, big house, its ground floor stuffed to bursting with furniture and cases: the stocks of the Kiel Landesmuseum. Almost buried under cases of paintings is a black concert grand. I open the lid over the keys, pick up a lovingly worked piece of embroidery and read the music of a theme from Humperdinck's *Hansel and Gretel*. The Platens explain: Humperdinck composed that wonderful opera on this piano. Weissenhaus will be our quarters when the end of the war comes. On 30 April we move in.

In the attics of the Weissenhaus manor four carpets are hung up for me as a substitute for four walls. A mattress, a box or two – and I have a comfortable 'room'.

The next day – it is 1 May 1945 – I switch on my radio to the first notes of the second movement of Bruckner's Seventh Symphony on the Hamburg radio. 'Very solemnly and very slowly' the tubas and violas join in. I think I know the recording: Furtwängler with the Berlin Philharmonic. It does me good to be listening to this symphony again after all this time.

Suddenly I am seized with suspicion. Why are they broadcasting Bruckner's Seventh today? When the movement ends I know the answer: in an emotional voice the speaker announces:

'It is reported from the Führer's headquarters that our Führer, Adolf Hitler, died this afternoon for Germany at his command post in the Reich Chancellery, fighting to his last breath against Bolshevism. On 30 April the Führer named Grand Admiral Dönitz as his successor.'

GEORGE MACDONALD FRASER
Quartered Safe Out Here
Dividing up a dead comrade's possessions

While Germany was collapsing into unconditional surrender in May 1945, the war against Japan in the Pacific and South-East Asia continued. George MacDonald Fraser (the author of the Flashman novels) was then serving as a private soldier in the 9th Battalion the Border Regiment, a unit of war-service soldiers from the English Lake District counties of Westmorland and Cumberland. A tough lot of hill farmers, fishermen and shipbuilders, they soldiered against the Japanese in a spirit of dour determination to get the war over and get home. Nine Section, a group of less than a dozen infantrymen, had no particular love for each other. They recognized, however, that the life of each individual depended upon the unconditional support of all the rest, and regarded the death of any one of them as a diminution of the group's chances of survival. Phlegmatic, unemotional, they found their own silent ways of mourning a comrade's death in battle. Here Fraser describes how they said their farewells to Tich Little, a Nine Section soldier who had died in battle with the Japanese. This author later became an officer of a Highland regiment in the war's aftermath. His memoir, published in the 1990s, of life as a private soldier in a remote and bitter campaign is one of the classic literary achievements of the Second World War, indeed of any war.

*   *   *

The aftermath was as interesting as the battle. Fiction and the cinema have led us to expect certain reactions from men in war, and the conventions of both demand displays of emotion, or a restraint which is itself highly emotional. I don't know what Nine Section felt, but whatever it was didn't show. They expressed no grief, or anger, or obvious relief, or indeed any emotion at all; they betrayed no symptoms of shock or disturbance, nor were they nervous or short-tempered. If they were quieter than usual that evening, well; they were dog-tired. Discussion of the day's events was limited to a brief reference to Gale's death, and to the prospects of the wounded: Steele had been flown out on a 'flying taxi', one of the tiny fragile monoplanes to which stretchers were strapped; it was thought his wound was serious. Parker was said to be in dock in Meiktila (and a few weeks later there were to be ironic congratulations when he returned to the section with a romantic star-shaped scar high on his chest; penicillin was a new marvel then).

Not a word was said about Tich Little, but a most remarkable thing happened (and I saw it repeated later in the campaign) which I have never heard of elsewhere, in fact or fiction, although I suspect it is as old as war.

Tich's military effects and equipment – not, of course, his private possessions, or any of his clothing – were placed on a groundsheet, and it was understood that anyone in the section could take what he wished. Grandarse took one of his mess-tins; Forster, his housewife [sewing and mending kit], making sure it contained only Army issue and nothing personal; Nixon, after long deliberation, took his rifle, an old Lee-Enfield shod in very pale wood (which surprised me, for it seemed it might make its bearer uncomfortably conspicuous); I took his pialla, which was of superior enamel, unlike the usual chipped mugs. Each article was substituted on the groundsheet with our own possessions – my old pialla, Forster's housewife, and so on – and it was bundled up for delivery to the quartermaster. I think everyone from the original section took something.

It was done without formality, and at first I was rather shocked, supposing that it was a coldly practical, almost ghoulish proceeding – people exchanging an inferior article for a better one, nothing more, and indeed that was the pretext. Nick worked the bolt, squinted along the sights, hefted the rifle, and even looked in its butt-trap [which housed the pull-through and oil bottle for cleaning the rifle] before nodding approval; Grandarse tossed his old mess-tin on to the groundsheet with a mutter about the booger's 'andle being loose. But of course it had another purpose: without a word said, everyone was taking a memento of Tich.

An outsider might have thought, mistakenly, that the section was unmoved by the deaths of Gale and Little. There was no outward show of sorrow, no reminiscences or eulogies, no Hollywood heart-searchings or phony philosophy. Forster asked 'W'ee's on foorst stag?'; Grandarse said 'Not me, any roads; Ah's aboot knackered,' and rolled up in his blanket; Nick cleaned Tich's rifle; I washed and dried his pialla; the new section commander – that young corporal who earlier in the day had earned the Military Medal – told off the stag roster; we went to sleep. And that was that. It was not callousness or indifference or lack of feeling for two comrades who had been alive that morning and were now names for the war memorial; it was just that there was nothing to be said.

It was part of war; men died, more would die, that was past, and what mattered now was the business in hand; those who lived would get on with it. Whatever sorrow was felt, there was no point in talking or brooding about it, much less in making, for form's sake, a parade of it. Better and healthier to forget it, and look to tomorrow.

The celebrated British stiff upper lip, the resolve to conceal emotion which is not only embarrassing and useless, but harmful, is just plain common sense.

But that was half a century ago. Things are different now, when the media seem to feel they have a duty to dwell on emotion, the more harrowing the better, and to encourage its indulgence. The cameras close on stricken families at funerals, interviewers probe relentlessly to uncover grief, pain, fear, and shock, know no reticence or even decency in their eagerness to make the viewers' flesh creep, and wallow in the sentimental cliché (victims are always 'innocent', relatives must be 'loved ones'). And the obscene intrusion is justified as 'caring' and 'compassionate' when it is the exact opposite.

The pity is that the public shapes its behaviour to the media's demands. The bereaved feel obliged to weep and lament for the cameras (and feel a flattering importance at their attention). Even young soldiers, on the eve of action in the Gulf [1991], confessed, under a nauseating inquisition designed to uncover their fears, to being frightened – of course they were frightened, just as we were, but no interviewer in our time was so shameless, cruel, or unpatriotic as to badger us into admitting our human weakness for public consumption, and thereby undermining public morale, and our own. In such a climate, it is not to be wondered at that a general should agonize publicly about the fears and soul-searchings of command – Slim and Montgomery and MacArthur had them, too, but they would rather have been shot than admit it. They knew the value of the stiff upper lip.

The damage that fashionable attitudes, reflected (and created) by television, have done to the public spirit, is incalculable. It has been weakened to the point where it is taken for granted that anyone who has suffered loss and hardship must be in need of 'counselling'; that soldiers will suffer from 'post-battle traumatic stress' and need psychiatric help. One wonders how Londoners survived the Blitz without the interference of unqualified, jargon-mumbling 'counsellors', or how an overwhelming number of 1940s servicemen returned successfully to civilian life without benefit of brain-washing. Certainly, a small minority needed help; war can leave terrible mental scars – but the numbers will increase, and the scars enlarge, in proportion to society's insistence on raising spectres which would be better left alone. Tell people they should feel something, and they'll not only feel it, they'll regard themselves as entitled and obliged to feel it.

It is a long way from the temple wood to Sheffield – and not only in miles. I knew a young Liverpudlian who, following the Hillsborough disaster [in April 1989, ninety-six Liverpool Football Club supporters were crushed to death during a game at the Hillsborough stadium in Sheffield], stayed away from work because, he said, of the grief he felt for those supporters of his team who had died on the terraces. He didn't know them, he hadn't been there, but he was too distressed to work. (Suppose Grandarse or the Battle of Britain pilots, with infinitely greater cause, had been too distressed to fight?) One shouldn't be too hard on the young man; he had been conditioned to believe that it was right, even proper, to indulge his emotions; he probably felt virtuous for having done so.

Fortunately for the world, my generation didn't suffer from spiritual hypochondria – but then, we couldn't afford it. By modern standards, I'm sure we, like the whole population who endured the war, were ripe for counselling, but we were lucky; there were no counsellors. I can regret, though, that there were no modern television 'journalists', transported back in time, to ask Grandarse: 'How did you *feel* when you saw Corporal Little shot dead?' I would have liked to hear the reply.

ERIC LOMAX
The Railway Man

The British and American opponents of the Japanese fought with the resolution they did because they knew the consequences of falling captive into that enemy's hands. During the 1930s, when the Japanese government fell progressively into the hands of ultra-nationalist military officers, recruits to the Japanese armed forces were subjected in training to a regime of deliberate brutalization. They were taught that to surrender to the enemy was worse than death and that prisoners taken from the enemy had, reciprocally, surrendered all right to humane and honourable treatment. Eric Lomax describes the consequences of falling into Japanese hands.

*   *   *

The guards conducted the five of us to the main guardroom where we were brusquely ordered to stand to attention, a few feet in front of the building and well away from any shade or protection from the sun. The guardroom was a flimsy three-sided wood and thatch structure, open in the front, with a table across the gap. A guard stood at attention on the side nearest the camp entrance; a few more were seated behind the table. Among them was a large, fat and rather elegantly dressed white-haired man, who now proceeded to address us in fluent American English. He ordered us forward. His attitude was aggressive, sneering and hostile as he checked our identities, making contemptuous references to Western duplicity and cowardice throughout the short procedure.

He ordered us back into the sun. There we stood beside a long ditch, neatly spaced like five telegraph poles along a road. The time was ten o'clock in the morning.

The morning and afternoon dragged on, every minute almost an hour. When you are forced to stand stiffly to attention in a blazing hot sun you have nothing to do but think; yet thought is a process that should be directed by the will, and under extreme stress thoughts spin away on their own, racing faster and faster like a machine out of control, one that has lost the touch of a human hand.

There was nothing we could do about it now: we stood there, knowing it was coming. The wretched little guardroom was no bigger than a domestic living room, and the few guards sprawling inside it or on guard behind us controlled the lives of several hundred men. So few to hold so many.

We stood for twelve hours with our backs to that hut. The nerves and flesh of the back become terribly sensitive and vulnerable when turned to an enemy. At any moment I expected to feel a rifle-butt on my spine, a bayonet thrust between my shoulder-blades. All we heard was their talk, their occasional rough laughter.

The intense heat of the sun, the irritation of flies and mosquitoes feeding on sweat, itching skin, the painful contraction of eyes against the light and even the fear of violent death had been superseded, by the evening, by the even more powerful sensation of a burning thirst. They gave us nothing to drink, all day, but they allowed us occasionally to go to the latrine. On one of these visits I regretfully disposed of my diary. The flimsy pages covered with neat notes on books, on grammar, on lists of collectable stamps fluttered into the stinking trench.

As dusk fell the five of us were moved into a closer and more compact group in front of the guardroom. The darkness came on with singular abruptness. We were lit by a weak light from behind us in the guardroom. A time signal was heard as a noisy party of Japanese and Koreans approached through the dark from the direction of the camp offices. They looked like NCOs, their uniforms dishevelled, one or two of them unsteady on their feet. All of them carried pick-helves. They stopped to talk to the guards, as though exchanging ideas about what to do with us.

Major Smith was called out in front of our line, and told to raise his arms right up over his head. His tall, gaunt figure, his thin arms held out like a scarecrow's, looked terribly weak and pitiful. He stood there on the edge of the circle of light. I thought for a moment

– a last gasp of hope – that this was the beginning of an advanced form of their endless standing to attention. A hefty Japanese sergeant moved into position, lifted his pick-handle, and delivered a blow across Smith's back that would have laid out a bull. It knocked him down, but he was trodden on and kicked back into an upright position. The same guard hit him again, hard. All the thugs now set to in earnest. Soon little could be seen but the rise and fall of pick-helves above the heads of the group and there were sickening thuds as blows went home on the squirming, kicking body, periodically pulled back on to its feet only to be knocked down again. Bill Smith cried out repeatedly that he was fifty years of age, appealing for mercy, but to no avail. The group of attackers seemed to move in concert with their crawling, bloodied victim into the darkness beyond the range of the miserable lighting from the guardroom, but the noises of wood on flesh continued to reach us from the dark of the parade ground.

They were using pickaxe-shafts: like solid, British Army-issue handles, and perhaps that is indeed what they were. The guards behind us did not move. There was no expectation that we ourselves would move, intervene, run away: merely the slack, contemptuous knowledge that we were trapped. That first blow: like a labourer getting into the rhythm of his job, then the others joining in, a confused percussive crescendo of slaps and thuds on flesh and bone. They kept kicking him, getting him up, putting him down – until he stopped moving altogether, unconscious or dead, I could not tell. Nor could I tell how long it all took. How does one measure such time? Blows had replaced the normal empty seconds of time passing, but I think it took about forty minutes to get him to lie still.

The gang came back out of the night. My special friend Morton Mackay was called forward. I was next in line. As they started on Mackay and the rain of fearful blows commenced I saw to the side another group of guards pushing a stumbling and shattered figure back towards the guardhouse. Smith was still alive; he was allowed to drop in a heap in the ditch beside the entrance.

Mackay went down roaring like a lion, only to be kicked up again; within a matter of minutes he was driven into the semi-darkness and out of the range of the lights, surrounded by the flailing pick-helves which rose and fell ceaselessly. I remember thinking that in the bad light they looked like the blades of a windmill, so relentless was their action. In due course Mackay's body was dragged along and dumped beside Smith's in the ditch.

The moments while I was waiting my turn were the worst of my life. The expectation is indescribable; a childhood story of Protestant martyrs watching friends die in agony on the rack flashed through my mind. To have to witness the torture of others and to see the preparations for the attack on one's own body is a punishment in itself, especially when there is no escape. This experience is the beginning of a form of insanity.

Then me. It must have been about midnight. I took off my spectacles and my watch carefully, turned and laid them down on the table behind me in the guardroom. It was almost as if I was preparing to go into a swimming-pool, so careful was the gesture of folding them and laying them down. I must have had to take a couple of steps backward to perform this neat unconscious manoeuvre. None of the guards made a move or said a word. Perhaps they were too surprised.

I was called forward. I stood to attention. They stood facing me, breathing heavily. There was a pause. It seemed to drag on for minutes. Then I went down with a blow that shook every bone, and which released a sensation of scorching liquid pain which seared through my entire body. Sudden blows struck me all over. I felt myself plunging downwards into an abyss with tremendous flashes of solid light which burned and agonized. I could identify the periodic stamping of boots on the back of my head, crunching my face into the gravel; the crack of bones snapping; my teeth breaking; and my own involuntary attempts to respond to deep vicious kicks and to regain an upright position, only to be thrown to the ground once more.

At one point I realized that my hips were being damaged and I remember looking up and seeing the pick-helves coming down towards my hips, and putting my arms in the way to deflect the blows. This seemed only to focus the clubs on my arms and hands.

I remember the actual blow that broke my wrist. It fell right across it, with a terrible pain of delicate bones being crushed. Yet the worst pain came from the pounding on my pelvic bones and the base of my spine. I think they tried to smash my hips. My whole trunk was brutally defined for me, like having my skeleton etched out in pain.

It went on and on. I could not measure the time it took. There are some things that you cannot measure in time, and this is one of them. Absurdly, the comparison that often comes to my mind is that torture was indeed like an awful job interview: it compresses time strangely, and at the end of it you cannot tell whether it has lasted five minutes or an hour.

I do know that I thought I was dying. I have never forgotten, from that moment onwards, crying out 'Jesus', crying out for help, the utter despair of helplessness. I rolled into a deep ditch of foul stagnant water which, in the second or two before consciousness was finally extinguished, flowed over me with the freshness of a pure and sweet spring.

I awoke and found myself standing on my feet. I do not recall crawling out of that ditch but the sun was already up. I was an erect mass of pain, of bloody contusions and damaged bones, the sun playing harshly on inflamed nerves. Smith and Slater were lying on the ground beside me, blackened, covered in blood and barely conscious. Mac and Knight were in a like state a few yards further away. We were only a few feet from the guardroom, close to the point where we had been standing the previous night. Slater was nearly naked; a pair of shorts and some torn clothing lay on the ground behind him, mudstained and bloodstained.

The guards simply ignored us. They stood in front of a barely moving, battered pile of human beings under the fierce sun and acted as though we were not there.

WILLIAM LAURENCE
Bombing Nagasaki

Even before the outbreak of the Second World War, scientists in Britain and America had become alarmed by the prospect of the enemies of Western democracy acquiring the knowledge necessary to make a nuclear weapon. At the war's outbreak, Britain succeeded in transferring from France to Britain the French government's stocks of heavy water, then an essential component of the process of separating the heavy from light elements of uranium, the fissile material from which scientists knew a nuclear warhead must be made. The British military nuclear research organization, codenamed Tube Alloys, was the recipient of the heavy-water stocks and used them in the research undertaken to advance development of an atomic bomb. Once the United States entered the war, in December 1941, however, Britain agreed with the government of the United States that all research into the development of nuclear weapons must be concentrated in America, where the first uranium chain reaction had been initiated at the University of Chicago by the emigré Italian scientist, Enrico Fermi, in 1941.

Unified under the umbrella of the so-called 'Manhattan Military District', the Allied nuclear weapons research programme succeeded by mid-1945 in creating two nuclear warheads, a heavy uranium bomb and a plutonium bomb. A test warhead was developed at Los Alamos and exploded in the New Mexico desert in July. On 6 August, the uranium bomb was dropped by an American B-29 bomber at Hiroshima. Three days later the plutonium bomb was exploded over Nagasaki. William Laurence's news report describes the operation, Laurence having flown on the mission in one of the supporting aircraft. Six days after the Nagasaki bombing, the Japanese Emperor announced Japan's unconditional surrender.

*   *   *

Atomic Bombing of Nagasaki Told by a Flight Member
By William L. Laurence

The first signs of dawn came shortly after 5 o'clock. Sergeant Curry, who had been listening steadily on his earphones for radio reports, while maintaining a strict radio silence himself, greeted it by rising to his feet and gazing out the window.

'It's good to see the day,' he told me. 'I get a feeling of claustrophobia hemmed in in this cabin at night.'

He is a typical American youth, looking even younger than his 20 years. It takes no mind-reader to read his thoughts.

'It's a long way from Hoopeston, Ill.,' I find myself remarking.

'Yep,' he replies, as he busies himself decoding a message from outer space.

'Think this atomic bomb will end the war?' he asks hopefully.

'There is a very good chance that this one may do the trick,' I assure him, 'but if not, then the next one or two surely will. Its power is such that no nation can stand up against it very long.'

This was not my own view. I had heard it expressed all around a few hours earlier, before we took off. To anyone who had seen this man-made fireball in action, as I had less than a month ago in the desert of New Mexico, this view did not sound overoptimistic.

By 5:50 it was real light outside. We had lost our lead ship [i.e. aircraft], but Lieutenant Godfrey, our navigator, informs me that we had arranged for that contingency. We have an assembly point in the sky above the little island of Yakoshima, southeast of Kyushu, at 9:10. We are to circle there and wait for the rest of our formation.

Our genial bombardier [bomb-aimer], Lieutenant Levy, comes over to invite me to take his front-row seat in the transparent nose of the ship and I accept eagerly. From that vantage point in space, 17,000 feet above the Pacific, one gets a view of hundreds of miles on all sides, horizontally and vertically. At that height the vast ocean below and the sky above seem to merge into one great sphere.

I was on the inside of that firmament, riding above the giant mountains of white cumulus clouds, letting myself be suspended in infinite space. One hears the whirl of the motors behind one, but it soon becomes insignificant against the immensity all around and is before long swallowed by it. There comes a point where space also swallows time and one lives through eternal moments filled with an oppressive loneliness, as though all life had suddenly vanished from the earth and you are the only one left, a lone survivor traveling endlessly through interplanetary space.

My mind soon returns to the mission I am on. Somewhere beyond these vast mountains of white clouds ahead of me there lies Japan, the land of our enemy. In about four hours from now one of its cities, making weapons of war for use against us, will be wiped off the map by the greatest weapon ever made by man. In one-tenth of a millionth of a second, a fraction of time immeasurable by any clock, a whirlwind from the skies will pulverize thousands of its buildings and tens of thousands of its inhabitants.

Our weather planes ahead of us are on their way to find out where the wind blows. Half an hour before target time we will know what the winds have decided.

Does one feel any pity or compassion for the poor devils about to die? Not when one thinks of Pearl Harbor and of the Death March on Bataan [in the Philippines; scene of the surrender of US and Philippine forces to the Japanese, April 1942].

Captain Bock informs me that we are about to start our climb to bombing altitude.

He manipulates a few knobs on his control panel to the right of him and I alternately watch the white clouds and ocean below me and the altimeter on the bombardier's panel. We reached our altitude at 9 o'clock. We were then over Japanese waters, close to their mainland. Lieutenant Godfrey motioned to me to look through his radar scope. Before me was the outline of our assembly point. We shall soon meet our lead ship and proceed to the final stage of our journey.

We reached Yakoshima at 9:12 and there, about 4,000 feet ahead of us, was The Great Artiste with its precious load. I saw Lieutenant Godfrey and Sergeant Curry strap on their parachutes and I decided to do likewise.

We started circling. We saw little towns on the coastline, heedless of our presence. We kept on circling, waiting for the third ship in our formation.

It was 9:56 when we began heading for the coastline. Our weather scouts had sent us code messages, deciphered by Sergeant Curry, informing us that both the primary target as well as the secondary were clearly visible.

The winds of destiny seemed to favor certain Japanese cities that must remain nameless. We circled about them again and again and found no opening in the thick umbrella of clouds that covered them. Destiny chose Nagasaki as the ultimate target.

We had been circling for some time when we noticed black puffs of smoke coming through the white clouds directly at us. There were fifteen bursts of flak in rapid succession, all too low. Captain Bock changed his course. There soon followed eight more bursts of flak, right up to our altitude, but by this time they were too far to the left.

We flew southward down the channel and at 11:33 crossed the coastline and headed straight for Nagasaki about 100 miles to the west. Here again we circled until we found an opening in the clouds. It was 12:01 and the goal of our mission had arrived.

We heard the prearranged signal on our radio, put on our arc-welder's glasses and watched tensely the manoeuvrings of the strike ship about half a mile in front of us. 'There she goes!' someone said.

Out of the belly of The Great Artiste what looked like a black object went downward.

Captain Bock swung around to get out of range, but even though we were turning away in the opposite direction, and despite the fact that it was broad daylight in our cabin, all of us became aware of a giant flash that broke through the dark barrier of our arc-welder's lenses and flooded our cabin with intense light.

We removed our glasses after the first flash, but the light still lingered on, a bluish-green light that illuminated the entire sky all around. A tremendous blast wave struck our ship and made it tremble from nose to tail. This was followed by four more blasts in rapid succession, each resounding like the boom of cannon fire hitting our plane from all directions.

Observers in the tail of our ship saw a giant ball of fire rise as though from the bowels of the earth, belching forth enormous white smoke rings. Next they saw a giant pillar of purple fire, 10,000 feet high, shooting skyward with enormous speed.

By the time our ship had made another turn in the direction of the atomic explosion the pillar of purple fire had reached the level of our altitude. Only about forty-five seconds had passed. Awe-struck, we watched it shoot upward like a meteor coming from the earth instead of from outer space, becoming ever more alive as it climbed skyward through the white clouds. It was no longer smoke, or dust, or even a cloud of fire. It was a living thing, a new species of being, born right before our incredulous eyes.

At one stage of its evolution covering millions of years in terms of seconds, the entity assumed the form of a giant square totem pole, with its base about three miles long, tapering off to about a mile at the top. Its bottom was brown, its center was amber, its top white. But it was a living totem pole, carved with many grotesque masks grimacing at the earth.

Then, just when it appeared as though the thing had settled down into a state of permanence, there came shooting out of the top a giant mushroom that increased the height of the pillar to a total of 45,000 feet. The mushroom top was even more alive than the pillar, seething and boiling in a white fury of sea foam, sizzling upward and then descending earthwards, a thousand Old Faithful geysers rolled into one.

It kept struggling in an elemental fury, like a creature in the act of breaking the bonds that held it down. In a few seconds it had freed itself from its gigantic stem and floated upward with tremendous speed, its momentum carrying it into the stratosphere to a height of about 60,000 feet.

But no sooner did this happen when another mushroom, smaller in size than the first one, began emerging out of the pillar. It was as though the decapitated monster was growing a new head. As the first mushroom floated off into the blue it changed its shape into a flower-like form, its giant petal curving downward creamy white outside, rose-colored inside. It still retained that shape when we last gazed at it from a distance of about 200 miles.

The New York Times, *9 September 1945*

C. W. BOWMAN
Red Thunder, Tropic Lightning

The war in Korea (1950–3) committed a fresh generation of Americans to combat, bitter in character and frustrating in its outcome. The conflict, endorsed as a legitimate peace-making operation by the United Nations, did not, however, engender domestic political dissent. The war in Vietnam, eventually involving half a million American troops between 1965 and 1972, divided the nation. Opponents objected that the war did not involve vital national interests and that, moreover, it was fought in a brutal way. Supporters of the war policy held that its prosecution was justified by America's commitment to preserve an anti-Communist government in South Vietnam, and was essential as a demonstration of the United States's determination to oppose the subversion and overthrow of free government throughout the Communist area of influence.

Opponents and supporters of the war effort all recognized that combat with the Viet Cong (Communist-led South Vietnamese guerrilla force and revolutionary army) and the North Vietnamese forces was of a particularly cruel and intense character. Never more so than in the struggle for control of the tunnel systems which the Communist forces had created all over South Vietnam, particularly in the region around Saigon. Tunnelling is a distinctively Asian style of warfare, practised by the Japanese, Chinese and Vietnamese. It confronted their Western opponents with demands on an individual soldier's courage to which nothing in their training or experience had previously exposed them. C. W. Bowman, an American serviceman who volunteered for the work of clearing tunnels of the enemy, describes the ordeal.

*　　*　　*

Tunnel warfare was another war in itself, an underground war. People asked me if I had a death wish: Why would anybody want to go down into the tunnel? I don't know. I was eighteen, and you're not going to die, at least you think you're not going to die. You're invincible. I guess it was because I had a good sixth sense or whatever you'd call it.

They had tunnels of all sizes. In some, you had to crawl on your hands and knees, in some, you could stand up and walk, and in some, you could almost drive a truck through. Mostly, I ran tunnels around Cu Chi and the Iron Triangle. Most of them were what I called 'hands-and-knees tunnels' because you had to crawl. Tunnels were cool inside – some went far underground – but you'd still be sweating. When you first entered, you'd look: most of the tunnels dropped straight down and took off on a 90-degree turn. That way, you couldn't see what was down the tunnel until you dropped down into it. You couldn't throw a grenade down there or tear gas or anything else because then you couldn't go in. A grenade would eat up the oxygen. So a lot of times, depending on how the tunnel was dug, it would be like dropping straight down into it. You had a flashlight in one hand and a pistol in the other. You never knew if you were going to come eyeball to eyeball with somebody right then. There were all kinds of little things you had to look for. There were snakes down there, and the V C [Viet Cong] planted nests of fire ants. Now, fire ants would build a nest out of leaves, and if you hit one, they would just pour out of those leaves. Once they got on you, the only way you could get them off was to pull them off: if you brushed them, they still hung on to you. They would draw blood or take little pieces of meat as you pulled them off, and it would burn like fire. I guess that's why they called them fire ants.

Usually, my friend Gary and I went in tunnels. It gets strange down there; it's quiet. It's cool, but the sweat is running off your body – rivers of sweat running down your back, dripping off your shoulders and arms, running over your nose, into your eyes. I never really got claustrophobia. But sometimes, you think you hear something up ahead, and your heart starts pounding: your chest hurts because your heart is pounding so hard. Sometimes, you could sense or feel there was something up ahead, but you couldn't really see it. You'd freeze in place, and you had to talk to your body to get it to move. Your arms and your legs feel about 1,000 pounds each, and at the same time, you feel so weak that you don't know if you had to if you could pull the trigger of your pistol or not. You want to back up and leave, and then you don't want to back up and leave because it's your job to go through the tunnel. The body sometimes feels like it's trying to tear itself in half. Part of it wants to go ahead because of the unknown, the challenge, but the other part, because of the fear and the unknown, wants to go back from where you came from. It wipes you out. When you come out of a tunnel, you are drained. You just have to sit down and pour some water over your head or whatever: take a break and regroup.

I've been in sandy tunnels, laterite tunnels. You'd get it all over you, and it would stick to you. We'd come up on trapdoors where we'd actually have to sit on the edge of the trapdoor, drop our legs through, and put our hands over our heads to narrow our shoulders so we could drop down. No telling what you'd find. We found medical equipment, surgical instruments, weapons, clothing, documents. Every tunnel was a little bit different. They had false floors in them booby-trapped with punji stakes, so you could fall through the floor and end up in a punji pit [full of sharpened bamboo spikes, often coated with poisonous or infective matter]. They might ambush you when you stuck your head up a trapdoor and stab you with a pike. There were false walls in the tunnel: you'd go through the tunnel, and they'd be on the other side of the wall watching you through a peephole, and they could open fire on you. I was fortunate

and lucky that a lot of this didn't happen to me. Still, Gary and I
did run into the VC or NVA [North Vietnamese Army] in the
tunnels. We had our shootouts. I've seen in books where people
had .22 or .38 pistols with silencers on them. But all we had were
.45s, and when you start firing up a tunnel with a .45 the concussion
damn near kills you. Gary and I both have come out of there with
nosebleeds, and I ruptured my eardrums at least once. So it's a
completely different world. And everything is fair game: if it moves,
you can shoot. As a matter of fact, you better shoot. You can't take
the time to say, 'Halt, who goes there?' You just kill whatever you
come up on. There is no place to run, no place to hide down there.
You come face to face with somebody, either they're dead or you're
dead.

I've seen guys break. They would go down into the tunnels, and
then one day – maybe they had a dream or something – and they
say, 'No, no, no.' The rule was, if you broke once, you never went
down a tunnel again. There were all kinds of ways some guys broke.
Some guys sat down and cried; some guys would drop down into
the tunnel, and as soon as they dropped down, they'd start shooting,
and that was it. Or you could just say, 'Hey, I've had it, I can't do
it any more.' We had a lieutenant who thought he could order people
down there. Once I went down a well, not a tunnel, and told Gary
to pull me back because there wasn't any oxygen. The lieutenant
called me all sorts of things, candyass and stuff. He told Gary to go
down, and Gary told him to kiss his ass. I loved it. The lieutenant
went crazy and was going to courtmartial us, but the captain laughed
at it. You'd find barracks-room types everywhere.

JAMES FENTON
The Fall of Saigon

The United States withdrew its forces from South Vietnam in
1972, trusting to a programme of 'Vietnamization' (strengthening the South Vietnamese army) to secure the independence
of the country after the signing of the Geneva Accords. In
1975 the Communist North Vietnamese government effectively
abrogated the Geneva agreement and initiated a full-scale military offensive into the South which, in a number of weeks,
overwhelmed the ARVN (Army of the Republic of Vietnam)
and established a Communist government in Saigon. James
Fenton, one of the most experienced correspondents of the
ten-year war, describes Saigon's last days before the Communist
take-over.

*　　*　　*

Early on the morning of 30 April, I went out of my hotel room to
be greeted by a group of hysterical Koreans. 'The Americans have
called off the evacuation!' said one. The group had been unable to
get into the [US] Embassy, had waited the whole night and had
now given up. Of all the nationalities to fear being stranded in
Saigon, the Koreans had most reason. I went up to breakfast in the
top-floor restaurant, and saw that there were still a few Jolly Green
Giants [troop-carrying helicopters] landing on the Embassy, but that
the group on the Alliance Française building appeared to have been
abandoned. They were still standing there on the roof, packed tight
on a set of steps. Looking up at the sky, they seemed to be taking
part in some kind of religious ritual, waiting for a sign. In the Brinks
building, the looting continued. A lone mattress fell silently from a
top-floor balcony.

There was one other group at breakfast – an eccentric Frenchman
with some Vietnamese children. The Frenchman was explaining to

the waiter that there had been some binoculars available the night before, and he wanted to use them again. The waiter explained that the binoculars belonged to one of the hotel guests.

'That doesn't matter,' said the Frenchman, 'bring them to me.'

The waiter explained that the binoculars were probably in the guest's room.

'Well go and get them then!' said the Frenchman. It seemed extraordinary that the Frenchman could be so adamant, and the waiter so patient, under the circumstances. I had orange juice and coffee, and noted that the croissants were not fresh.

Then I went to the American Embassy, where the looting had just begun. The typewriters were already on the streets. Outside there was a stink of urine from where the crowd had spent the night and several cars had been ripped apart. I did not bother to check what had happened to mine, but went straight into the Embassy with the looters.

The place was packed, and in chaos. Papers, files, brochures and reports were strewn around. I picked up one letter of application from a young Vietnamese student, who wished to become an Embassy interpreter. Some people gave me suspicious looks, as if I might be a member of the Embassy staff – I was, after all, the only one there with a white face – so I began to do a little looting myself, to show that I was entering into the spirit of the thing. Somebody had found a package of razor blades, and removed them all from their plastic wrappers. One man called me over to a wall-safe, and seemed to be asking if I knew the number of the combination. Another was hacking away at an air-conditioner, another dismantling a fridge.

On the first floor there was more room to move, and it was here I came across the Embassy library. I collected the following items: one copy of *Peace Is Not at Hand* by Sir Robert Thompson, one of the many available copies of *The Road from War* by Robert Shaplen, Barrington Moore's *Social Origins of Dictatorship and Democracy* (I had been meaning to read it for some time), a copy of a pacification report from 1972 and some Embassy notepaper. Two things I could not take (by now I was not just pretending to loot – I had become quite involved): a reproduction of an 1873 map of Hanoi, and a

framed quotation from Lawrence of Arabia, which read, 'Better to
let them do it imperfectly than to do it perfectly yourself, for it is
their country, their way, and your time IS short.' Nearby I found a
smashed portrait of President Ford, and a Stars and Stripes, mangled
in the dirt.

I found one room which had not yet been touched. There were
white chairs around a white table, and on the table the ashtrays were
full. I was just thinking how eerie it looked, how recently vacated,
when the lights went out. At once, a set of emergency lights,
photosensitively operated, turned themselves on above each doorway. The building was still partly working; even while it was being
torn to pieces, it had a few reflexes left.

From this room, I turned into a small kitchen, where a group of
old crones were helping themselves to jars of Pream powdered milk.
When they looked up and saw me, they panicked, dropped the
powdered milk and ran. I decided that it would be better to leave
the building. It was filling up so much that it might soon become
impossible to get out. I did not know that there were still some
marines on the roof. As I forced my way out of the building, they
threw tear-gas down on the crowd, and I found myself running hard,
in floods of tears.

Although the last helicopter was just now leaving, people still
thought there were other chances of getting out. One man came up
to me and asked confidentially if I knew of the alternative evacuation
site. He had several plausible reasons why he was entitled to leave.
Another man, I remember, could only shout, 'I'm a professor, I'm
a professor, I'm a professor,' as if the fact of his academic status
would cause the Jolly Green Giants to swoop down out of the sky
and whisk him away.

There was by now a good deal of activity on the streets. Military
trucks went to and fro across town, bearing loads of rice, and family
groups trudged along, bearing their possessions. As I finished writing
my Embassy story, the sirens wailed three times, indicating that the
city itself was under attack. I returned to the hotel roof to see what
was happening. The group on the Alliance Française building was
still there, still waiting for its sign. Across the river, but not far away,

you could see the artillery firing, and the battle lines coming closer. Then two flares went up, one red, one white. Somebody said that the white flare was for surrender. In the restaurant, the waiters sat by the radio. I asked them what was happening. 'The war is finished,' said one.

I looked down into the square. Almost at once, a waiter emerged from the Continental and began to hoist a French tricolour on the flagpole. There were groups of soldiers, apparently front-line troops, sitting down. From the battlefield across the river, the white flares began to go up in great numbers. Big Minh's broadcast had been heard – offering unconditional surrender – and in a matter of minutes the war would be well and truly over.

ANDY MCNAB
Bravo Two Zero

Iraq's illegal occupation of Kuwait in August 1990 caused the leading Western powers, with widespread assistance from Middle Eastern states, to mount a major operation, launched from bases in Saudi Arabia, to expel the Iraqi invaders and restore the legitimate Kuwaiti government. Among the forces engaged were elements of the British Special Air Service Regiment, which operated largely behind Iraqi lines. Sergeant Andy McNab (a pseudonym) was one of the most enterprising SAS soldiers in the campaign. Here he describes an engagement with Iraqi troops inside enemy lines.

*   *   *

People put down a fearsome amount of covering fire. You don't fire on the move. It slows you up. All you have to do is get forward, get down and get firing so that the others can move up. As soon as you get down on the ground your lungs are heaving and your torso is moving up and down, you're looking around for the enemy but you've got sweat in your eyes. You wipe it away, your rifle is moving up and down in your shoulder. You want to get down in a nice firing position like you do on the range, but it isn't happening that way. You're trying to calm yourself down to see what you're doing, but you want to do everything at once. You want to stop this heavy breathing so you can hold the weapon properly and bring it to bear. You want to get rid of the sweat so you can see your targets, but you don't want to move your arm to rub your eye because you've got it in the fire position and you want to be firing to cover the move of the others as they come forward.

I jumped up and ran forward another 15 metres – a far longer bound than the textbooks say you should. The longer you are up the longer you are a target. However, it is quite hard to hit a fast-moving man and we were pumped up on adrenalin. You're immersed in your own little world. Me and Chris running forward, Stan and Mark backing us up with the Minimi [light machine-gun]. Fire and manoeuvre. The others were doing the same, legging it forward. The rag-heads [Iraqis] must have thought we were crazy but they had put us in the situation and this was the only way out.

You could watch the tracer coming at you. You heard the burning, hissing sound as the rounds shot past or hit the ground and spun off into the air. It was scary stuff. There's nothing you can do but jump up, run, get down; jump up, run, get down. Then lie there panting, sweating, fighting for breath, firing, looking for new targets, trying to save ammo . . .

Once I had moved forward and started firing the Minimis stopped and they, too, bounded forward. The sooner they were up ahead the better, because of their superior fire-power.

The closer we got the more the Iraqis were flapping. It must have been the last thing they expected us to do. They probably didn't realize it was the last thing we wanted to do.

You're supposed to count your rounds as you're firing, but in practice it's hard to do. At any moment when you need to fire you should know how many are left, and change mags if you have to. Lose count and you'll hear a 'dead man click'. You pull the trigger and the firing pin goes forward but nothing happens. In practice, counting to thirty is unrealistic. What you actually do is wait for your weapon to stop firing, then press the button and let the mag fall, slap another straight on and off you go. If you are well drilled in this it's second nature and requires no mental action. It just happens. The Armalite [automatic rifle] is designed so that when you've stopped firing the working parts are to the rear, so you can slap another mag on and let the working parts go forward so that a round is taken into the breech. Then you fire again, at anything that moves.

We had got up to within 50 metres of them. The APC [armoured personnel carrier] nearest me started to retreat, gun still firing. Our rate of fire slowed. We had to husband the rounds.

The [Iraqi] truck was on fire. I didn't know if any of us was hit. There wouldn't have been a lot we could do about it anyway.

I couldn't believe that the APC was backing off. Obviously it was worried about the anti-armour rockets and knew the other one had been hit, but for it to withdraw was absolutely incredible. Some of the infantry ran with it, jumping into the back. They were running, turning, giving it good bursts, but it was a splendid sight. I fancied a cabby myself with my 66 [portable anti-tank rocket, shoulder-launched from a tube], and discovered that in the adrenalin rush I'd left it with my bergen [pack]. Wanker!

At the other end, Vince was up with Legs and still going forward. They were shouting to psych each other up. The rest of us put down covering fire.

Mark and Dinger stood up and ran forwards. They were concentrating on the APC ahead of them that they had hit with their 66s. They'd scored a 'mobility kill' – its tracks couldn't move, though it could still use its gun. They were putting in rounds hoping to shatter the gunner's prism. If I'd been in his boots I would have got out of the wagon and legged it, but then, he didn't know who he had pursuing him. They got up to the APC and found the rear doors still open. The Iraqis hadn't battened themselves down. An L2 grenade was lobbed in and exploded with its characteristic dull thud. The occupants were killed instantly.

We kept going forwards into the area of the trucks in four groups of two, each involved in its own little drama. Everybody was bobbing and moving with Sebastian Coe legs on. We'd fire a couple of rounds, then dash and get out of the way, then start again. We tried to fire aimed shots. You pick on one body and fire until he drops. Sometimes it can take as many as ten rounds.

There is a set of sights on the 203 [grenade launcher] but you don't always have time to set it up and fire. It was a case of just take a quick aim and get it off. The weapon 'pops' as it fires. I watched the bomb going through the air. There was a loud bang and showers of dirt. I heard screaming. Good. It meant they were bleeding, not shooting – and they'd become casualties that others now had to attend to.

We found ourselves on top of the position. Everybody who could do so had run away. A truck was blazing furiously ahead of us. A burnt-out APC smoked at the far-left extreme. Bodies were scattered over a wide area. Fifteen dead maybe, many more wounded. We disregarded them and carried on through. I felt an enormous sense of relief at getting the contact over with, but was still scared. There would be more to come. Anybody who says he's not scared is either a liar or mentally deficient.

SOURCES AND ACKNOWLEDGEMENTS

The following list is a combination of sources for the pieces in this anthology and acknowledgements. The author and the publishers are grateful for permission to reproduce this material. Every effort has been made to acknowledge all copyright owners prior to going to press. Penguin regret any inadvertent omissions or inaccuracies. These will be rectified at the earliest opportunity.

All the poetry is taken from *The Oxford Book of War Poetry*, edited by Jon Stallworthy, Oxford University Press, 1984. Those in copyright appear in the list below.

Thucydides, 'The Melian Dialogue'. From Thucydides, *The History of the Peloponnesian War*, edited in translation by Sir R. W. Livingstone, Oxford University Press, 1968, pp. 266–74. Reproduced by permission of the Trustees of Oxford University Press.

Thucydides, 'The Final Sea Battle'. From Thucydides, *The History of the Peloponnesian War*, edited in translation by Sir R. W. Livingstone, Oxford University Press, 1968, pp. 373–6. Reproduced by permission of the Trustees of Oxford University Press.

Xenophon, 'The Battle of Cunaxa and Death of Cyrus'. From Xenophon, *The Persian Expedition*, translated by Rex Warner, Penguin Books, 1949, Book I, pp. 47–51. © Rex Warner, 1949. Reproduced by permission of Penguin Books Ltd.

Xenophon, 'A Plundering Expedition'. From Xenophon, *The Persian Expedition*, translated by Rex Warner, Penguin Books, 1949, Book V, pp. 177–81. © Rex Warner, 1949. Reproduced by permission of Penguin Books Ltd.

Julius Caesar, 'First British Expedition'. From Julius Caesar, *The Gallic Wars*, translated by John Warrington, Everyman's Library, 1953, Book IV, pp. 65–8. Reprinted by permission of Everyman's Library, David Campbell Publishers Ltd.

Josephus, 'The Horrors of the Siege'. From Josephus, *The Jewish War*, translated by G. A. Williamson, revised by E. Mary Smallwood, Penguin Books, 1959, revised edition, 1969, pp. 294–302. © G. A. Williamson, 1959, 1969. Reproduced by permission of Penguin Books Ltd.

Usāmah Ibn-Munqidh, 'An Arab-Syrian Gentleman'. From Usāmah Ibn-Munqidh, *An Arab-Syrian Gentleman and Warrior in the Period of the Crusades*, translated by Philip K. Hitti, I. B. Tauris & Co. Ltd, 1987, pp. 101–5. Reproduced by permission of I. B. Tauris & Co. Ltd (UK) and Princeton University Press (USA).

Jean Froissart, 'The Battle of Crécy, 26 August 1346'. From Sir Jean Froissart, *Chronicles of England, France and Spain*, translated by John Bourchier, Lord Berners, 1523–5.

Jehan de Wavrin, 'A French Knight's Account of Agincourt'. From Jehan de Wavrin, *Chronicles, 1399–1422*, translated by Sir W. Hardy and E. Hardy, 1887.

Andrew Wheatcroft, 'The Fall of Constantinople'. From Andrew Wheatcroft, *The Ottomans: Dissolving Images*, Viking, 1993, pp. 12–21. © 1993 Andrew Wheatcroft. Reproduced by permission of Penguin Books Ltd.

Diego Hurtado de Mendoze, 'Such Botching, Disorder and Chaos'. From Diego Hurtado de Mendoze, *The War in Granada*, translated by Martin Shuttleworth, The Folio Society, 1982, pp. 188–91. Reproduced by permission of The Folio Society.

Francesco Balbi di Correggio, 'The Siege of Malta'. From Francesco Balbi di Correggio, *The Siege of Malta*, translated by Major Henry Alexander Balbi, The Folio Society, 1961, pp. 139–47. Reproduced by permission of The Folio Society.

Inga Clendinnen, 'Aztecs'. From Inga Clendinnen, *Aztecs*, Cambridge University Press, 1991, pp. 93–8 and 87–8. Reproduced by permission of the author and Cambridge University Press.

Father Paul Ragueneau, 'An Attack by Iroquois Warriors'. From Father Paul Ragueneau, in *Black Gown and Redskins: Adventures and Travels of the Early Jesuit Missionaries in North America, 1610–1791*, edited by Edna Kenton, Longmans, Green & Co., 1956, pp. 225–31. © Edna Kenton, 1956. Reproduced by permission of Penguin Books Ltd.

William Dunbar, 'Braddock at the Monongahela'. William Dunbar,

'Braddock at the Monongahela', unpublished manuscript, New York Public Library, Hardwicke Collection, vol. 136, doc. no. 6, pp. 186–8. Published by permission of The New York Public Library Manuscripts and Archives Division, Astor, Lenox and Tilden Foundations.

Anna Myers, 'The Revolution Remembered (1)'. From John C. Dann, *The Revolution Remembered: Eyewitness Accounts of the War for Independence*, University of Chicago Press, 1980, pp. 268–74.

Jacob Zimmerman, 'The Revolution Remembered (2)'. From John C. Dann, *The Revolution Remembered: Eyewitness Accounts of the War for Independence*, University of Chicago Press, 1980, pp. 285–8.

David Crockett, 'Davy Crockett'. From David Crockett, *A Narrative of the Life of David Crockett*, Carey, Hart & Co., 1834, pp. 106–13.

John D. Hunter, 'Captivity Among the Indians'. From John D. Hunter, *Captivity Among the Indians of North America*, Longman, Hurst, Rees, Orme & Browne, 1823, pp. 320–33.

Sergeant William Lawrence, 'Fugitive and Recruit 1804–6'. From *A Dorset Soldier: The Autobiography of Sergeant William Lawrence 1790–1869*, edited by Eileen Hathaway, Spellmount, 1993, pp. 17–19. Reproduced by permission of Spellmount Publishers.

Sergeant William Lawrence, 'Badajoz, March–April 1812'. From *A Dorset Soldier: The Autobiography of Sergeant William Lawrence 1790–1869*, edited by Eileen Hathaway, Spellmount, 1993, pp. 64–70. Reproduced by permission of Spellmount Publishers.

A Gentleman Volunteer, 'The Battle of Vitorio'. From *A Gentleman Volunteer: The Letters of George Hennell from the Peninsular War, 1812–1813*, edited by Michael Glover, Heinemann, 1979, pp. 87–93 (letter dated 29 June 1813). Reproduced by permission of Heinemann.

Helen Roeder, 'Captain Roeder'. From Helen Roeder, *The Ordeal of Captain Roeder*, Methuen, 1960, pp. 175–89. © Helen Roeder, 1960.

Victor Hugo, 'Russia 1812'. From *Imitations* by Robert Lowell. Copyright © 1959 by Robert Lowell. Copyright renewed © 1987 by Harriet, Sheridan and Caroline Lowell. Reprinted by permission of Faber and Faber Ltd (UK) and Farrar, Straus & Giroux, Inc. (USA).

Private Wheeler, 'The Letters of Private Wheeler'. From *The Letters of Private Wheeler*, edited by Captain B. H. Liddell Hart, The Windrush

Press, 1993, pp. 170–74. Reproduced by permission of David Higham Associates on behalf of the author's estate.

Duke of Wellington, 'Wellington's Waterloo Despatch'. Field Marshal the Duke of Wellington, official despatch after Waterloo, 19 June 1815, in General Muffling, *A Sketch of the Battle of Waterloo*, Gerard, 1883, pp. 65–77.

James Bodell, 'A Soldier's View of Empire'. From *A Soldier's View of Empire: The Reminiscences of James Bodell*, edited by Keith Sinclair, Bodley Head, 1982, pp. 41–2.

Henry Clifford, 'Clifford in the Crimea'. From *Henry Clifford V. C.: His Letters and Sketches from the Crimea*, Michael Joseph, 1956, pp. 69–76.

Colin Frederick Campbell, 'Letters from Camp'. From Colin Frederick Campbell, *Letters from Camp*, Richard Bentley & Son, 1894, pp. 87–95.

W. H. Fitchett, 'The Relief of Lucknow'. From W. H. Fitchett, *The Tale of the Great Mutiny*, Smith, Elder & Co., 1906, pp. 214–18.

Lieutenant-Colonel Fremantle, 'The Fremantle Diary'. From *The Fremantle Diary, Being the Journal of Lieutenant-Colonel James Arthur Lyon Fremantle, Coldstream Guards, on his three months in the Southern States*, edited by Walter Lord, Little, Brown & Co., 1954, pp. 205–9. Reprinted by permission of Sterling Lord Literistic Inc., USA. Copyright © 1954 by Walter Lord.

Elizabeth B. Custer, 'General Custer'. From Elizabeth B. Custer, *Boots and Saddles, or Life in Dakota with General Custer*, Sampson Low, Marston, Searle & Rivington, 1885, pp. 286–90.

Rudyard Kipling, 'Tommy'. From *Rudyard Kipling's Verse: Definitive Edition*, Hodder and Stoughton and Doubleday, 1940. Reproduced by permission of A. P. Watt Ltd for The National Trust for Places of Historic Interest or Natural Beauty (UK).

Isandhlwana and Rorke's Drift, 'Four eyewitness accounts'. From Frank Emery, *The Red Soldier: Letters from the Zulu Wars*, Hodder & Stoughton, 1977, pp. 82–3, pp. 85–6, pp. 87–91 and pp. 126–30. © 1977 Frank Emery.

Sir Henry Newbolt, 'Vitaï Lampada'. From *Selected Poems of Henry Newbolt*, Hodder and Stoughton, 1981. Reproduced by permission of Peter Newbolt.

Stephen Graham, 'How the news of war came to a village on the Chinese

frontier'. From Stephen Graham, *Russia and the World*, Cassell, 1915. ©
the Estate of Stephen Graham, 1915.

Erwin Rommel, 'Infantry Attacks'. From General Field Marshal Erwin
Rommel, *Infantry Attacks*, new introduction by Manfred Rommel,
Greenhill Books, 1990, pp. 10–13.

Compton Mackenzie, 'Gallipoli Memories'. From Compton Mackenzie,
Gallipoli Memories, Cassell, 1929, pp. 123–5. Reproduced by permission
of Weidenfeld & Nicolson Ltd.

Sidney Rogerson, 'Twelve Days'. From Sidney Rogerson, *Twelve Days*,
1930, reissued Gliddon Books, 1988, pp. 85–99.

John Glubb, 'A Soldier's Diary of the Great War'. From John Glubb, *Into
Battle: A Soldier's Diary of the Great War*, Cassell, 1977, pp. 67–9, 153 and
185–8. Reproduced by permission of Weidenfeld & Nicolson Ltd.

Brigadier-General E. L. Spears, 'Prelude to Victory'. From Brigadier-
General E. L. Spears, CB, CBE, MC, *Prelude to Victory*, Introduction
by The Rt Hon. Winston S. Churchill, PC, MP, Jonathan Cape, 1939,
pp. 395–401 and 490–3. Copyright Brigadier-General E. L. Spears,
1939.

W. B. Yeats, 'An Irish Airman Foresees His Death'. From *Collected Poems*,
copyright 1919 by Macmillan Publishing Co., renewed 1937 by Bertha
Georgie Yeats. Reproduced with permission of A. P. Watt on behalf
of Michael B. Yeats (UK) and Scribner/Simon & Schuster Inc. (USA).

Robert Graves, 'Goodbye To All That'. From Robert Graves, *Goodbye To
All That*, 1929, revised edition 1957, Penguin Books Ltd, 1960, pp. 154–
9. Reproduced by permission of A. P. Watt Ltd on behalf of the Trustees
of the estate of Robert Graves/Copyright Trust.

Ernest Hemingway, 'Wounded (1)'. From Ernest Hemingway, *Selected Letters
1917–1961*, edited by Carlos Baker, Granada Publishing Ltd, 1981, pp. 13–
16. Reproduced by permission of Charles Scribner's Sons, Macmillan
Publishing.

Siegfried Sassoon, 'The Hero'. From *Collected Poems of Siegfried Sassoon*, ©
1918, 1920 by E. P. Dutton & Co. Copyright 1936, 1946, 1947, 1948
by Siegfried Sassoon. Used by kind permission of George Sassoon
(UK) and of Viking Penguin, a division of Penguin Putnam Inc. (USA).

Gerald Uloth, 'Riding to War'. From Gerald Uloth, *Riding to War*, Monks
Publishing, 1993, pp. 52–5. Reproduced by permission of A. C. Uloth.

Siegfried Sassoon, 'The General'. From *Collected Poems of Siegfried Sassoon*, © 1918, 1920 by E. P. Dutton & Co. Copyright 1936, 1946, 1947, 1948 by Siegfried Sassoon. Used by kind permission of George Sassoon (UK) and of Viking Penguin, a division of Penguin Putnam Inc. (USA).

Isaac Babel, 'Red Cavalry'. From Isaac Babel, *Collected Stories*, translated by Walter Morison, Associated Book Publishers, pp. 37–9. Reproduced by permission of Penguin Putnam Inc. (USA).

Wilfred Owen, 'Anthem for Doomed Youth'. From *The Complete Poems of Wilfred Owen*, edited by Jon Stallworthy, 1963. © 1963 by Chatto & Windus. Reproduced by permission of Chatto & Windus (UK) and New Directions Publishing Corporation (USA).

John Masters, 'Bugles and a Tiger'. From John Masters, *Bugles and a Tiger: A Personal Adventure*, Michael Joseph, 1956, pp. 78–84. © John Masters, 1956. Reproduced by permission of Penguin Books Ltd.

Alexander Stahlberg, 'Bounden Duty (1)'. From Alexander Stahlberg, *Bounden Duty: The Memoirs of a German Officer 1932–45*, translated by Patricia Crampton, Brassey's UK Ltd, 1990, pp. 67–73. English translation © 1990 Patricia Crampton. First published in German under the title *Die Verdammte Pflicht: Erinnerungen 1932 bis 1945*. © 1987 Verlag Ullstein GmbH Berlin–Frankfurt/Main.

Tim Bishop, 'One Young Soldier'. From Tim Bishop, *One Young Soldier*, edited by Bruce Shand, Michael Russell, 1993, pp. 10–12. Reproduced by permission of Michael Russell Publishers Ltd.

George Orwell, 'Wounded (2)'. From George Orwell, *Homage to Catalonia*, Secker & Warburg, 1938, pp. 133–5. Reproduced by permission of Secker & Warburg Ltd and A. M. Heath and Co. Ltd on behalf of Mark Hamilton as the Literary Executor of the estate of the late Sonia Brownell Orwell.

Bruce Shand, 'Previous Engagements'. From Bruce Shand, *Previous Engagements*, Michael Russell, 1990, pp. 29–32. Reproduced by permission of Michael Russell Publishers Ltd

Winston Churchill, 'BBC broadcast, London, 19 May 1940'. From *The Speeches of Winston Churchill*, edited and introduced by David Cannadine, Penguin Books Ltd, 1989, pp. 151–4. Previously published as *Blood, Toil, Tears and Sweat* by Cassell Publishers, 1989. Reproduced by permission of Cassell Publishers.

Stuart Milner-Barry, 'Codebreakers'. From F. H. Hinsley and Alan Stripp, *Codebreakers: The Inside Story of Bletchley Park*, Oxford University Press, 1993, pp. 89–99. Reproduced by permission of Oxford University Press.

Peter Cremer, 'U-333'. From Peter Cremer, *U-333: The Story of a U-Boat Ace*, translated by Lawrence Wilson, The Bodley Head, 1984, pp. 154–9. © Peter Cremer and translation © Lawrence Wilson, 1984.

Hugh Dundas, 'Flying Start'. From Hugh Dundas, *Flying Start: A Fighter Pilot's War Years*, Century, 1988, pp. 67–70. Reproduced by permission of Century (UK) and St Martin's Press, Inc. (USA).

George Kennard, 'Loopy'. From George Kennard, *Loopy: The Autobiography of George Kennard*, Leo Cooper, 1990, pp. 46–51. © George Kennard, 1990. Reproduced by permission of Pen & Sword Books Ltd.

Studs Terkel, 'The Good War'. From Paul Edwards, in Studs Terkel, *The Good War*, The New Press and Hamish Hamilton, 1984, pp. 561–73. © 1984 *The Good War* by Studs Terkel. Reproduced by permission of The New Press, 450 West 41 St, New York, NY 10036 (USA) and Penguin Books Ltd (UK).

Obituary, 'David Stirling'. From *The Daily Telegraph Second Book of Obituaries: Heroes and Adventurers*, edited by Hugh Massingberd, Macmillan, 1996, pp. 154–60. Reproduced by permission of Macmillan Publishers Ltd (UK) and the *Daily Telegraph* Plc (USA).

Keith Douglas, 'Aristocrats'. From *The Complete Poems of Keith Douglas*, edited by Desmond Graham, Oxford University Press, 1978. © Marie J. Douglas, 1978. Reproduced by permission of Faber and Faber Ltd.

Alexander Stahlberg, 'Bounden Duty (2)'. From Alexander Stahlberg, *Bounden Duty: The Memoirs of a German Officer 1932–45*, translated by Patricia Crampton, Brassey's UK Ltd, 1990, pp. 207–11. English translation © 1990 Patricia Crampton. First published in German under the title *Die Verdammte Pflicht: Erinnerungen 1932 bis 1945*. © 1987 Verlag Ullstein GmbH Berlin–Frankfurt/Main.

Geyr von Schweppenburg, 'On the Other Side of the Hill'. From General Leo, Freiherr Geyr von Schweppenburg, 'On the Other Side of the Hill', 1964, in *Articles of War: The Spectator Book of World War II*, edited by Fiona Glass and Philip Marsden-Smedley, Grafton Books, 1989, pp. 311–17. Reproduced by permission of Fiona Glass and Philip Marsden-Smedley.

Marie-Louise Osmont, 'Normandy Diary'. From *The Normandy Diary of Marie-Louise Osmont*, Discovery Communication Inc./Random House, 1994, pp. 40–4. © Marie-Louise Osmont, 1994.

Ernie Pyle, 'Battle and Breakout in Normandy'. From Ernie Pyle, 'Battle and Breakout in Normandy', wire-service copy, 1 July 1944, in *Reporting World War II*, Part Two: *American Journalism 1944–1946*, The Library of America, Penguin Books, 1995, pp. 194–6. Reproduced by permission of Scripps Howard Foundation.

David Smiley, 'Albanian Assignment'. From David Smiley, *Albanian Assignment*, Chatto & Windus, 1984, pp. 60–3. © David Smiley, 1984. Reproduced by permission of Maggie Noach Literary Agency on behalf of the author.

Franklin Lindsay, 'Beacons in the Night'. From Franklin Lindsay, *Beacons in the Night: With the OSS and Tito's Partisans in Wartime Yugoslavia*, Stanford University Press, 1993, pp. 131–4. Reproduced by permission of the publishers, Stanford University Press. © 1993 by the Board of Trustees of the Leland Stanford Junior University.

Alexander Stahlberg, 'Bounden Duty (3)'. From Alexander Stahlberg, *Bounden Duty: The Memoirs of a German Officer 1932–45*, translated by Patricia Crampton, Brassey's UK Ltd, 1990, pp. 394–9. English translation © 1990 Patricia Crampton. First published in German under the title *Die Verdammte Pflicht: Erinnerungan 1932 bis 1945*. © 1987 Verlag Ullstein GmbH Berlin–Frankfurt/Main.

George MacDonald Fraser, 'Quartered Safe Out Here'. From George MacDonald Fraser, *Quartered Safe Out Here*, Harvill/Collins, 1992, pp. 87–90. Reproduced by permission of Curtis Brown Ltd on behalf of George MacDonald Fraser. Copyright © George MacDonald Fraser, 1992.

Eric Lomax, 'The Railway Man'. From Eric Lomax, *The Railway Man: A POW's Searing Account of War, Brutality and Forgiveness*, Jonathan Cape, 1995, pp. 116–21. © Eric Lomax, 1995. Reproduced by permission of Jonathan Cape (UK) and W. W. Norton and Co., Inc. (USA).

William Laurence, 'Bombing Nagasaki'. William L. Laurence, 'Atomic Bombing of Nagasaki Told by a Flight Member', *New York Times*, 9 September 1945, in *Reporting World War II*, Part Two: *American Journalism 1944–1946*, The Library of America, 1995, pp. 768–72. © 1946 by The New York Times. Reprinted by permission.

C. W. Bowman, 'Red Thunder, Tropic Lightning'. From C. W. Bowman, 'Red Thunder, Tropic Lightning', in Eric M. Bergerud, *Red Thunder, Tropic Lightning*, Westview Press, 1993, pp. 190–2. Reproduced by permission of Westview Press.

James Fenton, 'The Fall of Saigon'. From James Fenton, 'The Fall of Saigon', in *Granta 15*, Granta Publications, 1985, pp. 80–2. © James Fenton, 1985. Reproduced by permission of the Peters, Fraser and Dunlop Group Ltd on behalf of the author.

Andy McNab, 'Bravo Two Zero'. From Andy McNab, *Bravo Two Zero*, Corgi, a division of Transworld Publishers Ltd, 1993, pp. 118–21. © Andy McNab 1993. All rights reserved.

INDEX OF AUTHORS

INDEX